Interrelationships Between Corals and Fisheries

CRC
MARINE BIOLOGY
SERIES

The late Peter L. Lutz, Founding Editor
David H. Evans, Series Editor

PUBLISHED TITLES

Biology of Marine Birds
E.A. Schreiber and Joanna Burger

Biology of the Spotted Seatrout
Stephen A. Bortone

Early Stages of Atlantic Fishes: An Identification Guide for the Western Central North Atlantic
William J. Richards

Biology of the Southern Ocean, Second Edition
George A. Knox

Biology of the Three-Spined Stickleback
Sara Östlund-Nilsson, Ian Mayer, and Felicity Anne Huntingford

Biology and Management of the World Tarpon and Bonefish Fisheries
Jerald S. Ault

Methods in Reproductive Aquaculture: Marine and Freshwater Species
Elsa Cabrita, Vanesa Robles, and Paz Herráez

Sharks and Their Relatives II: Biodiversity, Adaptive Physiology, and Conservation
Jeffrey C. Carrier, John A. Musick, and Michael R. Heithaus

Artificial Reefs in Fisheries Management
Stephen A. Bortone, Frederico Pereira Brandini, Gianna Fabi, and Shinya Otake

Biology of Sharks and Their Relatives, Second Edition
Jeffrey C. Carrier, John A. Musick, and Michael R. Heithaus

The Biology of Sea Turtles, Volume III
Jeanette Wyneken, Kenneth J. Lohmann, and John A. Musick

The Physiology of Fishes, Fourth Edition
David H. Evans, James B. Claiborne, and Suzanne Currie

Interrelationships Between Corals and Fisheries
Stephen A. Bortone

Interrelationships Between Corals and Fisheries

EDITED BY **Stephen A. Bortone**

CRC Press is an imprint of the
Taylor & Francis Group, an **informa** business

This book was produced with funding assistance by the Gulf of Mexico Fishery Management Council under award number NA11NMF4410063 from the Coral Reef Conservation Program, National Oceanic and Atmospheric Administration, U.S. Department of Commerce. The statements, findings, conclusions, and recommendations are those of the editor and authors and do not necessarily reflect the views of the Gulf of Mexico Fishery Management Council, the National Oceanic and Atmospheric Administration or the Department of Commerce.

CRC Press
Taylor & Francis Group
6000 Broken Sound Parkway NW, Suite 300
Boca Raton, FL 33487-2742

First issued in paperback 2019

ISBN-13: 978-1-4665-8830-1 (hbk)
ISBN-13: 978-0-367-37854-7 (pbk)

Library of Congress Cataloging-in-Publication Data

Interrelationships between corals and fisheries / edited by Stephen A. Bortone.
pages cm. -- (CRC marine biology series ; 16)
Includes bibliographical references and index.
ISBN 978-1-4665-8830-1
1. Coral reefs and islands--Monitoring. 2. Coral reef ecology. 3. Corals--Habitat. 4. Coral reef fishes--Monitoring. I. Bortone, Stephen A., editor of compilation.

QE565.I58 2014
551.42'4--dc23 2014006681

Visit the Taylor & Francis Web site at
http://www.taylorandfrancis.com

and the CRC Press Web site at
http://www.crcpress.com

Dedication

This volume is dedicated to recently deceased colleagues in fisheries: Karen M. Burns (Mote Marine Laboratory and the Gulf of Mexico Fishery Management Council), Richard L. Leard (Gulf of Mexico Fishery Management Council), Tom McIlwain (University of Southern Mississippi), Vernon Minton (Alabama Department of Conservation and Natural Resources), Russell Nelson (fisheries consultant), and Larry B. Simpson (Gulf States Marine Fisheries Commission). Each played a significant role toward achieving sustainability of world fisheries.

Contents

Preface

This book is the result of an idea stimulated by a request for proposals from the U.S. National Oceanographic and Atmospheric Agency's (NOAA) Coral Reef Conservation Program (CRCP). In partial fulfillment of a grant to the Gulf of Mexico Fishery Management Council, a workshop (Interrelationships between Coral Reefs and Fisheries) was held in Tampa, Florida on 20–22 May, 2013, to discuss current and emerging threats as well as challenges and opportunities for managing corals and associated fisheries. In brief, the workshop aimed to answer four questions:

1. What are the "essential" features of the interrelationships between corals and fisheries?
2. If negative trends in coral condition continue, should we adjust management plans for coral-associated species?
3. What adjustments can be made or new techniques developed that will enhance our management approach to both corals and fisheries?
4. What more do we need to know to answer these questions?

The workshop served as an information baseline toward a better understanding of how corals and the consequences of coral condition influence fish populations, especially as these relate to the management of those populations. Workshop participants were asked to consider possible future states for coral reef fisheries, supply information on the specific linkages between corals and various fish species, and provide evidence that fish respond negatively to coral loss.

The presentations and abstracts from this workshop are freely available on the Gulf Council FTP site (http://www.gulfcouncil.org/about/ftp.php) in the "Coral/Workshop/Presentations" folder. Audio/video recordings are archived and sections can be made available upon request to the Gulf Council (info@gulfcouncil.org). This book is chiefly a compilation of content from presentations offered during the workshop but modified as a result of interactions and discussions with colleagues. Not all workshop presenters opted to prepare a chapter for inclusion here and many chapter authors chose to add additional materials not included in the workshop. Several papers were not specifically presented at the workshop but were developed in line with the central theme of this book.

All chapters were peer reviewed by national and international experts in corals and fisheries, but the content of the chapters remains the responsibility of the chapter authors. It is hoped that these chapters, written by selected world experts, will add to the collective knowledge on this subject and serve environmental scientists and resource managers alike.

Stephen A. Bortone
Tampa, Florida

Acknowledgments

This book was produced with funding assistance by the Gulf of Mexico Fishery Management Council under award number NA11NMF4410063 from the Coral Reef Conservation Program, National Oceanic and Atmospheric Administration, U.S. Department of Commerce. The statements, findings, conclusions, and recommendations are those of the editor and authors and do not necessarily reflect the views of the Gulf of Mexico Fishery Management Council, the National Oceanic and Atmospheric Administration, or the Department of Commerce. The staff at the Coral Reef Conservation Program were particularly helpful in providing guidance for the project. I especially thank Liz Fairey, Jennifer Koss, and Jen Schull.

Staff from the Gulf of Mexico Fishery Management Council were instrumental in every aspect of this project. I thank John Froeschke, Cathy May Readinger, Carrie M. Simmons, and especially Mark Mueller, who coordinated and managed the workshop.

Organizational direction for the theme of the workshop was provided by many individuals, but especially by the members of the Gulf of Mexico Fishery Management Council's Special Coral Scientific and Statistical Committee including Thomas Cuba, Richard Dodge, Walter C. Jaap, Judith Lang, Rob Ruzicka, George Schmahl, and Andrew N. Shepard. I also thank Graciela Garcia-Moliner (Caribbean Fishery Management Council) and Anna Martin (South Atlantic Fishery Management Council) for their valuable input to the organization of the workshop.

I thank the following reviewers for providing expert comments on the substance and scientific merit of individual chapters: Peter Auster, Michael Burton, Angela B. Collins, Nick Farmer, John Froeschke, Virginia Garrison, Walter C. Jaap, Ken Lindeman, Charles Messing, Margaret Miller, Henry Mullins, Dave Reed, Andrea Quattrini, Caroline Rogers, George Sedberry, and Carrie M. Simmons.

This book depended on the cooperation and diligence of the chapter authors. I thank them for their professional approach and material as well as the timeliness of their responses to my, at times, nagging but, hopefully, encouraging correspondence.

Lastly, I thank the workshop participants and attendees. Their collective comments, discussions, and ideas truly provided the foundation of this work.

Editor

Stephen A. Bortone recently retired as executive director of the Gulf of Mexico Fishery Management Council and now serves as a principal in an environmental consulting firm (Osprey Aquatic Sciences, Inc.) specializing in fisheries and other aquatic sciences. Formerly he was director of the Minnesota Sea Grant College Program located in Duluth, where he was professor of biology at the University of Minnesota Duluth. Also, he held positions as director of the marine laboratory at the Sanibel-Captiva Conservation Foundation in Sanibel, Florida, director of environmental science at the Conservancy of Southwest Florida, and director of the Institute for Coastal and Estuarine Research while professor of biology at the University of West Florida. Dr. Bortone received his BS degree (biology) from Albright College in Reading, PA, his MS degree (biological sciences) from Florida State University, Tallahassee, and his PhD (marine science) from the University of North Carolina at Chapel Hill in 1973.

For the past 45 years, he has conducted research on fisheries and the life history of aquatic organisms, especially fishes, chiefly in the southeastern United States and the Gulf of Mexico. He has published over 160 scientific articles on broad aspects of the aquatic sciences, but mostly on fisheries. In addition, he has edited four books on the aquatic sciences—seagrasses, biology of the spotted sea trout, artificial reefs in fisheries management and estuarine indicators—all with CRC Press.

Dr. Bortone has traveled widely while conducting his research and teaching activities. He has served as visiting scientist at the Johannes Gutenberg University (Mainz, Germany) and conducted extensive field surveys with colleagues from La Laguna University in the Canary Islands. He was Mary Ball Washington Scholar at University College Dublin, Ireland. He has received numerous teaching and research awards, including the titles "fellow" from the American Institute of Fishery Research Biologists, "certified fisheries professional" from the American Fisheries Society, and "certified senior ecologist" from the Ecological Society of America.

Contributors

William S. Arnold
Southeast Regional Office
National Marine Fisheries Service
St. Petersburg, Florida

Stephen A. Bortone
Osprey Aquatic Sciences, Inc.
Tampa, Florida

Sandra Brooke
Florida State University Coastal and Marine Laboratory
St. Teresa, Florida

Angela B. Collins
Fish and Wildlife Research Institute
Florida Fish and Wildlife Conservation Commission
St. Petersburg, Florida

Andrew David
Southeast Fisheries Science Center
National Marine Fisheries Service
Panama City, Florida

Stephanie Farrington
Harbor Branch Oceanographic Institute
Florida Atlantic University
Fort Pierce, Florida

John Froeschke
Gulf of Mexico Fishery Management Council
Tampa, Florida

Stacey Harter
Southeast Fisheries Science Center
National Marine Fisheries Service
Panama City, Florida

Mark A. Hixon
Department of Biology
University of Hawai'i
Honolulu, Hawaii

Thomas F. Hourigan
Deep Sea Coral Research and Technology Program
National Oceanographic and Atmospheric Agency
Silver Spring, Maryland

Walter C. Jaap
Lithophyte Research LLC
St. Petersburg, Florida

Pierre Kleiber
Western Pacific Fishery Management Council
Honolulu, Hawaii

Anders Knudby
Department of Geography
Simon Fraser University
Burnaby, Canada

Janet A. Ley
Fish and Wildlife Research Institute
Florida Fish and Wildlife Conservation Commission
St. Petersburg, Florida

Anna Martin
South Atlantic Fishery Management Council
North Charleston, South Carolina

Carole C. McIvor
Southeast Ecological Science Center
U.S. Geological Survey
St. Petersburg, Florida

Mark Mueller
Gulf of Mexico Fishery Management Council
Tampa, Florida

David Naar
College of Marine Science
University of South Florida
St Petersburg, Florida

Ivan Nagelkerken
Southern Seas Ecology Laboratory
The University of Adelaide
Adelaide, Australia

John C. Ogden
Department of Integrative Biology
University of South Florida
Tampa, Florida

Simon J. Pittman
National Centers for Coastal Ocean Science
National Oceanic and Atmospheric Administration
Silver Spring, Maryland

and

Marine Institute
Plymouth University
Plymouth, United Kingdom

John K. Reed
Harbor Branch Oceanographic Institute
Florida Atlantic University
Fort Pierce, Florida

Steve W. Ross
Center for Marine Science
University of North Carolina at Wilmington
Wilmington, North Carolina

Rob Ruzicka
Fish and Wildlife Research Institute
Florida Fish and Wildlife Conservation Commission
St. Petersburg, Florida

Marlowe Sabater
Western Pacific Fishery Management Council
Honolulu, Hawaii

Peter F. Sale
Institute for Water, Environment and Health
United Nations University
Hamilton, Canada

Paul W. Sammarco
Louisiana Universities Marine Consortium
Chauvin, Louisiana
and
Department of Oceanography and Coastal Sciences
Louisiana State University
Baton Rouge, Louisiana

Andrew N. Shepard
Gulf of Mexico University Research Collaborative
Florida Institute of Oceanography
St. Petersburg, Florida

Carrie M. Simmons
Gulf of Mexico Fishery Management Council
Tampa, Florida

chapter one

Introduction

The need to discern the relationships between corals and fisheries

Stephen A. Bortone

Contents

Overview

This chapter sets the tone and describes the reasons behind the effort that all authors have put into this book. I begin with an overview of the various stages of fisheries management practiced in the United States and provide a summary of the current state of management thought. The present status of world fisheries is also examined along with the general condition of, and trends in, coral health. Arguments are presented to explain why it is timely to examine the interrelationships between corals and fisheries. Lastly, there is an overview of the presentations offered herein, and the contribution that each makes toward elucidating important features of the interrelationships.

Status of fisheries

The world's wild capture fisheries attained a plateau after a long but steady increase in landings prior to the 1970s, with a more gradual increase from 1970 to 1990 (Figure 1.1). In recognition of this trend, and in anticipation of a leveling off and because of other factors, the U.S. Congress adopted the Fishery Conservation and Management Act in 1976. This act was subsequently modified and known as the Magnuson-Stevens Fishery Management Act, with its most recent iteration being in (NOAA 2007). The Magnuson-Stevens Act helped herald a new style of fisheries management that is slowly being adopted as a worldwide model to help attain and maintain fisheries at sustainable levels. Basically, the Magnuson-Stevens Act established national standards (a total of 10 in its most recent authorization) that gave direction to the National Marine Fisheries Service and the eight Regional Fishery Management Councils. These 10 standards are aimed at managing U.S. fisheries in a sustainable manner for the benefit of the nation.

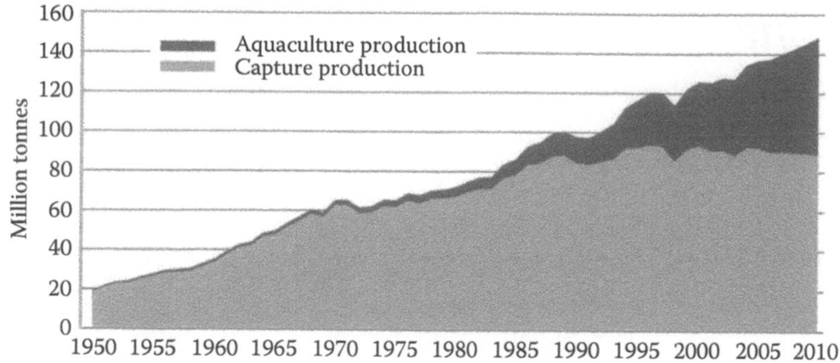

Figure 1.1 World capture fisheries and aquaculture production from 1950 to 2010. (From FAO (Food and Agriculture Organization). 2012. The state of world fisheries and aquaculture 2012. Rome: FAO of the United Nations.)

There have been at least three general stages or phases that can be used to characterize the main thinking with regard to managing fish in the United States. Prior to the act's implementation, the general attitude toward fishery management was one that incorporated the "boundless sea" concept. This presumed that the natural resources in the world's seas were limitless and little or no regulation was necessary other than natural "supply and demand" market forces. Examination of Figure 1.1 shows the folly of this attitude toward these resources as overcapitalization of fishing fleets and excessive demands for fish helped lead to the plateau effect witnessed in the trends of wild capture fisheries. Between the 1970s and 1990s and after adoption and implementation of the provisions of the Magnuson-Stevens Act, the second phase of fisheries ethics guided U.S. fisheries policy within the exclusive economic zone (EEZ) of U.S. federal waters, which extends approximately 200 nautical miles (370 km) out from U.S. shores. During this second phase, the management attitude toward sustainable fisheries was that management decisions would be made with consideration of an "environmental capacity" for each fishery. In other words, each stock had a determinable limit and management decisions should be based on attaining or maintaining that limit.

The result of this management strategy, although theoretically sound, was not particularly effective as many fisheries continued to decline or did not increase to former levels of abundance at rates sufficient for the benefit of the nation. The overall effectiveness of managing to an environmental capacity was less than hoped for because of several factors. Some of these were the inexactness of our understanding of ecological processes, a lack of appreciation of the importance of natural variation in these process, and the assumption that environmental capacity was constant. We know now that environmental capacity is ever-changing (perhaps even trending). Of course many other factors were at play, but recognition of the effect of these and other factors led to the adoption of a "precautionary approach" to fisheries management. Consequently, the reauthorized Magnuson-Stevens Act of 1996 incorporated several changes that helped establish this new precautionary approach. Among these changes was the establishment of a direction to end overfishing in all fisheries in U.S. federal waters by a fixed time (i.e., by 2011 but with some exceptions). Additionally, the new precautionary approach called for an increase in the role of science in the decision-making process to determine the acceptable biological catch, to improve recreational data collection, and to more fully integrate economics and social values into the decision-making process when reducing overfishing.

It was out of this shift in approach toward fisheries management, and the recognition that our environment may not have the presumed capacity that it once had, that a more specific appreciation for the ecosystem emerged. Some of this appreciation can be witnessed in the direction to the Regional Fishery Management Councils and the National Marine Fisheries Service to give consideration to ecosystem management. Previously, management was a species- (or stock-) specific endeavor. Annual catch limits were determined for each species under the management regime of the appropriate management agency with little consideration given to the effects of these management decisions on other species or the environment.

A coral perspective

Concomitant with recognition of the importance of the ecosystem in improving fishery management, the National Oceanic and Atmospheric Administration (NOAA) instituted an aggressive program to attain improved management of corals and coral habitat through its Coral Reef Conservation Program. This program "supports grants and cooperative agreements in the following categories: State and Territory Coral Reef Management; State and Territory Coral Reef Ecosystem Monitoring; Coral Reef Ecosystem Research; Projects to Improve or Amend Coral Reef Fishery Management Plans; Domestic Coral Reef Conservation; and International Coral Reef Conservation."

Some of the urgency for advancing the development of this effort on behalf of corals was due to incoming information that indicated corals were globally declining in extent and relative health (Figure 1.2). Many federal agencies began to take the ecosystem-based management approach, but in waters with coral-based communities, this undertaking largely fell under the auspices of the federal coral program in the United States. Clearly, managing corals *per se* for sustainability is an inexact science under our current abilities to manage ecosystems. This is even more apparent when one considers the enormous global pressures that corals are experiencing with regard to habitat loss, global climate change,

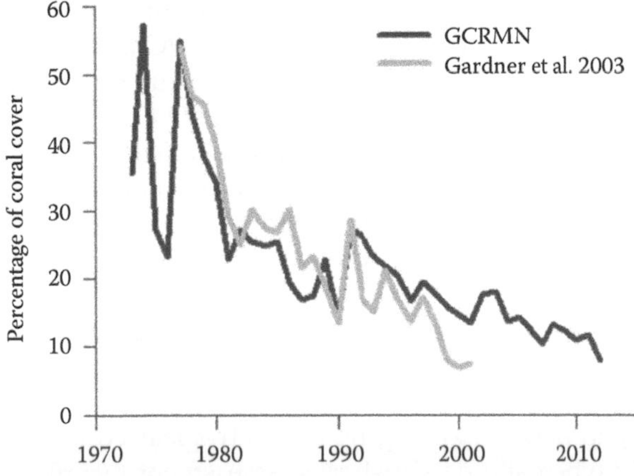

Figure 1.2 **(See color insert)** Decline in percentage of coral cover on Caribbean coral reefs from 1963 to the present based on data compiled by the IUCN (i.e., the International Union for the Conservation of Nature's Tropical Americas Coral Reef Resilience Workshop) with yearly averages weighted by the area surveyed per study and compared to Gardner et al. (2003) (yearly averages weighted by the inverse of a study's sample variance). GCRMN, Global Coral Reef Monitoring Network. Credit: © IUCN (International Union for the Conservation of Nature).

increases in coral disease, and physical pressures owing to increased contact with humans through recreation and development.

Coral and fisheries interrelationships

Herein lies the premise and impetus for the workshop mentioned in the Preface, the chapters that follow, and the idea behind the entire book. Today we witness fisheries that are below their projected capacity or fisheries that are unable to attain capacity at a sufficiently beneficial rate. We also know that corals are declining in their extent as a habitat. Moreover, our current state of knowledge gives us some insight into the importance of the ecosystem in helping sustain the communities of organisms of which it is comprised. Linking trends in coral reef habitat with the ability of this habitat to support the fisheries associated with it does not require a great increase in our ability to analyze the situation. It is intuitive that if corals continue to decline in the ecological services they provide, their associated fisheries will suffer.

The chapters that follow address issues associated with this premise through a variety of approaches.

Initially, an overall appreciation is offered of the importance of the interrelationship between corals and fisheries and the ecological services that both provide (Chapter 2, Sale and Hixon). The next two chapters help describe the current state of coral fisheries management from the standpoint of two of the Regional Fishery Management Councils (Chapter 3, Simmons et al.; Chapter 4, Martin). Following these perspectives, the extensive continental areas off the southeastern portion of the United States and the Gulf of Mexico are examined to demonstrate the current extent of our understanding of continental shelf areas, especially in deep water (Chapter 5, Reed et al.; Chapter 6, Jaap et al.; Chapter 7, Sammarco; Chapter 8, Hourigan). These presentations are significant in that they describe the current distribution and condition of corals and information we must acquire if we are to fully understand the interrelationships between corals and fisheries.

After these are two chapters that highlight the special situations that occur in some parts of the world, especially with regard to the significant roles that both corals and fisheries play relative to other ecosystem features. Shepard (Chapter 9) examines the current state of understanding regarding coral ecosystem exposure to an unexpected, large-scale perturbation such as an oil spill, while Ley (Chapter 10) explores the ecological services linkages between mangroves with both corals and fisheries.

Reef fishes associated with corals are special in several respects. Fish associated with these fisheries tend to be data poor; compared to most high seas, large-scale fisheries in the northern hemisphere (e.g., Atlantic cod and halibut species), many tropical fisheries are not dominated by one of a few species as the diversity of fish species in tropical waters is high. Additionally, the high species diversity, a feature of coral-associated fisheries, leads to the relative rarity of many species in landings. Also, corals occur in less developed parts of the world and consequently they have received less attention and directed research than fisheries off more developed nations. Managing fish in tropical, coral reef-associated fisheries has been problematic when standard fisheries assessment techniques are used. Sabater and Kleiber (Chapter 11) address aspects of this important issue and suggest an expansion in approaches used to determine stock status among data-poor fish species typically associated with coral reefs.

The availability of global information systems (GIS) has revolutionized the study of coral reefs. The accuracy afforded the scientific community, as well as the ability to more readily incorporate multiple factors into data analyses, has brought a higher level of data sophistication and utility to the forefront. Pittman and Knudby (Chapter 12) and Mueller et al. (Chapter 13) demonstrate the incorporation of GIS into the evaluation of coral reefs. Their work also indicates how future research questions can be addressed. Continuing with the theme of incorporating new information, Ogden et al. (Chapter 14) take a multilayered approach toward the future management of coral reef and reef fish fisheries by summarizing much of the current knowledge on the complexities of their interactions.

Lastly, Bortone (Chapter 15) offers some suggestions for research and management activities that will help direct efforts toward resolving questions regarding the future management of corals, their associated fisheries, and the ecosystem services they provide.

Onward

The chapter authors have offered their unique perspectives on issues related to understanding the relationship between corals and fisheries. The body of work offered here will help ensure the future successful management of both resources and their ecosystem components. While the examples offered are chiefly from the southeastern United States, it is understood that the features related to the comanagement of these valued ecosystem components are global with global consequences. The net effect of this collection of perspectives will help direct future efforts toward fully understanding and appreciating the difficulties that lie before the resource management community.

References

Gardner TA, Cote IM, Gill JA, Grant A, Watkinson AR. 2003. Long-term region-wide declines in Caribbean corals. Science. 301:958–960.

FAO (Food and Agriculture Organization). 2012. The state of world fisheries and aquaculture 2012. Rome: FAO of the United Nations.

NOAA (National Oceanic and Atmospheric Administration). 2007. Magnuson-Steven Fishery Conservation and Management Act Reauthorized. Available from: http://www.nmfs.noaa.gov/msa2007/index.html. Accessed 10 March, 2014.

chapter two

Addressing the global decline in coral reefs and forthcoming impacts on fishery yields

Peter F. Sale and Mark A. Hixon

Contents

An immediate problem for one fifth of humanity

LandScan data (Bright et al. 2012) reveal that, worldwide, 1.36 billion people now live within 100 km of a tropical coastline—that is, almost one fifth of the global population live on a narrow strip that is just 7% of the available land. These people occupy 9 megacities (10 million inhabitants or larger), but also live in numerous much smaller settlements. The average population density (145 people/km^2) of this strip is twice that of inland populations, with coastal populations growing more rapidly (Sale et al. in press). Most of these people live in developing countries and many depend on their immediate coastal waters for much of their animal protein food, other resources, and their livelihoods. Only the wealthy are sufficiently well connected to global markets that they do not depend directly on their local coastal waters. Coastal fisheries provide both food and livelihoods for millions of poor people along these tropical coasts (Donner and Potere 2007).

Coral reefs are one of the principal habitats of these coastal waters, and while there are many coastal fisheries over nearby seagrass beds, sand flats, and rocky hardbottom habitats, the fish taken are mostly ecologically associated closely with coral reefs (Ogden and Gladfelter 1983). Whether this close ecological connection means that fishery production is heavily dependent on the status of nearby coral reef habitat is not yet clear.

In developing countries, coastal fisheries are principally artisanal and small scale (Newton et al. 2007, Cochrane et al. 2011). Fish are caught from shore and from small boats using nets, traps, and spears, as well as by hand (some of these "fish" are shallow-water invertebrates taken from reef flats at low tide). Use of chemicals or dynamite to catch fish is generally illegal, but still widespread and chronically damaging (Fox and Caldwell 2006). Numerous families depend on these fisheries for the majority of the protein in their diets and major portions of the catch never enter a commercial market. Coastal fisheries are generally open and only weakly regulated, and evidence of overfishing is widespread (Friedlander and DeMartini 2002, Robbins et al. 2006, DeMartini et al. 2008, Sandin et al. 2008, Stallings 2009).

Coral reefs are currently experiencing severe anthropogenic impacts of several types, and worldwide, coral reef habitat is becoming seriously degraded (Pandolfi et al. 2003, 2005, Bellwood et al. 2004, Birkeland 2004, Wilson et al. 2006). Recent data from Australia (De'ath et al. 2012), the Caribbean (Perry et al. 2013), and elsewhere (Carpenter et al. 2008, Wilkinson 2008) show substantial reductions in coral cover and in coral growth rates, and increases in the rate of reef erosion over the last couple of decades. Over the last 27 years, 50% of cover on the Great Barrier Reef has been lost (De'ath et al. 2012); loss of cover across the Caribbean is even more severe (Jackson et al. 2013). The impacts responsible include overfishing, inappropriate coastal development, pollution with nutrients, heavy metals, and other chemicals, siltation, and coral mining, all of which are essentially local in their effects. Causes also include ocean warming, sea-level rise, increased intensity of storms, and more variable precipitation leading to more extreme flooding and siltation, all of which are due to anthropogenic climate change and cannot be mitigated by local effort (Maina et al. 2011). Finally, the release of CO_2 into the atmosphere, in addition to causing climate change, is resulting in increased rates of transfer of CO_2 to ocean waters, which in turn leads to the reduction in pH termed ocean acidification (Feely et al. 2004, Sabine et al. 2004, Doney et al. 2009). Each of these impacts has effects on corals and on coral reefs that lead to reef decline (Hoegh-Guldberg et al. 2007, Pandolfi et al. 2011, Meissner et al. 2012), with likely deleterious effects on artisanal reef fisheries and the social systems they support (Munday et al. 2008, Cinner et al. 2012, 2013, Sale et al. in press).

Decline of coral reefs

Coral reefs are now declining at such a rate, and on such a geographical scale, that it is no longer possible to dismiss claims the loss will be total by 2050 (Hoegh-Guldberg et al. 2007, Sale 2008, 2011). Coral reefs, as we knew them during the 1960s, will no longer exist in 2050 unless we institute far more effective management of locally acting threats. Simultaneously, we should take real steps to reduce the global emissions of CO_2 and other greenhouse gases, because if we do not, even the best-managed coral reefs (with respect to local threats) may very well fail to survive the warming and acidification. Given that it has now been 16 years since the global epidemic of mass coral bleaching in 1997–1998, and our emissions of CO_2 continue to rise steeply while local management of coral reefs remains largely ineffective, optimism about our ability to stem the decline of coral reefs is very difficult to muster.

There are several reasons for the decline of coral reefs (Sale 2008, 2011), all related to the dependence of reef ecosystems on the survival and health of a group of ecologically delicate species, the corals themselves. The rocky structure of coral reefs with its complex topography is the result of calcification by a broad range of organisms, especially the corals (Kinsey and Davies 1979, Barnes and Chalker 1990). Erosional forces, some also biogenic, modify the accreted structure, and prevailing patterns of water flow further shape the resulting reef, so that a stable coral reef is a dynamic equilibrium between processes of growth in which new rock is created as calcified skeletons, and processes of erosion in which the rocky structure is broken down by a variety of bioeroders, by storms, by changes in sea level, and by the routine effects of wave action and currents.

Bioeroders range from minute sponges and worms to 1 m long parrotfishes, and their rasping, burrowing, and dissolving activities gradually reduce consolidated limestone rock into sand (Glynn 1997). A single, 1 m long, green humphead parrotfish (*Bolbometopon muricatum*) can remove 5 tons of structural reef carbonates per year, almost half of it living coral (Bellwood et al. 2003). Cyclonic storms break up living coral, shift rubble and sediments, scour and abrade corals and other sessile species, and generally wear down coral reefs, while also throwing rock and sediment up onto reef flats to add to or form new islands of up to 3–4 m elevation above high tide levels. Changes in sea level, whether due to global shifts, subsidence, or elevation of underlying rock, can lift a reef above sea level resulting in rapid death, lower it modestly allowing it to flourish, or take it deeper at rates too quick for compensation by reef-building processes, causing it to be drowned. Normal wave action and currents have continual low-level erosional impacts on a reef (Glynn 1997).

The modern corals (order Scleractinia) have been present since the early Mesozoic period and are likely derived from earlier rugose corals dominant in the Paleozoic era (Veron et al. 1996). Most Scleractinia contain symbiotic single-celled dinoflagellates (zooxanthellae) within their tissues that greatly increase the capacity of the coral to calcify. Zooxanthellate Scleractinia necessarily are restricted to shallow water by the photosynthetic requirements of the algae, but some azooxanthellate forms occur in deep water where they build slow-growing reefs. We know a lot less about the ecology of deepwater reefs; this chapter deals with the shallow-water reefs built by tropical zooxanthellate corals, while recognizing that fishery species may also depend on deepwater reefs.

Zooxanthellate corals are limited to the tropics by temperature, and to clear, oceanic, low-nutrient waters. Further, because they are restricted to shallow waters, they typically occur relatively close to coastlines—coastlines that are increasingly occupied by human settlements. The present-day global decline in coral reefs began at different times in different locations as humans interacted with reefs (Bell et al. 2006). Inappropriate coastal development has caused increased siltation and pollution, changes to patterns of water flow, and physical destruction of inshore reefs. Agricultural and industrial activities well inland have contributed to coastal pollution and nutrient enrichment of coastal waters. In some places, corals are mined for building materials, lime, and the curio trade. Overfishing, or fishing using inappropriate methods, has led to some direct loss of corals and substantial trophic shifts in reef ecosystems (Hughes et al. 2007). The latter are most damaging when important grazers and bioeroders such as the larger parrotfishes (Scaridae) are harvested. Paradoxically, bioerosion by parrotfishes, which directly destroys reef structure, also creates bare rocky surfaces that favor coral settlement and growth (Bellwood et al. 2003). Removal of herbivores more generally reduces controls on algal growth and can result in a phase shift from a coral-dominated to an algae-dominated system (Hughes 1994, Mumby 2006, 2009, Hughes et al. 2007, Ledlie et al. 2007). The latter tend to be less diverse, and less productive of fishery species, although primary production may be as high as, or higher than, that of a coral-dominated system.

Given the generally weak legal structures governing activities in most coastal waters, this suite of human impacts is present to varying degrees along most tropical coasts. As human populations have grown, these impacts have intensified, contributing to a general downward spiral in reef condition. Technical solutions are available to remedy all of these local impacts on coral reefs, but, as is often the case, societal, cultural, and political factors impede their application (Bell et al. 2006, Lotze et al. 2006). How does a fishery manager reduce fishing effort, or prohibit taking of certain species, in an artisanal fishery in which a major portion of the catch never enters an open market and in a community where there are few other sources of employment? How does a government control coastal development and the release of nutrients from upstream agricultural districts, when the human population is growing, poverty is rife, and efforts to feed and house people deplete all available resources?

In recent decades, the global effects of greenhouse gases have added to the burden on coral reefs. Rising sea surface temperatures have meant that periods of unusually warm water have become more prevalent, and since corals live close to their thermal maxima, these warm episodes have led to mass bleaching (Baker et al. 2008). When conditions causing the bleaching last for three or more weeks, extensive coral mortality can result. Ocean acidification appears to have already progressed far enough that it is measurably slowing the rates of growth of at least some coral species, thereby slowing the regenerative capacity of reefs (Pandolfi et al. 2011, Meissner et al. 2012). With increased coral mortality and slowed regeneration, any past equilibrium has shifted toward reef decline (Hoegh-Guldberg et al. 2007).

De'ath et al. (2009) recently reported the results of analysis of 328 cores from colonies of *Porites* spp. collected from 69 sites across the Great Barrier Reef. These cores showed a 14.2% decrease in rate of calcification and a 13.3% slowdown in growth (extension) rate since 1990. Both the extent and the abruptness of the changes are greater than any over the past 400 years. The authors attributed this regionwide decline to a combination of growing thermal stress and decreasing ocean pH—the capacity of corals of this genus to calcify was being measurably reduced. Tanzil et al. (2009) showed very similar results in a smaller study for corals at Phuket, Thailand. Subsequent work elsewhere has confirmed these trends while revealing important interspecific variation (Manzello 2010, Friedrich et al. 2012), and it remains unclear whether the data for individual genera of coral scale-up to a measurable decline in rate of reef accretion. In any case, the percentage cover of live coral is declining rapidly in many regions (Carpenter et al. 2008, Wilkinson 2008, De'ath et al. 2012, Jackson et al. 2013, Perry et al. 2013).

Effects of coral declines on reef fishes

Fishes are a highly diverse and vital component of the coral reef ecosystem. Some 8000 species of fish are associated with coral reefs (Bellwood et al. 2012), and reef fishes have a broad range of life histories, body sizes, and trophic roles, ranging from tiny prey to large apex predators. Energy flow upward through trophic levels including fishes ultimately supports reef fisheries that provide a major source of protein for associated human communities (Polunin and Roberts 1996). Consequently, it is highly likely that the decline of coral reefs will lead to loss of fishery productivity in tropical coastal waters (Bellwood et al. 2004, Cinner et al. 2012).

Living corals provide both food and living space for fishes, as well as settlement cues for incoming larvae (Kingsford et al. 2002, Gerlach et al. 2007, Munday et al. 2010). The food provided by corals includes not only the coral polyps themselves (supplying food for corallivores), but also a broad variety of prey that inhabit corals (supplying food for herbivores, invertivores, piscivores, and omnivores). The living space provided by the coral structure includes actual home sites (settlement, nursery, or permanent) as well as spawning/nesting sites, cleaning stations, refuges from strong currents, and especially, spatial refuges from predation (Graham and Nash 2013, Hixon 2014).

Recent major empirical reviews have demonstrated that reef fishes are closely associated with living corals. Coker et al. (2014) reported 93 species of coral inhabited by fishes, mainly branching species in the genera *Acropora* and *Porites*. They listed 39 fish families that inhabit live coral habitats, the species richness of which is dominated by smaller forms, including damselfishes (Pomacentridae), gobies (Gobiidae), wrasses (Labridae), cardinalfishes (Apogonidae), and butterflyfishes (Chaetodontidae). Many species inhabit corals only as juveniles (Coker et al. 2014), with a trend for larger fish to be less associated with coral, likely as individuals shift from spatial prey refuges to size refuges from predation (Alvarez-Filip et al. 2011). Consequently, there is often a strong positive relationship between reef structural complexity and fish density, biomass, and species richness (Gratwicke and Speight 2005, Graham and Nash 2013), as well as between coral species richness and fish species richness (Belmaker et al. 2008, Belmaker 2009, Messmer et al. 2011), and more complex reefs support more trophic levels (Alvarez-Filip et al. 2011). A recent modeling study using an Ecopath with Ecosim model parameterized for a rich coral reef system in Indonesia predicts a direct link between loss of coral biomass and loss of the capacity to produce fishery species, such that a loss of 50% of coral would reduce fishery production capacity by 30% (Sale et al. in press).

It is not surprising, therefore, that fish abundance and perhaps species richness decline substantially as coral reefs degrade, a pattern increasingly well documented in a variety of regions, including the Pacific (Sano 2000, Booth and Beretta 2002, Jones et al. 2004, Feary et al. 2007, Holbrook et al. 2008, Wilson et al. 2010a,b), the Atlantic/Caribbean (Acosta-Gonzalez et al. 2013), the Indian Ocean (Garpe et al. 2006, Graham et al. 2006), and globally (Wilson et al. 2006, Pratchett et al. 2008, 2011). These losses may undergo time lags: dead yet structurally intact corals may initially support most fishes except live-coral specialists, but then gradually erode into rubble inhabited by relatively few species (Garpe et al. 2006, Pratchett et al. 2008). Time lags may also occur as recruitment fails, reducing replenishment of local populations (Graham et al. 2007, McCormick et al. 2010). In some cases, there may be substantial shifts in species composition and/or evenness of reef fishes despite few changes in overall fish abundance and species richness (Lindahl et al. 2001, Bellwood et al. 2006, Cheal et al. 2008). In any case, the threat of extirpation and even extinction now looms large for coral reef fishes, especially obligate corallivores (Graham et al. 2011). This threat will accelerate through a variety of mechanisms as the oceans continue to warm and acidify (Munday et al. 2008).

Delaying coral reef decline

As noted earlier, the locally acting stressors that help degrade coral reefs are all amenable to local intervention, and the technologies for intervening successfully are largely well understood (Bell et al. 2006, Lotze et al. 2006). In most cases, the primary need is simply enhanced awareness of potential problems before they arise. For example, it is relatively easy to plan coastal improvements with due recognition of existing patterns of freshwater runoff and awareness that disrupting these patterns can kill nearby reefs. Similarly, if patterns of water flow are known, it can be obvious that sediment or pollutants released into the water at particular locations will be deposited on downstream reefs, smothering or poisoning them. While straightforward conceptually, the successful application of environmental insights during major coastal development projects depends on a rigorous environmental impact assessment (EIA) process backed up by regulatory structures that are strong enough to encourage compliance instead of violation followed by the willing payment of assessed penalties. Rigorous EIA procedures and a culture of compliance with environmental regulations are not hallmarks of developing countries, and are often bypassed in the more advanced countries when political will is weak, profit margins consequent on development are high, or the value of coastal marine ecosystems is poorly appreciated by stakeholders (Wanless and Maier 2007, Lindeman et al. 2010).

Addressing chronic, and growing, coastal pollution due to agricultural and industrial activities far inland can be more difficult. The jurisdiction of agencies responsible for the management of coastal areas seldom extends inland, and the economic value of the upland activities may be such that a collaborative approach to encourage agriculture and industry to act in the interests of a distant coastal ocean will achieve very little change in behavior. Stable political systems, and an electorate well informed on the economic value of sustainably managed coastal ecosystems, are essential if any remission of pollution is to be gained. Our difficulty in the developed world in restoring coastal dead zones, such as those at the mouth of the Mississippi River or in Chesapeake Bay, shows how difficult it is to apply common sense to the management of nutrients and pollutants on a watershed-wide scale (Diaz and Rosenberg 2008, Doney 2010).

In developing countries, the chronic overfishing of reefs, including frequent use of inappropriate methods, is a particularly vexing stressor on coral reefs (Cinner et al. 2009). These problems are due to the fact that large populations of poorly educated people adopt fishing as the employment of last resort, and because coastal fisheries, to the extent they are managed at all, are managed as open-access fisheries. The obvious solution would be a reduction in fishing effort. However, sustained social, economic, and political effort over many years would first be needed to create other employment opportunities, to encourage fishers to stop fishing, and to enforce a no-take status in networks of marine protected areas. Such sustained effort is difficult to achieve when most fishery management agencies are poorly resourced by governments that have often given up on the idea that creation of alternative employment is even possible (Bell et al. 2006, Sale et al. in press).

The development of open aquaculture in pens and on subtidal leases could generate alternative employment while subsidizing food production from traditional capture fisheries (Naylor and Burke 2005). However, too often, inadequately managed aquaculture development leads directly to new coastal pollution, degradation of coastal ecosystems such as mangroves, which are important as nursery habitat for capture fishery species, and acrimonious conflict among stakeholder groups—not a way of solving stresses due to overfishing (Goldburg and Naylor 2005).

Overriding all such efforts to mitigate local stressors is the near universal tendency for environmental (including fishery) management to be fragmented among agencies and between tiers of government, with few agencies able to build a cadre of technically well-informed staff, or a holistic view of the problems and solutions. While simple technical fixes are available for most locally acting stressors, culture, tradition, politics, and failure of leadership ensure that the solutions are seldom implemented or sustained (Bell et al. 2011, Sale et al. in press).

Globally acting stressors are becoming progressively more intense. This is due both to the environmental consequences of greenhouse gas emissions, and to the increasing pressure on coastal fisheries caused by growing worldwide demand for fishery products as communities grow and become more affluent (Halpern et al. 2008, Mora et al. 2013). These stressors cannot be addressed by local interventions alone. It might be possible to reduce demand and thereby reduce fishing effort in a community by closing opportunities for export trade, but such barriers are usually pierced by enterprising individuals willing to violate the law. However, the effects of globally acting stressors can be reduced in some cases by addressing local impacts effectively. For example, it has been shown both on the Great Barrier Reef (Wooldridge 2009) and in the Florida Keys (Wagner et al. 2010) that the extent of coral bleaching during major warming events was less severe where waters contained lower nutrient levels. By reducing the strength of locally acting stressors, it should thus become possible to "buy time" for coral reefs, enabling them to better withstand stresses such as warming. Ultimately, however, unless we act globally to reduce greenhouse gas emissions, coral reefs as we know them are going to largely disappear (Pandolfi et al. 2003, Carpenter et al. 2008).

Adaptation to coral reef loss

The decline and ultimate loss of coral reefs will not bring an end to coastal fisheries, but catches are likely to change in composition, and the capacity of reefs to produce fishery species also seems likely to fall. Indeed, the abundance of reef fish is already clearly declining, severely in some cases (Jones et al. 2004, Paddack et al. 2009, Stallings 2009). While humans will probably find ways to adapt as the situation changes, just as we have been adapting to changing yields as coral reefs are degraded, the substantial decline of fishery production capacity on a global scale will become a serious issue for food security in future decades (Garcia and Rosenberg 2010). Food insecurity, driven by growing human populations, expanding affluence, and worsening impacts of climate change on food production, is already increasing and will continue to do so. Experts differ in the degree of optimism they display. In a recent review, Wheeler and von Braun (2013) report that the positive effects on plant growth due to increased CO_2 in the atmosphere are now known to be less than had been anticipated, and that food production is going to fall most severely across the tropics as climate change advances. They quote Knox et al. (2012), who stated that yields across Africa are predicted by 2050 to change by –17% (wheat), –5% (maize), –15% (sorghum), and –10% (millet), and across South Asia by –16% (maize) and –11% (sorghum). While their emphasis is on agricultural crops, Wheeler and von Braun conclude that committed climate change requires that we begin to adapt to global food insecurity in the next 20–30 years.

Under these circumstances, any actions that can improve the likelihood of tropical coastal fisheries remaining productive should be pursued aggressively. Obviously, a far higher priority should be being given to maintaining the sustainability of present-day coastal fisheries, using recognized technical solutions: strong prohibitions on the use of inappropriate fishing methods such as dynamite and chemicals, provision of alternative livelihoods and reduction of fishing effort, strengthening and effective enforcement of regulations governing coastal pollution and habitat destruction, and the creation of networks of marine protected areas to reduce fishing effort while enhancing fishery and reef conservation. Beyond these actions, there should be continued research on the cues used by reef fishes to return to reef habitat at the end of larval life, and consideration of methods for enhancing the availability of such cues, or of preserving those cues as reefs degrade. With the right cues in place, it may become possible to enhance settlement of fishes to a site where they can be protected during the first critical weeks or months (see Almany and Webster 2006), thereby enhancing ultimate yields.

With further reef decline, the loss of living reefs could be mitigated by providing artificial reefs, built specifically to provide the habitat attributes needed by fish. How best to do this is not yet clear, but reefs should be carefully designed not simply thrown together using whatever surplus materials or vehicles happen to be at hand (Bohnsack and Sutherland 1985). It is sometimes asserted that reef fishes are drawn only to the three-dimensional physical structure of reefs, so that only physical structure is important in artificial reef design. In reality, as shown in the section on Effects of coral declines on reef fishes, the fishes that inhabit coral reefs are dependent on far more than structural shelter; the high productivities of coral reef ecosystems provide the food source for reef fishes, as well as the remainder of the entire reef food web of which fishes are one part. While relatively few fishes directly consume corals, the living reef provides the broad variety of algal and invertebrate prey that supports the productivity and diversity of reef fishes. Therefore, artificial reefs intended to mitigate coral reef loss must be designed with the benefit of detailed comparisons with living reefs (Carr and Hixon 1997), and especially with the goal of increasing local production rather than merely aggregating fish from surrounding habitats (Grossman et al. 1997).

What is called for, in fact, is the kind of ecosystem-based management of fisheries that has long been promoted, but taken to a new level where those facilities or services that degraded coral reefs can no longer deliver are provided in order to facilitate the continued production of fishery species (McClanahan et al. 2011, Aswani et al. 2012, Sale et al. in press). The good news in an otherwise bleak future is that these possible steps, available if we decide not to just "muddle through," do not require global agreement: a local community can unilaterally decide to improve its fisheries, to sustain its reefs, or to replace lost reefs with artificial structures. The possibility of effective local action needs to be promulgated widely, in the hope that some local communities will show the wisdom that most of us, at least to now, appear to have lacked (McClanahan et al. 2008). To put this last point more bluntly, what is needed is a dramatic improvement in management performance, addressed to all locally acting stressors with the aim of sustaining or improving fishery yield; at present, most nations show little inclination to embark on this journey.

An immediate effort to improve all aspects of the management of tropical coastal waters, including management of fisheries, could be a major boost to tropical coastal ecosystems. This approach would best equip us to cope with reef degradation due to greenhouse gas emissions while we make a major global effort to reduce those emissions. Billions of people depending on tropical coastal seas will be severely affected if the largely piecemeal, unsustained, and otherwise ineffective management of coastal waters is allowed to continue.

References

Acosta-Gonzales G, Rodriguez-Zaragoza FA, Hernandez-Landa RC, Arias-Gonzalez JE. 2013. Additive diversity partitioning of fish in a Caribbean coral reef undergoing shift transition. PLoS ONE. 8(6):e65665.

Almany GR, Webster MS. 2006. The predation gauntlet: Early post-settlement mortality in coral reef fishes. Coral Reefs. 25:19–22.

Alvarez-Filip L, Gill JA, Dulvy NK. 2011. Complex reef architecture supports more small-bodied fishes and longer food chains on Caribbean reefs. Ecosphere. 2(10):118. Available from: http://dx.doi.org/10.1890/ES11-00185.1 via the Internet. Accessed 1 February, 2014.

Aswani S, Christie P, Muthiga NA, Mahon R, Primavera JH, Cramer LA, Barbier EB, Granek EF, Kennedy CJ, Wolanski E, et al. 2012. The way forward with ecosystem-based management in tropical contexts: Reconciling with existing management systems. Mar Policy. 36:1–10.

Baker AC, Glynn PW, Riegl B. 2008. Climate change and coral reef bleaching: An ecological assessment of long-term impacts, recovery trends and future outlook. Estuar Coast Shelf Sci. 80:435–471.

Barnes DJ, Chalker BE. 1990. Calcification and photosynthesis in reef-building corals and algae. *In*: Dubinsky Z, editor. Coral reefs. Ecosystems of the world. Volume 25. Amsterdam: Elsevier. p. 109–131.

Bell JD, Johnson JE, Hobday AJ. 2011. Vulnerability of tropical Pacific fisheries and aquaculture to climate change. Noumea: Secretariat of the Pacific Community.

Bell JD, Ratner BD, Stobutzki I, Oliver J. 2006. Addressing the coral reef crisis in developing countries. Ocean Coast Manage. 49:976–985.

Bellwood DR, Hoey AS, Choat JH. 2003. Limited functional redundancy in high diversity systems: Resilience and ecosystem function on coral reefs. Ecol Lett. 6:281–285.

Bellwood DR, Hughes TP, Folke C, Nyström M. 2004. Confronting the coral reef crisis. Nature. 429:827–833.

Bellwood DR, Hughes TP, Hoey AS. 2006. Sleeping functional group drives coral-reef recovery. Curr Biol. 16:2434–2439.

Bellwood DR, Renema W, Rosen BR. 2012. Biodiversity hotspots, evolution and coral reef biogeography: A review. *In*: Gower D, Johnson KG, Richardson J, Rosen BR, Williams ST, Rüber L, editors. Biotic evolution and environmental change in southeast Asia. London: Cambridge University Press. p. 216–245.

Belmaker J. 2009. Species richness of resident and transient coral-dwelling fish responds differentially to regional diversity. Glob Ecol Biogeogr. 18:426–436.

Belmaker J, Ziv Y, Shashar N, Connolly SR. 2008. Regional variation in the hierarchical partitioning of diversity in coral-dwelling fishes. Ecology. 89:2829–2840.

Birkeland C. 2004. Ratcheting down the coral reefs. BioScience. 54:1021–1027.

Bohnsack JA, Sutherland DL. 1985. Artificial reef research: A review with recommendations for future priorities. Bull Mar Sci. 37:11–39.

Booth DJ, Beretta GA. 2002. Changes in a fish assemblage after a coral bleaching event. Mar Ecol Prog Ser. 245:205–212.

Bright EA, Coleman PR, Rose AN, Urban ML; LandScan 2011 [Internet]. Oak Ridge, TN: Oak Ridge National Laboratory; 2012. Available from: http://www.ornl.gov/landscan/ via the Internet. Accessed 1 February, 2014.

Carpenter KE, Abrar M, Aeby G, Aronson RB, Banks S, Bruckner A, Chiriboga A, Cortés J, Delbeek JC, DeVantier L, et al. 2008. One-third of reef-building corals face elevated extinction risk from climate change and local impacts. Science. 321:560–563.

Carr MH, Hixon MA. 1997. Artificial reefs: The importance of comparisons with natural reefs. Fisheries. 22(4):28–33.

Cheal AJ, Wilson SK, Emslie MJ, Dolman AM, Sweatman H. 2008. Responses of reef fish communities to coral declines on the Great Barrier Reef. Mar Ecol Prog Ser. 372:211–223.

Cinner JE, Huchery C, Darling ES, Humphries AT, Graham NAJ, Hicks CC, Marshall N, McClanahan TR. 2013. Evaluating social and ecological vulnerability of coral reef fisheries to climate change. PLoS ONE. 8(9):e74321. Available from: http://dx.doi.org/10.1371/journal.pone.0074321 via the Internet. Accessed 1 February, 2014.

Cinner JE, McClanahan TR, Daw TM, Graham NAJ, Maina J, Wilson SK, Hughes TP. 2009. Linking social and ecological systems to sustain coral reef fisheries. Curr Biol. 19:206–212.

Cinner JE, McClanahan TR, Graham NAJ, Daw TM, Maina J, Stead SM, Wamukota A, Brown K, Bodin O. 2012. Vulnerability of coastal communities to key impacts of climate change on coral reef fisheries. Glob Environ Chang. 22:12–20.

Cochrane KL, Andrew NL, Parma AM. 2011. Primary fisheries management: A minimum requirement for provision of sustainable human benefits in small-scale fisheries. Fish Fish. 12:275–288.

Coker DJ, Wilson SK, Pratchett MS. 2014. Importance of live coral habitat for reef fishes. Rev Fish Biol Fisher. 24:89–126.

De'ath G, Fabricius KE, Sweatman H, Puotinenb M. 2012. The 27-year decline of coral cover on the Great Barrier Reef and its causes. Proc Natl Acad Sci U S A. 109:17995–17999.

De'ath G, Lough JM, Fabricius KE. 2009. Declining coral calcification on the Great Barrier Reef. Science. 323:116–119.

DeMartini EE, Friedlander AM, Sandin SA, Sala E. 2008. Differences in fish-assemblage structure between fished and unfished atolls in the northern Line Islands, central Pacific. Mar Ecol Prog Ser. 365:199–215.

Diaz RJ, Rosenberg R. 2008. Spreading dead zones and consequences for marine ecosystems. Science. 321:926–929.

Doney SC. 2010. The growing human footprint on coastal and open-ocean biogeochemistry. Science. 328:1512–1516.

Doney SC, Fabry VJ, Feely RA, Kleypas KA. 2009. Ocean acidification: The other CO_2 problem. Ann Rev Mar Sci. 1:169–192.

Donner SD, Potere D. 2007. The inequity of the global threat to coral reefs. BioScience. 57:214–215.

Feary DA, Almany GR, Jones GP, McCormick MI. 2007. Coral degradation and the structure of tropical reef fish communities. Mar Ecol Prog Ser. 333:243–248.

Feely RA, Sabine CL, Lee K, Berelson W, Kleypas J, Fabry VJ, Millero FJ. 2004. Impact of anthropogenic CO_2 on the $CaCO_3$ system in the oceans. Science. 305:362–366.

Fox HE, Caldwell RL. 2006. Recovery from blast fishing on coral reefs: A tale of two scales. Ecol Appl. 16:1631–1635.

Friedlander AM, DeMartini EE. 2002. Contrasts in density, size, and biomass of reef fishes between the northwestern and the main Hawaiian Islands: The effects of fishing down apex predators. Mar Ecol Prog Ser. 230:253–264.

Friedrich T, Timmerman A, Abe-Ouchi A, Bates NR, Chikamoto MO, Church MJ, Dore JE, Gledhill DK, González-Dávila M, Heinemann M, et al. 2012. Detecting regional anthropogenic trends in ocean acidification against natural variability. Nat Clim Chang. 2:167–171.

Garcia SM, Rosenberg AA. 2010. Food security and marine capture fisheries: Characteristics, trends, drivers and future perspectives. Philos Trans Roy Soc B Biol Sci. 365:2869–2880.

Garpe KC, Yahya SAS, Lindahl U, Öhman MC. 2006. Long-term effects of the 1998 coral bleaching event on reef fish assemblages. Mar Ecol Prog Ser. 315:237–247.

Gerlach G, Atema J, Kingsford MJ, Black KP, Miller-Sims V. 2007. Smelling home can prevent dispersal of reef fish larvae. Proc Natl Acad Sci U S A. 104:858–863.

Glynn PW. 1997. Bioerosion and coral reef growth: A dynamic balance. *In*: Birkeland C, editor. Life and death of coral reefs. New York: Chapman and Hall. p. 69–98.

Goldburg R, Naylor R. 2005. Future seascapes, fishing, and fish farming. Front Ecol Environ. 3:21–28.

Graham NAJ, Chabanet P, Evans RD, Jennings S, Letourneur Y, MacNeil MA, McClanahan TR, Őhman MC, Polunin NVC, Wilson SK, et al. 2011. Extinction vulnerability of coral reef fishes. Ecol Lett. 14:341–348.

Graham NAJ, Nash KL. 2013. The importance of structural complexity in coral reef ecosystems. Coral Reefs. 32:315–326.

Graham NAJ, Wilson SK, Jennings S, Polunin NVC, Bijoux JP, Robinson J. 2006. Dynamic fragility of oceanic coral reef ecosystems. Proc Natl Acad Sci U S A. 103:8425–8429.

Graham NAJ, Wilson SK, Jennings S, Polunin NVC, Robinson J, Bijoux JP, Daw TM. 2007. Lag effects in the impacts of mass coral bleaching on coral reef fish, fisheries, and ecosystems. Conserv Biol. 21:1291–1300.

Gratwicke B, Speight MR. 2005. The relationship between fish species richness, abundance and habitat complexity in a range of shallow tropical marine habitats. J Fish Biol. 66:650–667.

Grossman GD, Jones GP, Seaman WJ. 1997. Do artificial reefs increase regional fish production? A review of existing data. Fisheries. 22(4):17–23.

Halpern BS, Walbridge S, Selkoe KA, Kappel CV, Micheli F, D'Agrosa C, Bruno JF, Casey KS, Ebert C, Fox HE, et al. 2008. A global map of human impact on marine ecosystems. Science. 319:948–952.

Hixon MA. 2014. Predation: Piscivory and the ecology of coral-reef fishes. *In*: Mora C, editor. Ecology and conservation of fishes on coral reefs: The functioning of an ecosystem in a changing world. Cambridge, UK: Cambridge University Press.

Hoegh-Guldberg O, Mumby PJ, Hooten AJ, Steneck RS, Greenfield P, Gomez E, Harvell CD, Sale PF, Edwards AJ, Caldeira K, et al. 2007. Coral reefs under rapid climate change and ocean acidification. Science. 318:1737–1742.

Holbrook SJ, Schmitt RJ, Brooks AJ. 2008. Resistance and resilience of a coral reef fish community to changes in coral cover. Mar Ecol Prog Ser. 371:263–271.

Hughes TP. 1994. Catastrophes, phase-shifts, and large-scale degradation of a Caribbean coral reef. Science. 265:1547–1551.

Hughes TP, Rodrigues MJ, Bellwood DR, Ceccarelli D, Hoegh-Guldberg O, McCook L, Moltschaniwskyj N, Pratchett MS, Steneck RS, Willis B, et al. 2007. Phase shifts, herbivory, and the resilience of coral reefs to climate change. Curr Biol. 17:360–365.

Jackson J, Donovan M, Cramer K, Lam V, editors. 2013. Status and trends of Caribbean coral reefs: 1970–2012. Washington, DC: Global Coral Reef Monitoring Network, c/o International Union for the Conservation of Nature, Global Marine and Polar Program.

Jones GP, McCormick MI, Srinivasan M, Eagle JV. 2004. Coral decline threatens fish biodiversity in marine reserves. Proc Natl Acad Sci U S A. 101:8251–8253.

Kingsford MJ, Leis JM, Shanks A, Lindeman KC, Morgan SG, Pineda J. 2002. Sensory environments, larval abilities and local self-recruitment. Bull Mar Sci. 70(1):309–340.

Kinsey DW, Davies PJ. 1979. Carbon turnover, calcification and growth in coral reefs. *In*: Trudinger PA, Swaine DJ, editors. Biogeochemical cycling of mineral-forming elements. Studies in Environmental Science 3. Amsterdam: Elsevier. p. 131–162.

Knox J, Hess T, Daccache A, Wheeler T. 2012. Climate change impacts on crop productivity in Africa and South Asia. Environ Res Lett. 7(3):034032. Available from: http://dx.doi.org/10.1088/1748-9326/7/3/034032 via the Internet. Accessed 1 February, 2014.

Ledlie MH, Graham NAJ, Bythell JC, Wilson SK, Jennings S, Polunin NVC, Hardcastle J. 2007. Phase shifts and the role of herbivory in the resilience of coral reefs. Coral Reefs. 26:641–653.

Lindahl U, Ohman MC, Schelten CK. 2001. The 1997/1998 mass mortality of corals: Effects on fish communities on a Tanzanian coral reef. Mar Pollut Bull. 42:127–131.

Lindeman KC, Gibson HT, Yu H. 2010. Participatory climate adaptation in coastal Florida: Increasing roles for water-users and independent science. Proc Gulf Caribb Fish Inst. 62:7–11.

Lotze HK, Lenihan HS, Bourque BJ, Bradbury RH, Cooke RG, Kay MC, Kidwell SM, Kirby MX, Peterson CH, Jackson JBC, et al. 2006. Depletion, degradation, and recovery potential of estuaries and coastal seas. Science. 312:1806–1809.

Maina J, McClanahan TR, Venus V, Ateweberhan M, Madin J. 2011. Global gradients of coral exposure to environmental stresses and implications for local management. PLoS ONE. 6(8):e23064. Available from: http://dx.doi.org/10.1371/journal.pone.0023064 via the Internet. Accessed 1 February, 2014.

Manzello DP. 2010. Coral growth with thermal stress and ocean acidification: Lessons from the eastern tropical Pacific. Coral Reefs. 29:749–758.

McClanahan TR, Graham NAJ, MacNeil MA, Muthiga NA, Cinner JE, Bruggemann JH, Wilson SK. 2011. Critical thresholds and tangible targets for ecosystem-based management of coral reef fisheries. Proc Natl Acad Sci U S A. 108:17230–17233.

McClanahan TR, Hicks CC, Darling ES. 2008. Malthusian overfishing and efforts to overcome it on Kenyan coral reefs. Ecol Appl. 18:1516–1529.

McCormick MI, Moore JAY, Munday PL. 2010. Influence of habitat degradation on fish replenishment. Coral Reefs. 29:537–546.

Meissner KJ, Lippmann T, Gupta AS. 2012. Large-scale stress factors affecting coral reefs: Open ocean sea surface temperature and surface seawater aragonite saturation over the next 400 years. Coral Reefs. 31:309–319.

Messmer V, Jones GP, Munday PL, Holbrook SJ, Schmitt RJ, Brooks AJ. 2011. Habitat biodiversity as a determinant of fish community structure on coral reefs. Ecology. 92:2285–2298.

Mora C, Frazier AG, Longman RJ, Dacks RS, Walton MM, Tong EJ, Sanchez JJ, Kaiser LR, Stender YO, Anderson JM, et al. 2013. The projected timing of climate departure from recent variability. Nature. 502:183–187.

Mumby PJ. 2006. The impact of exploiting grazers (Scaridae) on the dynamics of Caribbean coral reefs. Ecol Appl. 16:747–769.

Mumby PJ. 2009. Phase shifts and the stability of macroalgal communities on Caribbean coral reefs. Coral Reefs. 28:761–773.

Munday PL, Dixson DL, McCormick MI, Meekan MG, Ferrari MCO, Chivers DP. 2010. Ocean acidification alters larval behaviour and impairs recruitment to reef fish populations. Proc Natl Acad Sci U S A. 107:12930–12934.

Munday PL, Jones GP, Pratchett MS, Williams AJ. 2008. Climate change and the future for coral reef fishes. Fish Fish. 9:261–285.

Naylor R, Burke M. 2005. Aquaculture and ocean resources: Raising tigers of the sea. Ann Rev Environ Res. 30:185–218.

Newton K, Côté IM, Pilling GM, Jennings S, Dulvy NK. 2007. Current and future sustainability of island coral reef fisheries. Curr Biol. 17:655–658.

Ogden JC, Gladfelter EH. 1983. Coral reefs, seagrass beds and mangroves: Their interaction in the coastal zones of the Caribbean. UNESCO Rep Mar Sci. 23:133.

Paddack MJ, Reynolds JD, Aguilar C, Appeldoorn RS, Beets J, Burkett EW, Chittaro PM, Clarke K, Esteves R, Fonseca AC, et al. 2009. Recent region-wide declines in Caribbean reef fish abundance. Curr Biol. 19:590–595.

Pandolfi JM, Bradbury RH, Sala E, Hughes TP, Bjorndal KA, Cooke RG, McArdle D, McClenachan L, Newman MJH, Paredes G, et al. 2003. Global trajectories of the long-term decline of coral reef ecosystems. Science. 301:955–958.

Pandolfi JM, Connolly SR, Marshall DJ, Cohen AL. 2011. Projecting coral reef futures under global warming and ocean acidification. Science. 333:418–422.

Pandolfi JM, Jackson JBC, Baron N, Bradbury RH, Guzman HM, Hughes TP, Kappel CV, Micheli F, Ogden JC, Possingham HP, et al. 2005. Are U.S. coral reefs on the slippery slope to slime? Science. 307:1725–1726.

Perry CT, Murphy GN, Kench PS, Smithers SG, Edinger EN, Steneck RS, Mumby PJ. 2013. Caribbean-wide decline in carbonate production threatens coral reef growth. Nat Commun. 4:1402. Available from: http://dx.doi.org/10.1038/ncomms2409 via the Internet. Accessed 1 February, 2014.

Polunin NVC, Roberts CM. 1996. Reef fisheries. London: Chapman and Hall.

Pratchett MS, Hoey AS, Wilson SK, Messmer V, Graham NAJ. 2011. Changes in biodiversity and functioning of reef fish assemblages following coral bleaching and coral loss. Diversity. 3:424–452.

Pratchett MS, Munday PL, Wilson SK, Graham NAJ, Cinner JE, Bellwood DR, Jones GP, Polunin NVC, McClanahan TR. 2008. Effects of climate-induced coral bleaching on coral-reef fishes: Ecological and economic consequences. Oceanogr Mar Biol Ann Rev. 46:251–296.

Robbins WD, Hisano M, Connolly SR, Choat JH. 2006. Ongoing collapse of coral-reef shark populations. Curr Biol. 16:2314–2319.

Sabine CL, Feely RA, Gruber N, Key RM, Lee K, Bullister JL, Wanninkhof R, Wong CS, Wallace DWR, Tilbrook B, et al. 2004. The oceanic sink for anthropogenic CO_2. Science. 305:367–371.

Sale PF. 2008. Management of coral reefs: Where we have gone wrong and what we can do about it. Mar Pollut Bull. 56:805–809.

Sale PF. 2011. Our dying planet: An ecologist's view of the crisis we face. Berkeley, CA: University of California Press.

Sale PF, Agardy T, Ainsworth CH, Feist BE, Bell JD, Christie P, Hoegh-Guldberg O, Mumby PJ, Feary DA, Saunders MI, et al. In press. Transforming management of tropical coastal seas to cope with challenges of the 21st century. Glob Environ Chang.

Sandin SA, Smith JE, DeMartini EE, Dinsdale EA, Donner SD, Friedlander AM, Konotchick T, Malay M, Maragos JE, Obura D, et al. 2008. Baselines and degradation of coral reefs in the Northern Line Islands. PLoS ONE. 3(2):e1548. Available from: http://dx.doi.org/10.1371/journal.pone.0001548 via the Internet. Accessed 1 February, 2014.

Sano M. 2000. Stability of reef fish assemblages: Responses to coral recovery after catastrophic predation by *Acanthaster planci*. Mar Ecol Prog Ser. 198:121–130.

Stallings CD. 2009. Fishery-independent data reveal negative effect of human population density on Caribbean predatory fish communities. PLoS ONE. 4(5):e5333. Available from: http://dx.doi.org/10.1371/journal.pone.0005333 via the Internet. Accessed 1 February, 2014.

Tanzil JTI, Brown BE, Tudhope AW, Dunne RP. 2009. Decline in skeletal growth of the coral *Porites lutea* from the Andaman Sea, South Thailand, between 1984 and 2005. Coral Reefs. 28:519–528.

Veron JEN, Odorico DM, Chen CA, Miller DJ. 1996. Reassessing evolutionary relationships of scleractinian corals. Coral Reefs. 15:1–9.

Wagner DE, Kramer P, van Woesik R. 2010. Species composition, habitat, and water quality influence coral bleaching in Southern Florida. Mar Ecol Prog Ser. 408:65–78.

Wanless HR, Maier KL. 2007. An evaluation of beach renourishment sands adjacent to reefal settings, southeast Florida. Southeast Geol. 45:25–42.

Wheeler T, von Braun J. 2013. Climate change impacts on global food security. Science. 341:508–513.

Wilkinson C. 2008. Status of coral reefs of the world: 2008. Townsville, Australia: Global Coral Reef Monitoring Network and Reef and Rainforest Research Centre.

Wilson SK, Depczynski M, Fisher R, Holmes TH, O'Leary RA, Tinkler P. 2010a. Habitat associations of juvenile fish at Ningaloo Reef, Western Australia: The importance of coral and algae. PLoS ONE. 5(12):e15185. Available from: http://dx.doi.org/10.1371/journal.pone.0015185 via the Internet. Accessed 1 February, 2014.

Wilson SK, Fisher R, Pratchett MS, Graham NAJ, Dulvy NK, Turner RA, Cakacaka A, Polunin NVC. 2010b. Habitat degradation and fishing effects on the size structure of coral reef fish communities. Ecol Appl. 20:442–451.

Wilson SK, Graham NAJ, Pratchett MS, Jones GP, Polunin NVC. 2006. Multiple disturbances and the global degradation of coral reefs: Are reef fishes at risk or resilient? Glob Chang Biol. 12:2220–2234.

Wooldridge SA. 2009. Water quality and coral bleaching thresholds: Formalising the linkage for the inshore reefs of the Great Barrier Reef, Australia. Mar Pollut Bull. 58:745–751.

chapter three

Distribution and diversity of coral habitat, fishes, and associated fisheries in U.S. waters of the Gulf of Mexico

Carrie M. Simmons, Angela B. Collins, and Rob Ruzicka

Contents

Introduction

This chapter summarizes current knowledge on coral reef areas, artificial habitats, and fishes in the Gulf of Mexico. We focus on shallow- to mid-water (<90 m) coral reef habitats within the Gulf of Mexico Fishery Management Council (Gulf Council) jurisdiction. Many of these areas have supported fisheries for multiple species in the families Lutjanidae and Epinephelidae since the 1800s. The Gulf Council, since its formation in 1976, has interacted with the National Oceanic and Atmospheric Administration (NOAA) Office of National Marine Sanctuaries, to recognize some of these habitats as important ecological resources deserving protection (http://sanctuaries.noaa.gov). We describe Gulf Council current management efforts and provide an updated literature review of these reef systems and their associated fishes. Some areas within the Gulf of Mexico have been extensively studied, while others have only preliminary or qualitative baseline data. Overall, there has been a lack of consistent research effort and monitoring of reef systems throughout the Gulf of Mexico, which limits our understanding of current conditions. This gap in knowledge creates many challenges for monitoring and management. Ultimately the larger goal of this chapter is to demonstrate the close association between many of the reef fish species managed by the Gulf Council and the coral reef habitats within the Gulf of Mexico. Currently, 31 reef fish species and over 142 scleractinian coral species are managed under the Gulf Council's respective fishery management plans for the Gulf of Mexico (GMFMC 2011). Reef habitats in the Gulf of Mexico provide lifetime living quarters for many managed reef fish species, and these habitats also serve as areas for feeding and refuge for other pelagic, more transient species. Even benthic species with strong site fidelity, such as gag (*Mycteroperca microlepis*) and Atlantic goliath grouper (*Epinephelus itajara*), are capable of moving long distances between reef habitats, defined as rocky limestone outcroppings, ledges, or artificial structures (McGovern et al. 2005, Addis et al. 2007, Collins 2009), demonstrating the connectivity of habitats over a broad geographic region. The health and persistence of these reef systems are important for productive and sustainable fisheries.

Management efforts in the Gulf of Mexico

The Gulf Council has designated areas of importance for both coral and fish since 1982 (Figure 3.1). Areas may be classified as sanctuaries, reserves, habitat areas of particular concern (HAPCs), or any combination of these designations. The Gulf Council continues to work with the National Marine Sanctuaries Program toward the common goal of resource conservation.

In 1982, the Florida Middle Grounds was designated as the northernmost hermatypic coral community in the Gulf of Mexico and thus a coral HAPC (GMFMC and SAFMC

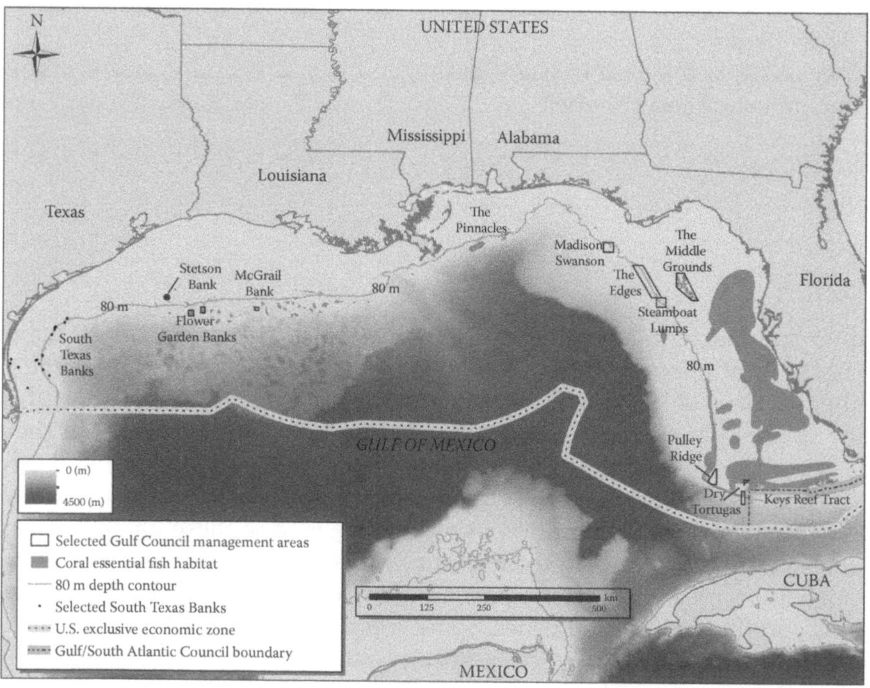

Figure 3.1 **(See color insert)** Map of the Gulf of Mexico exclusive economic zone (yellow dashed line) depicting Gulf Council designated fishery management areas (black boxes) and other known areas of importance with shallow and mesophotic corals. The orange area is the 2005 coral essential fish habitat, the thin gray line is the 80 m depth contour, and the blue dashed line indicates the jurisdictional boundary between the Gulf Council and the South Atlantic Fishery Management Council.

1982; Table 3.1). This designation prohibited coral harvest in the Florida Middle Grounds and prohibited the following gear types: bottom longlines, trawls, dredges, pots, and traps (GMFMC and SAFMC 1982). The Madison-Swanson and Steamboat Lumps marine reserves occur west of the Florida Middle Grounds and provide valuable reef structure that supports multiple species of economically important reef fishes (Smith et al. 2006, Coleman et al. 2011). Information on the invertebrate community is limited due to depth, but these areas have been identified as critical spawning areas for multiple species of grouper. Madison-Swanson and Steamboat Lumps were established as reserves in 2000 and have restricted fishing regulations (GMFMC 1999) in an effort to rebuild the gag (*M. microlepis*) population by protecting spawning aggregations (Smith et al. 2006). Madison-Swanson and Steamboat Lumps reserves are closed to all fishing from 1 November through 30 April, the primary spawning season for multiple grouper species in this region including gag (*M. microlepis*). Bottom fishing remains prohibited throughout the remainder of the year, but surface trolling is allowed (GMFMC 2003). In 2008, the Gulf Council established a new reserve between Madison-Swanson and Steamboat Lumps, called "The Edges," which is also closed to all fishing from 1 January through 30 April (GMFMC 2008).

Table 3.1 Current Management Efforts and Year Designated as a Sanctuary, Habitat Area of Particular Concern (HAPC), and/or Reserve in the Gulf of Mexico

Area	Sanctuary	HAPC	Reserve	Reference
Alderdice Bank		2006		GMFMC (2005)
Bouma Banks		2006		GMFMC (2005)
Fathom 29		2006		GMFMC (2005)
Florida Middle Grounds*		1984		GMFMC and SAFMC (1982), GMFMC (2005)
East Flower Garden*	1992	2006		GMFMC (2005)
West Flower Garden*	1992	2006		GMFMC (2005)
Geyer Bank		2006		GMFMC (2005)
Jakkula Bank		2006		GMFMC (2005)
McGrail Bank*		2006		GMFMC (2005)
MacNeil Bank		2006		GMFMC (2005)
Madison-Swanson*		2006	2000	GMFMC (1999, 2003, 2005, 2008)
Pulley Ridge*		2006		GMFMC (2005)
Rankin Bright Bank		2006		GMFMC (2005)
Rezak-Sidner Bank		2006		GMFMC (2005)
Steamboat Lumps*			2000	GMFMC (1999, 2003, 2008)
Stetson*	1996	2006		GMFMC (2005)
Sonnier		2006		GMFMC (2005)
The Edges*			2009	GMFMC (2008)
Tortugas North*		2006	2002	GMFMC (2001, 2005)
Tortugas South*		2006	2002	GMFMC (2001, 2005)

*An asterisk indicates areas that have fishing regulations or restrictions. Please see text for specific fishing regulations and restrictions by area. The effective date is listed under each type of management effort, which may differ from the corresponding reference document date.

In the southeastern Gulf of Mexico, the Tortugas North and South Ecological Reserves were established as a cooperative effort between the Gulf Council and Florida Keys National Marine Sanctuary Program (GMFMC 2001). A portion of the North Reserve is in Florida state waters, while the entire South Ecological Reserve is in federal waters. The Tortugas Ecological Reserves are closed to all fishing, and anchoring is prohibited. These areas are within the Gulf Council's jurisdiction, but adjacent areas later incorporated in Dry Tortugas National Park have varying use designations and regulations that are outside the Gulf Council's jurisdiction, and therefore are not discussed in this chapter. The Gulf Council designated these two marine reserves as HAPCs to be consistent with previous habitat and coral areas including other sanctuaries and reserves in the Gulf of Mexico, even though no additional regulations were established (GMFMC 2005).

In the western Gulf of Mexico, the East and West Flower Garden Banks were designated as a sanctuary in 1992 and nearby Stetson Bank was designated as a sanctuary in 1996 (Table 3.1). Hook and line fishing is the only type of fishing permitted within the Flower Garden Bank Sanctuary, and anchoring is prohibited.

In 2005, the Gulf Council completed a review of all currently closed areas and considered several new areas for HAPC classification and gear restrictions (GMFMC 2005). Thirteen additional areas were designated as HAPCs in 2006, and four existing reserves or sanctuaries were given HAPC status (Table 3.1). Eleven of the areas designated as a sanctuary, reserve, HAPC, or any combination of these designations have fishing or gear restrictions in place to protect these habitats and their associated fishes (Table 3.1).

The identification and continued monitoring and management of these habitats enables the conservation of regions that contribute disproportionately more to the ecological and economic value of the Gulf of Mexico. Not all habitats are created equally, and it has been repeatedly demonstrated that high quality, healthy habitat is crucial to successful fisheries (Dayton et al. 1995). Reef fishing and associated industries (commercial and recreational) contribute significantly to the economy of the Gulf States (Bell et al. 1983, Jaap 1984, Weninger and Waters 2003), so the identification and conservation of these areas has significant socioeconomic implications in addition to simply protecting a natural resource.

Eastern Gulf of Mexico

Pulley Ridge is the deepest (30–40 m) known zooxanthellate coral reef in the United States, and was designated as a HAPC in 2006. Anchoring is prohibited and fishing is restricted to hook and line fishing only (i.e., no traps, trawls, or bottom longlines allowed). The area is suspected to be an important site for spawning snappers and groupers, and could potentially serve as a source of larval fish and invertebrate recruits for the Florida Keys National Marine Sanctuary; however, the depth and distance from shore have inhibited detailed scientific investigations to date. Pulley Ridge is located off southwest Florida and is a unique deep hermatypic coral habitat with high relief up to 10 m (Harter et al. 2008). Over 10 species of stony coral have been recorded, the most dominant of which belong to the family Agariciidae. Coral cover may be as high as 60% in some locations (Jarrett et al. 2005, Hine et al. 2008). Many shelf-edge reefs support important spawning aggregations of snappers and groupers (Koenig et al. 2000, Coleman et al. 2004), so it is not unlikely that Pulley Ridge may also serve as an important site for these species. The hermatypic coral growth in this area is supported by the Loop Current that supplies Pulley Ridge with warm, clear oceanic water and limits stressors associated with closer proximity to land

(Jarrett et al. 2005). At least 60 species of reef fish have been identified during limited surveys at Pulley Ridge (Halley et al. 2003), and commercially and recreationally important species observed included hogfish (*Lachnolaimus maximus*), scamp (*Mycteroperca phenax*), black grouper (*Mycteroperca bonaci*), red grouper (*Epinephelus morio*), and sand tilefish (*Malacanthus plumieri*).

Moving north along the West Florida Shelf, hard bottom consists primarily of relic shorelines with low to moderate relief (2–8 m) limestone ledges (Smith 1976, Hine et al. 2008). Parker et al. (1983) estimated that the eastern Gulf of Mexico has between 40,000 and 50,000 km^2 of reef habitat. Low winter temperatures and seasonally high turbidity in nearshore waters of the West Florida Shelf limit significant growth of reef-building corals; however, a wide variety of scleractinian corals do occur. Up to 141 different coral species have been identified in various studies e.g., (Jaap et al. 1989, Colella et al. 2008, Cairns et al. 2009) and the area supports a significant diversity of fishes (101 species; Smith 1976).

The Florida Middle Grounds are over 125 km from land and offer a unique high-relief region for coral settlement and growth. This area is considered the northernmost coral reef community in the Gulf of Mexico with 22 identified scleractinian coral species (Coleman et al. 2005, Hine et al. 2008) and over 97 reef fish species (Hopkins et al. 1977), including 21 epinephelids. Gag (*M. microlepis*) and scamp (*M. phenax*) are abundant, along with multiple species of economically important snappers and jacks (Hopkins et al. 1977). The Florida Middle Grounds have historically supported productive commercial and recreational reef fish fisheries (Futch and Torpey 1966, Austin 1970), but their depth (45 m) and distance from shore allow for reduced fishing pressure compared to reefs closer to shore (Hine et al. 2008). For example, Collins and McBride (2011) found that hogfish in this region were larger, older, and changed sex at a size almost twice the length of hogfish collected closer to shore. This carbonate high-relief habitat with individual banks reaching approximately 12–15 m in height is unique to the West Florida Shelf where most of the hardbottom habitat does not exceed 2–8 m in relief (Hine et al. 2008). These geological features combined with clear oceanic waters and moderate current patterns make the Florida Middle Grounds a suitable habitat for reef fishes that are otherwise rare on the West Florida Shelf (Smith and Ogren 1974).

Three areas (Madison-Swanson, Steamboat Lumps, and The Edges) occur off the "Big Bend" region of Florida and are deepwater shelf-edge habitats that have historically served as important commercial and recreational grouper fishing grounds (Schirripa and Legault 1997, Coleman et al. 2011). Little is known about the benthic community composition of these reefs; however, they have been documented as spawning aggregation sites for multiple species of grouper (gag [*M. microlepis*], scamp [*M. phenax*], and red grouper [*E. morio*]) as well as red snapper (*Lutjanus campechanus*) (Coleman et al. 2011). Madison-Swanson and Steamboat Lumps were established as marine reserves to protect spawning grouper (Coleman et al. 1996), and are currently closed year round to reef fish fishing. Gag (*M. microlepis*) are protogynous hermaphrodites (i.e., fish are initially female, then eventually transition to male) that aggregate to spawn, and heavy fishing pressure is believed to have depleted the population of males (Hood and Schlieder 1992, Coleman et al. 1996). Males especially display strong site fidelity to spawning aggregation sites for several years (Coleman et al. 2011), and the proportion of male gag (*M. microlepis*) in the Gulf of Mexico has shown a decline from 17% in the 1980s to less than 5% in recent years (Coleman et al. 1996). The area known as "The Edges" is located between Madison-Swanson and Steamboat Lumps along the same bathymetric break (80 m), and recently was seasonally closed to all fishing to protect a larger portion of spawning fishes and further assist with rebuilding the gag (*M. microlepis*) population.

Central Gulf of Mexico

Natural reefs are rare in the north central Gulf of Mexico (Pensacola, Florida, to Pass Cavalla, Texas) and are estimated to cover only 3.3% of the area (Parker et al. 1983). There is little to no vertical relief as the inner shelf is composed of sand, mud, and silt (Ludwick 1964, Kennicutt et al. 1995, Dufrene 2005). The majority of reef fishing occurs on artificial structures due to limited natural habitat. Artificial structures have historically been used for reef fish enhancement and to increase fishing opportunities (Gallaway et al. 2009, Shipp and Bortone 2009, Kim et al. 2011). In 2012, when the Gulf Council considered the designation of oil and gas platforms as essential fish habitat, the estimated cover of artificial structures and substrates in state and federal waters of the Gulf of Mexico was 21.1 km^2 (GMFMC 2013). This estimate included active oil and gas platforms, all permitted artificial structure deployments, and any documented shipwrecks or other obstructions identified throughout the Gulf of Mexico (Figure 3.2).

It is well documented that several reef fish species such as gray triggerfish (*Balistes capriscus*) and red snapper (*L. campechanus*) recruit, feed, and reproduce around artificial substrates (Kurz 1995, Redman and Szedlmayer 2009, Szedlmayer 2011, Simmons and Szedlmayer 2011, 2012). Artificial reefs attract and provide long-term residences for juvenile and adult Atlantic goliath grouper (*E. itajara*), which may maintain residence at specific sites for years (Collins unpublished data), with the exception of brief forays during the spawning season. Several historically important spawning sites for Atlantic goliath grouper (*E. itajara*) are also at artificial reefs (typically deep shipwrecks; Sadovy and Eklund 1999), and individuals may travel over 100 km to these sites during the spawning season (Collins unpublished data). Pelagic fishes such as jacks (Carangidae) also aggregate at artificial structures and may remain at a site for extended periods (Brown et al. 2011). Sammarco

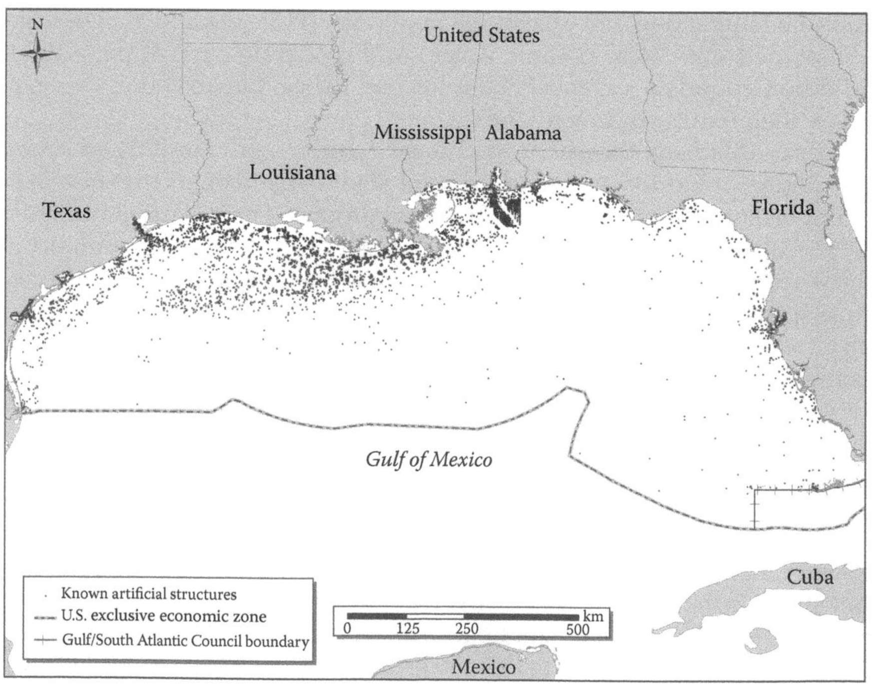

Figure 3.2 Known locations of artificial structures in state and federal waters of the Gulf of Mexico.

et al. (2004) found evidence that oil and gas platforms have facilitated the expansion of coral populations in the western Gulf of Mexico, acting as "stepping stones" for larval settlement in areas otherwise devoid of hard substrate. Hickerson et al. (2008) suggested that while oil and gas platforms provide structure for motile and sessile marine organisms, there is a concern that these artificial structures may also act as vectors for the spread of invasive and exotic species such as orange cup coral (*Tubastraea coccinea*). Invasive scorpaenids, including red lionfish (*Pterois volitans*) and devil firefish (*Pterois miles*), have also been associated with artificial and natural reefs in the Gulf of Mexico (Aguilar-Perera and Tuz-Suluband 2010, Collins personal observation) and have been reported along the Gulf coast as far north as Louisiana (nas.er.usgs.gov). As these scorpaenids expand throughout the Gulf of Mexico, it is likely that natural and artificial reefs will serve as refuge and habitat for lionfish recruits.

Western Gulf of Mexico

Three categories of banks in the northwestern Gulf of Mexico were identified based on geological criteria and are classified as: (1) midshelf banks with carbonate reef caps, (2) shelf-edge or outer-shelf bedrock banks with carbonate reef caps, and (3) reefs growing on relic carbonate shelf (Rezak et al. 1990). The first two categories include the banks off east Texas and Louisiana, and the third category includes the South Texas Banks (Rezak et al. 1990; Table 3.2). The hard banks of the northwestern Gulf of Mexico have been commercially fished for red snapper (*L. campechanus*) since the 1890s (Camber 1955), but were better described after increasing interest from the oil and gas industry during energy exploration along the continental shelf (Rezak et al. 1985, 1990).

Midshelf banks

Six midshelf banks have been identified in the northwestern Gulf of Mexico: Claypile, Sonnier, Stetson, Fishnet, Coffee Lump, and 32 Fathom (Table 3.2). Sonnier Bank is described as a salt dome structure that rises from depths of 80 m or less and have relief of 4–50 m (Rezak et al. 1990, Weaver et al. 2006). Three of the midshelf banks (Claypile, Sonnier, and Stetson) peak at depths shallow enough to include the *Millepora*-sponge zone described by Rezak et al. (1985, 1990). Of these six midshelf banks, most research regarding habitat and fish diversity has been performed on Stetson and Sonnier Banks (Rezak et al. 1985, 1990, Dennis and Bright 1988, Hickerson et al. 2008, DeBose et al. 2013). Sonnier Bank was designated as a HAPC in 2006.

Sonnier Bank is composed of eight separate banks associated with a single salt dome. These banks are characterized as being in the *Millepora*-sponge zone described by Rezak et al. (1985) with three primary genera of coral: *Stephanocoenia* sp., *Millepora* spp., and *Agaricia* spp. (Rezak et al. 1985, Weaver et al. 2006). Other corals identified later in surveys by the Minerals Management Service (MMS) include ten-ray star coral (*Madracis decactis*), yellow pencil coral (*Madracis mirabilis*), encrusting star coral (*Madracis pharensis*), lesser starlet coral (*Siderastea radians*), great star coral (*Montastraea cavernosa*), and fragile saucer coral (*Agarcia fragilis*) (Weaver et al. 2006).

Stetson Bank is one of the more studied midshelf banks in the western Gulf of Mexico and was included within the Flower Garden Banks National Marine Sanctuary in 1996. It is comprised of claystone and siltstone outcroppings and pinnacle formations, pushed up within 17 m of the sea surface at some points (Rezak et al. 1985, Hickerson et al. 2008, DeBose et al. 2013). It is located approximately 50 km north of the Flower Garden Banks

Table 3.2 Location and Size of Offshore Reef and Banks Identified in the Gulf of Mexico

Geographic area	Coral	Fish	Area (km^2)	References
Eastern Gulf of Mexico				
Pulley Ridge (restricted area)	10	60	250	Halley et al. (2005)
Florida Middle Grounds	22	170	645	Smith et al. (1975), Hopkins et al. (1977), Coleman et al. (2005)
Madison-Swanson		64	213	Coleman et al. (2004, 2011), David and Gledhill (2010)
Steamboat Lumps		53	193	Coleman et al. (2004), David and Gledhill (2010)
The Edges			316	Jaap et al. (1989)
Tortugas North, federal portion	53		41.4*	Jaap et al. (1989)
Tortugas South	28		187.2*	Tresslar (1974), Fandino (1977), Jaap et al. (1989)
Western Gulf of Mexico				
Midshelf banks				
Claypile		76	7.5	Dennis and Bright (1988), Department of Interior (MMS) (2008)
Sonnier	9	77	0.41	Rezak et al. (1985), Dennis and Bright (1988), Weaver et al. (2006), Department of Interior (MMS) (2008), Schmahl and Hickerson (2006), Schmahl et al. (2008)
Stetson		76	1.1	Dennis and Bright (1988), Department of Interior (MMS) (2008), Schmahl et al. 2008
Fishnet			1.0	Rezak et al. (1990), Department of Interior (MMS) (2008)
Coffee Lump			18.9	Rezak et al. (1990), Department of Interior (MMS) (2008)
32 Fathom			1.8	Rezak et al. (1990), Department of Interior (MMS) (2008)
Shelf-edge banks				
Alderdice	9	95	7.6	Dennis and Bright (1988), Weaver et al. (2006), Department of Interior (MMS) (2008)
Bouma		95	13.5	Dennis and Bright (1988), Department of Interior (MMS) (2008)
Bright	11	95	13.8	Rezak et al. (1985), Dennis and Bright (1988), Department of Interior (MMS) (2008), Schmahl and Hickerson (2006)
Diaphus		95	0.88	Dennis and Bright (1988), Department of Interior (MMS) (2008)
East Flower Garden	21	85	57.1	Bright et al. (1984), Dennis and Bright (1988), Department of Interior (MMS) (2008), Hickerson et al. (2008), Schmahl et al. (2008)

Table 3.2 (Continued) Location and Size of Offshore Reef and Banks Identified in the Gulf of Mexico

Geographic area	Coral	Fish	Area (km^2)	References
Elvers		95	0.55	Dennis and Bright (1988), Department of Interior (MMS) (2008)
Ewing		95	19.0	Dennis and Bright (1988), Department of Interior (MMS) (2008)
Rankin			6.3	Department of Interior (MMS) (2008)
Geyer	5	95	17.0	Rezak et al. (1985), Department of Interior (MMS) (2008), Schmahl and Hickerson (2006)
McGrail (formerly 18 Fathom)	9	78	2.5	Rezak et al. (1985), Dennis and Bright (1988), Weaver et al. (2006), Department of Interior (MMS) (2008), Schmahl and Hickerson (2006), Schmahl et al. 2008
Macneal			5.3	Department of Interior (MMS) (2008)
Jakkula		95	2.1	Dennis and Bright (1988), Department of Interior (MMS) (2008)
Rezak-Sidner		95	11.2	Dennis and Bright (1988), Department of Interior (MMS) (2008)
Sackett		95	2.2	Dennis and Bright (1988), Department of Interior (MMS) (2008)
West Flower Garden	21	85	71.7	Bright et al. (1984), Department of Interior (MMS) (2008), Hickerson et al. (2008), Schmahl et al. 2008
Applebaum			0.59	Department of Interior (MMS) (2008)
Sweet			0.18	Department of Interior (MMS) (2008)
Parker			12.7	Department of Interior (MMS) (2008)
South Texas Banks				
Aransas		66	0.51	Dennis and Bright (1988), Nash (2013)
Baker		66	1.39	Dennis and Bright (1988), Nash (2013)
Dream		66	2.07	Dennis and Bright (1988), Nash (2013)
Hospital Rock		66	2.41	Dennis and Bright (1988), Nash (2013)
North Hospital		66	1.42	Dennis and Bright (1988), Nash (2013)
South Baker		66	0.21	Dennis and Bright (1988), Nash (2013)
Small Adam			0.07	Nash (2013)
Big Adam			0.51	Nash (2013)
Blackfish ridge			1.36	Nash (2013)
Mysterious			3.64	Nash (2013)
Harte (formerly SE Mysterious)			0.37	Nash (2013)
Seabree			1.03	Nash (2013)
Southern Bank Test			1.02	Nash (2013)
Steamer			0.21	Nash (2013)

Notes: Number of scleratinian coral and reef fish species reported is based on the literature referenced. The asterisk indicates the size of the management area versus the physical bank or reef feature.

and is comparatively closer to land than other midshelf banks, so it is likely under greater fishing pressure than areas located further offshore. Stetson Bank is not considered a true accreting coral reef, but is described as a low diversity coral community with prominent sponge fauna (Hickerson et al. 2008). At Stetson Bank, the dominant coral is fire coral (*Millepora alcicornis*), a hydrozoan, which accounts for up to ~30% of the biotic cover. At least 10 other species of zooxanthellate corals have been identified for this area, including larger reef-building species such as *Montastraea* spp., but they comprise a much smaller proportion of the benthic cover (Hickerson et al. 2008, Schmahl et al. 2008).

The midshelf banks have a diverse number of fish species (Dennis and Bright 1988, Weaver et al. 2006). Sonnier Bank has multiple species managed by the Gulf Council that are commonly observed in deeper areas, including red snapper (*L. campechanus*), gray snapper (*Lutjanus griseus*), blackfin snapper (*Lutjanus buccanella*), vermilion snapper (*Rhomboplites aurorubens*), scamp (*M. phenax*), yellowmouth grouper (*Mycteroperca interstitialis*), greater amberjack (*Seriola dumerili*), gray triggerfish (*B. capriscus*), and cobia (*Rachycentron canadum*) (Weaver et al. 2006). Dennis and Bright (1988) noted that both red snapper (*L. campechanus*) and vermilion snapper (*R. aurorubens*) were common on the midshelf banks, and the most common grouper observed was the rock hind (*Epinephelus adscensionis*).

Shelf-edge banks

There are 18 defined shelf-edge banks in the northwestern Gulf of Mexico (Table 3.2). The best known and most studied of these are the East and West Flower Garden Banks (Bright et al. 1984, Rezak et al. 1985, 1990, Hickerson et al. 2008, 2012), located on the continental shelf edge 160 km south of the Texas–Louisiana border in 100–140 m of water. The West Flower Garden Bank is larger (77 km^2) than the East Flower Garden Bank (66 km^2) and the two banks are separated by approximately 22 km (Bright et al. 1984, Hickerson et al. 2008, 2012). Both are salt domes formed during the Jurassic period by the intrusion of salt plugs. Approximately 10,000 years later, coral polyps were carried north and settled to form coral reefs (Bright et al. 1984, Rezak et al. 1990, Cancelmo 2008). The Flower Garden Banks are often considered the northernmost coral reefs in the continental United States (Cancelmo 2008, Hickerson et al. 2008, 2012), although it could be argued that the Florida Middle Grounds should be awarded that status. Either way, the Flower Garden Banks display more characteristics of "true coral reefs" because they are accreting reefs and contain a greater assemblage of the principal reef-building species than do other locations in the Gulf of Mexico. They are located far enough offshore to maintain tropical characteristics including clear, nutrient-poor water and moderate temperatures ranging from 18°C to 32°C. These environmental characteristics are more similar to the oligotrophic properties of Caribbean waters than the eutrophic, turbid nearshore coastal areas in the Gulf of Mexico, which are subject to high turbidity and cold winter temperatures that limit reef-building corals (Bright et al. 1984, Cancelmo 2008).

The Flower Garden Banks contain one hydrocoral and 21 scleractinian corals, of which 18 are considered to be reef-building hermatypic coral (Bright et al. 1984, Hickerson et al. 2008). The Flower Garden Banks are considered relatively isolated from other coral reefs, being 690 km from the Campeche Banks off Mexico and approximately 1200 km from the Florida Keys reef tract (Hickerson et al. 2008). Due to their northern location and geographic distance from source populations, the Flower Garden Banks have lower diversity than other coral reefs in the south Atlantic. Interestingly, the community structure and dominant coral species most closely resemble the coral reefs of Bermuda (Bright et al. 1984).

On both the East and West Flower Garden Banks, more than 50% of the hard substratum is occupied by live coral (Bright et al. 1984, Hickerson et al. 2012, Schmahl et al. 2008). Boulder star coral (*Orbicella annularis*) is the most abundant species, covering 27%–40% of the hard substratum (Bright et al. 1984, Hickerson et al. 2008). The next most abundant coral species on the Flower Garden Banks are the symmetrical brain coral (*Diploria strigosa*), boulder brain coral (*Colpophyllia natans*), great star coral (*Montastrea cavernosa*), fire coral (*M. alcicornis*), and mustard hill coral (*Porites astreoides*), which together cover 20%–25% of the hard bottom (Bright et al. 1984, Hickerson et al. 2008). Only sparse patches of sand have been recorded between the Flower Garden Banks healthy hard coral cover (Bright et al. 1984).

Bank structure, depth, and location are significant determining factors for reef fish community structure and diversity. The most abundant families recorded during diver surveys for this region were Labridae, Pomacentridae, Epinephelidae, and Scaridae (Hickerson et al. 2008). Lutjanids important to Gulf of Mexico fisheries such as red snapper (*L. campechanus*) and vermilion snapper (*R. aurorubens*) were observed in relatively low numbers during diver surveys (Hickerson et al. 2008), but these species are regularly caught near the boundaries of the Flower Garden Banks (Levesque 2011). The largest economically important fish recorded during diver surveys within the sanctuary in 2004 and 2005 were black grouper (*M. bonaci*; 90 cm fork length [FL]), tiger grouper (*Mycteroperca tigris*; 90 cm FL), dog snapper (*Lutjanus jocu*; 83 cm FL), and yellowmouth grouper (*M. interstitialis*; 50 cm FL). The most frequently sighted groupers from 1995 through 2005 included yellowmouth grouper (*M. interstitialis*), tiger grouper (*M. tigris*), and black grouper (*M. bonaci*) (Pattengill-Semmens 2007), underscoring the value of this habitat to economically important reef fishes. In addition, two federally protected species were also recorded at the Flower Garden Banks in 2006: Nassau grouper (*Epinephelus striatus*) and Atlantic goliath grouper (*E. itajara*) (Hickerson et al. 2008).

After the Flower Garden Banks, McGrail Bank has the shallowest crest of the shelf-edge banks in the northwest Gulf of Mexico and is the only other bank in this area where high numbers of reef-building coral have been documented (Rezak et al. 1985, Schmahl and Hickerson 2006). McGrail was designated a HAPC in 2006. At least nine coral species and 78 fish species occur here, and recent remotely operated vehicle (ROV) surveys revealed a community dominated (~30% cover) by blushing star coral (*Stephanocoenia intersepta*) and planktivorous fishes (Rezak et al. 1985, Schmahl and Hickerson 2006, Weaver et al. 2006). Schmahl and Hickerson (2006) also reported multiple adult and juvenile marbled grouper (*Dermatolepis inermis*) during submersible and ROV surveys at this site, indicating the area could be important for this relatively uncommon grouper.

Alderdice Bank is unique in that it is the only shelf-edge bank known to display outcrops of basalt with the underlying salt dome (Rezak and Tieh 1984, Weaver et al. 2006). This bank has a high-profile structure that attracts large numbers of fish. Planktivores were the most dominant reef fishes observed during submersible surveys and comprised over 94% of the total fishes recorded (Weaver et al. 2006); however, other more economically important species were also well represented and included vermilion snapper (*R. aurorubens*), gray snapper (*L. griseus*), yellowmouth grouper (*M. interstitialis*), greater amberjack (*S. dumerili*), and almaco jack (*Seriola rivoliana*). Marbled grouper (*D. inermis*) have also been observed in this area (Weaver et al. 2006).

South Texas Banks

Fourteen banks off south Texas have been identified and named. These banks have unusual topographic relief compared to the surrounding area of flat outer continental shelf

(Carsey 1950, Nash et al. 2013). Most of the available information on the South Texas Banks is geological, and little is known about the community structure (Dennis and Bright 1988, Nash et al. 2013). The banks off south Texas are associated relic carbonate shelf instead of the salt domes from which the mid- and outer-shelf banks were formed (Rezak et al. 1990). Described Texas Banks are located at depths of 60–95 m, typically cresting between 55 and 82 m, with an average relief of 10–12 m (Rezak et al. 1990). Minimal hard substrate has been noted to extend above the nepheloid layer (Dennis and Bright 1988). Due to the depths (>56 m) of the carbonate crests, coral reef formation is not expected in this region and typical coral reef finfish species have not been observed (Dennis and Bright 1988, Rezak et al. 1990). These banks are known for their abundances of snappers, such as vermilion snapper (*R. aurorubens*), and are commonly referred to as "the snapper-banks" by fishermen (Nash et al. 2013). These banks also hold groupers, greater amberjacks (*S. dumerili*), almaco jack (*S. rivoliana*), blue runner (*Caranx crysos*), and great barracuda (*Sphyraena barracuda*), but the overall species richness of reef fish (66 species) is relatively low (Dennis and Bright 1988). The Southern Bank is the most studied bank and it is believed that other South Texas Banks have similar reef fish communities (Nash et al. 2013). The shallowest reef in the South Texas Banks is located in state waters (Nash et al. 2013).

Connectivity on the northwestern Gulf Banks

High-resolution bathymetry mapping has revealed structural connectivity between the hardbottom banks and reefs in the northwestern Gulf of Mexico (Hickerson et al. 2008). The connectivity of these regions is further supported by fish tagging studies. Manta rays have been documented to utilize multiple banks throughout the region as habitat for feeding and breeding (Hickerson et al. 2008). Some tagging studies have demonstrated the capacity of reef fish to travel tens to hundreds of kilometers between reef habitats. Even for species that are sedentary with high site fidelity, long distance movements are possible and likely can contribute to increased gene flow within the population. For example, in a study by McGovern et al. (2005), 23% of tagged gag (*M. microlepis*) were recaptured over 185 km from their original tagging location. Similarly, Atlantic goliath grouper (*E. itajara*) will travel over 100 km to reach spawning aggregations, and then return to their original "home" site (Pina-Amargós and González-Sansón 2009, Collins unpublished data). Managed areas may thus act as critical areas of refuge or reproduction for species in transit in addition to providing source populations for larval recruits to more exploited habitats.

The northern Gulf has less reef fish diversity than the more tropical and subtropical reefs of the southern Gulf of Mexico. Dennis and Bright (1988) attributed the limited number of reef fish species in the northwestern Gulf of Mexico to reduced habitat diversity and area, as well as increased distance from source populations. In addition, low reef fish species richness was found in association with a nepheloid layer as described in several of the banks (Dennis and Bright 1988, Rezak et al. 1990). Although the eastern Gulf of Mexico possesses 10 times the amount of available hardbottom habitat, the northwestern Gulf of Mexico is closer to the source populations of tropical Mexico (Dennis and Bright 1988). This has been offered as an explanation for the absence or rarity in the eastern Gulf of Mexico of several reef fish species that are found in the northwestern Gulf. Another possibility is the massive volumes of fresh water discharging from the Mississippi River which creates a natural physical barrier between eastern and western populations of the Gulf of Mexico (Rezak et al. 1985,1990). Additional information regarding the population connectivity of fishes and invertebrates within these hardbottom communities of the Gulf of Mexico would assist management efforts in defining essential coral and fish habitats.

Historically important reef fisheries

Hardbottom habitats have supported reef fisheries within the Gulf of Mexico for centuries (Camber 1955, Futch and Torpey 1966, Smith and Ogren 1974). The oceanographic circulation and currents in the Gulf of Mexico coupled with the geographic distribution of the hardbottom banks detailed above provide settlement opportunities for planktonic larvae from tropical lower latitudes (Smith and Ogren 1974). The resulting coral and associated invertebrate communities create oases in areas otherwise devoid of structure, providing a settlement base for successively higher trophic levels. These diverse epibenthic communities provide food, refuge, and reproductive opportunities for multiple species of fish, and it has been well established that these habitats are critical for successful fisheries (Dayton et al. 1995, Koenig et al. 2000).

The commercial and recreational reef fish fisheries associated with these habitats have an important socioeconomic role in coastal communities throughout the Gulf States. For example, from 2005 through 2009, the Gulf of Mexico reef fish commercial fleet landed over 77 million pounds of seafood, and grouper (primarily red grouper [*E. morio*] and gag [*M. microlepis*]) accounted for approximately $108 million of industry revenue (Solis et al. 2013). Commercial landings data for reef fish species in the Gulf of Mexico demonstrate the productivity and biological contribution of reef habitats in the southeast United States (NOAA Fisheries 2013). At the peak of the commercial red snapper (*L. campechanus*) fishery (1960s), landings averaged over 11 million pounds per year (Figure 3.3). Red snapper (*L. campechanus*) commercial landings in the Gulf of Mexico declined steadily after 1970, but have averaged approximately 3 million pounds annually since 2009. Incidentally, the decline in red

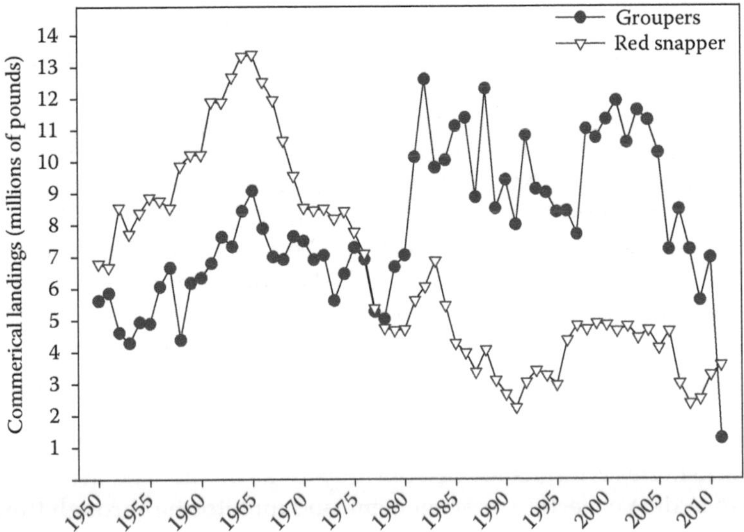

Figure 3.3 Commercial landings from the Gulf of Mexico in millions of pounds whole weight of red snapper and the most commonly landed groupers (gag, black grouper, scamp, red grouper, yellowmouth grouper, yellowedge grouper) from 1950 through 2011. Grouper landings were not separated by species until 1986. (From NOAA Fisheries, http://www.st.nmfs.noaa.gov/commercial-fisheries/commercial-landings/annual-landings/index, 2013.)

snapper (*L. campechanus*) landings coincided with an increase in Gulf grouper commercial landings (NOAA Fisheries 2013). Commercial grouper landings peaked in the Gulf of Mexico in the 1980s and 1990s, but have steadily declined since then (Figure 3.3). These declines in landings are due to a combination of factors including increased management, fishing restrictions, and decreased fish populations. Commercial fishing grounds for grouper in the northeastern Gulf of Mexico historically overlapped important spawning aggregation sites for gag (*M. microlepis*) and scamp (*M. phenax*) (Schirripa and Legault 1997), which resulted in the formation of marine reserves at Madison-Swanson and Steamboat Lumps to help protect the largest individuals from fishing mortality. The Sustainable Fisheries Act of 1996 (an amendment to the Magnuson-Stevens Fishery Conservation and Management Act) also spurred increased regulations and restrictions directed at protecting reef fish and their habitats. The reauthorization of the Magnuson-Stevens Act in 2006 provided an additional incentive to state and federal governments to ensure sustainable fisheries.

The remoteness of many of the banks in the Gulf of Mexico discussed herein kept them relatively isolated from intense fishing pressure until the 1960s. Nonetheless, trawlers and vertical line (bandit) fishermen have exploited these areas periodically since the 1800s. The Flower Garden Banks and the Florida Middle Grounds have been fished for red snapper (*L. campechanus*) since at least 1883 (Futch and Torpey 1966) and both areas remain a productive source for commercial and recreational fisheries. Distance from shore can often provide a type of refuge, and as fishing mortality decreases, the maximum size of the fish in the area may increase (Allman and Goetz 2009, Collins and McBride 2011, Levesque and Richardson 2011). Although fishing effort at sites farther offshore generally tends to be restricted to larger and fewer vessels (Levesque and Richardson 2011), advances in technology (side scan sonar, geographic information systems, outboard engines) continue to increase the capability to safely travel longer distances and better detect and target fish and their habitats. Landings data and regulatory requirements for commercial fisheries allow for relatively effective tracking of commercial reef fishing efforts. Recreational fishing effort is more difficult to quantify, but is believed to be increasing annually throughout the Gulf of Mexico (Levesque and Richardson 2011).

The National Marine Fisheries Service (NMFS) Office of Sustainable Fisheries updates its Status of U.S. Fisheries Report to Congress on a quarterly basis utilizing the most current stock assessment information (NOAA NMFS 2013). Currently, Congress has listed four reef fish species managed by the Gulf Council as overfished and in a rebuilding plan: red snapper (*L. campechanus*), gag (*M. microlepis*), greater amberjack (*S. dumerili*), and gray triggerfish (*B. capriscus*). Perhaps even more concerning is the lack of data for other species, resulting in inability to gauge the status and trends of the stock and thus an "unknown" classification for important reef fish such as hogfish (*L. maximus*), cubera snapper (*Lutjanus cyanopterus*), mutton snapper (*Lutjanus analis*), and members of the deepwater grouper complex. Without data to classify stock status, management efforts are difficult to justify, which often results in inaction until the population is in significant decline. Thus the persistence and continued success of these fisheries is intimately tied not only to the quality and availability of suitable habitat, but also to data collection, management, and monitoring through time.

Summary

This review identified commercially and recreationally important regions in the Gulf of Mexico and described their associated coral reef and fish communities. Basic information is still unavailable for many of these areas due to remoteness and deep depths, which underscores the need for long-term, or at a minimum, periodic assessments of coral communities within the Gulf of Mexico (Asch and Turgeon 2003). As new information is published on artificial and natural reefs of varying sizes and complexity, the role of artificial habitats in fisheries and management will become more apparent. This baseline information is necessary in order to track changes through time and gauge the response of these areas to environmental and anthropogenic perturbations, such as hurricanes, harmful algal blooms, oil spills, and fishing activities. The designation of HAPCs, reserves, and sanctuaries throughout the Gulf of Mexico demonstrates federal and state recognition of the value of these systems. Continued monitoring of well-described areas will assist managers in tracking the health of these systems through time. Additionally, directed research in regions where little information exists is imperative for successful management and increasing our understanding of the connectivity and ecological contributions of reef habitats within the Gulf of Mexico (Kenchington 1990, Jackson et al. 2012).

Acknowledgments

We thank our reviewers for helping to mold this book chapter, as well as J. Froeschke and M. Mueller for their help making maps. We also thank S. Bortone for inspiring us to write this book chapter.

References

Addis DT, Patterson WF III, Dance MA. 2007. Site fidelity and movement of reef fishes tagged at unreported artificial reef sites off NW Florida. Proc 60th Gulf Carib Fish Inst. 60:297–304.

Aguilar-Perera A, Tuz-Suluband A. 2010. Non-native, invasive red lionfish (*Pterois volitans* [Linnaeus, 1758]: Scorpaenidae), is first recorded in the southern Gulf of Mexico, off the Northern Yucatan Peninsula, Mexico. Aquat Invasions. 5(Suppl. 1):S9–S12.

Allman RJ, Goetz LA. 2009. Regional variation in the population structure of gray snapper, *Lutjanus griseus*, along the West Florida Shelf. Bull Mar Sci. 84(3):315–330.

Asch RG, Turgeon DD. 2003. Detection of gaps in the spatial coverage of coral reef monitoring projects in the US Caribbean and Gulf of Mexico. Revista de Biologia Tropical. 51(Suppl. 4):127–140.

Austin, H. 1970. Florida Middle Ground. Int Poll Bull. 2(2):71–72.

Bell FW, Sorensen PE, Leeworthy VR. 1983. The economic impact and valuation of saltwater recreational fisheries in Florida. Government Reports, Announcements and Index, National Technical Information Service (NTIS), US Department of Commerce 83(3):413.

Bright TJ, Kraemer GP, Minnery GA, Viada ST. 1984. Hermatypes of the Flower Garden Banks, northwestern Gulf of Mexico: A comparison to other western Atlantic reefs. Bull Mar Sci. 34(3):461–476.

Brown H, Benfield MC, Keenan SF, Powers SP. 2011. Movement patterns and home ranges of a pelagic carangid fish, *Caranx crysos*, around a petroleum platform complex. Mar Ecol Prog Ser. 403:205–218.

Cairns SD, Jaap WC, Lang JC. 2009. Scleractinia (Cnidaria) of the Gulf of Mexico. *In:* Felder DL, Camp DK, editors. Gulf of Mexico origin, waters, and biota. Volume 1, Biodiversity. Corpus Christi, TX: Texas A&M University Press. p. 333–348.

Camber CI. 1955. A survey of the red snapper fishery of the Gulf of Mexico, with special reference to the Campeche Banks. Florida Board of Conservation Marine Research Laboratory Technical Series No. 12.

Cancelmo J. 2008. Texas coral reefs. 1st ed. College Station: Texas A&M University Press.

Carsey JB. 1950. Geology of Gulf coastal area and continental shelf. AAPG Bull. 34:361–385.

Colella M, Richardson B, Johnson D, Ruzicka RR, Callahan MK, Bertin M, Wheaton JW, Schmitt S, Sheridan N. 2008. Assessment of population and community structure of sessile macro invertebrates following a benthic mortality event in the eastern Gulf of Mexico. Saint Petersburg, FL: Fish and Wildlife Research Institute, Florida Fish and Wildlife Conservation Commission.

Coleman FC, Baker PB, Koenig CC. 2004. A review of Gulf of Mexico marine protected areas. Fisheries. 29:10–21.

Coleman FC, Dennis G, Jaap W, Schmahl GP, Koenig C, Reed S, Beaver C. 2005. Status and trends of the Florida Middle Grounds. Technical Report to the Gulf of Mexico Fisheries Management Council, Tampa, FL.

Coleman FC, Koenig CC, Collins LA. 1996. Reproductive styles of shallow-water groupers (Pisces: Serranidae) in the eastern Gulf of Mexico and the consequences of fishing spawning aggregations. Environ Bio Fish. 47:129–141.

Coleman FC, Scanlon KM, Koenig CC. 2011. Groupers on the edge: Shelf edge spawning habitat in and around marine reserves of the northeastern Gulf of Mexico. Prof Geogr. 63(4):456–474.

Collins AB. 2009. A preliminary assessment of the abundance and size distribution of Goliath grouper *Epinephelus itajara* within a defined region of the central eastern Gulf of Mexico. Proc Gulf Carib Fish Inst. 61:184–190.

Collins AB, McBride RS. 2011. Demographics by depth: Spatially-explicit life history dynamics of a protogynous reef fish. Fish Bull. 109:232–242.

David A, Gledhill C. 2010. Reef fish observations in two marine protected areas in the northeastern Gulf of Mexico during 2010. Report to the Gulf of Mexico Fishery Management Council.

Dayton PK, Thrush SF, Agardy MT, Hofman RJ. 1995. Environmental effects of marine fishing. Aquat Cons Mar Fresh Eco. 5:205–232.

DeBose JL, Nuttall MF, Hickerson EL, Schmahl GP. 2013. A high-latitude coral community with an uncertain future: Stetson Bank, northwestern Gulf of Mexico. Coral Reefs. 32:255–267.

Dennis GD, Bright TJ. 1988. Reef fish assemblages on hard bank in the northwestern Gulf of Mexico. Bull Mar Sci. 43(2):280–307.

Department of Interior (MMS). 2008. Leasing activities information: Western and central Gulf of Mexico topographic features stipulation map package for oil and gas leases in the Gulf of Mexico. Available from: http://www.boem.gov/uploadedFiles/topo_features_package.pdf via the Internet. Accessed 28 January, 2014.

Dufrene TA. 2005. Geological variability and Holocene sedimentary record on the northern Gulf of Mexico inner to mid-continental shelf. MS Thesis, Louisiana State University, Baton Rouge, LA.

Fandino SV. 1977. Algunos studios sobre las madrepores del arrecife "La Blanquilla," Veracruz, Mexico. MS Thesis, Universidad Autonoma Metropolitana, Mexico.

Futch CR, Torpey JM. 1966. The red snapper, a valuable marine resource. Florida, Board of Conservation Marine Laboratory, Salt Water Fisheries Leaflet 4, p. 4. Available from: http://research.myfwc.com/engine/download_redirection_process.asp?file=ls4%5F3552%2Epdf&objid=38094&dltype=publication via the Internet. Accessed 28 January, 2014.

Gallaway BJ, Szedlmayer ST, Gazey WJ. 2009. A life history review for red snapper in the Gulf of Mexico with an evaluation of the importance of offshore petroleum platforms and other artificial reefs. Rev Fish Sci. 17:48–67.

GMFMC (Gulf of Mexico Fishery Management Council). 1999. Regulatory amendment to the reef fish fishery management plan to set 1999 gag/black grouper management measures (revised). Tampa, FL: Gulf of Mexico Fishery Management Council. Available from: http://gulfcouncil.org/Beta/GMFMCWeb/downloads/RF%20RegAmend%20-%201999-08.pdf via the Internet. Accessed 28 January, 2014.

GMFMC (Gulf of Mexico Fishery Management Council). 2001. Final generic amendment addressing the establishment of Tortugas Marine Reserves in the following fishery management plans of the Gulf of Mexico: Coastal migratory pelagics of the Gulf of Mexico and South Atlantic, coral and coral reefs, red drum, reef fish, shrimp, spiny lobster, and stone crab. Tampa, FL: Gulf of Mexico Fishery Management Council. Available from: http://www.gulfcouncil.org/Beta/GMFMCWeb/downloads/TORTAMENwp.pdf via the Internet. Accessed 28 January, 2014.

GMFMC (Gulf of Mexico Fishery Management Council). 2003. Final amendment 21 to the reef fish fishery management plan including regulatory impact review, initial regulatory flexibility analysis, and environmental assessment. Tampa, FL: Gulf of Mexico Fishery Management Council. Available from: http://www.gulfcouncil.org/Beta/GMFMCWeb/downloads/Amend21-draft%203.pdf via the Internet. Accessed 28 January, 2014.

GMFMC (Gulf of Mexico Fishery Management Council). 2005. Final generic amendment number 3 for addressing essential fish habitat requirements, habitat areas of particular concern, and adverse effects of fishing in the following fishery management plans of the Gulf of Mexico: Shrimp, red drum, reef fish, coastal migratory pelagics in the Gulf of Mexico and South Atlantic, stone crab, spiny lobster, and coral and coral reefs of the Gulf of Mexico. Tampa, FL: Gulf of Mexico Fishery Management Council. Available from: http://www.gulfcouncil.org/Beta/GMFMCWeb/downloads/FINAL3_EFH_Amendment.pdf via the Internet. Accessed 28 January, 2014.

GMFMC (Gulf of Mexico Fishery Management Council). 2008. Final amendment 30B to the reef fish fishery management plan. Tampa, FL: Gulf of Mexico Fishery Management Council. Available from: http://www.gulfcouncil.org/Beta/GMFMCWeb/downloads/Final%20Amendment%2030B%2010_10_08.pdf via the Internet. Accessed 28 January, 2014.

GMFMC (Gulf of Mexico Fishery Management Council). 2011. Final generic annual catch limits/accountability measures amendment for the Gulf of Mexico fishery management council's red drum, reef fish, shrimp, coral and coral reefs fishery management plans, including environmental impact statement, regulatory impact review, regulatory flexibility analysis, and fishery impact statement. Tampa, FL: Gulf of Mexico Fishery Management Council. Available from: http://www.gulfcouncil.org/docs/amendments/Final%20Generic%20ACL_AM_Amendment-September%209%202011%20v.pdf via the Internet. Accessed 28 January, 2014.

GMFMC (Gulf of Mexico Fishery Management Council). 2013. Generic amendment 4: Fixed petroleum platforms and artificial reefs and essential fish habitat: Options paper. Tampa, FL: Gulf of Mexico Fishery Management Council. Available from: http://www.gulfcouncil.org/docs/amendments/Artificial%20Reefs%20as%20EFH%20Amendment.pdf via the Internet. Accessed 28 January, 2014.

GMFMC and SAFMC (Gulf of Mexico Fishery Management Council and South Atlantic Fishery Management Council). 1982. Fishery management plan for coral and coral reefs in the Gulf of Mexico and South Atlantic Fishery Management Councils. Tampa, FL: Gulf of Mexico Fishery Management Council; Charleston, SC: South Atlantic Fishery Management Council. Available from: http://www.gulfcouncil.org/Beta/GMFMCWeb/downloads/Coral%20FMP.pdf via the Internet. Accessed 28 January, 2014.

Halley RB, Dennis GP, Weaver D, Coleman F. 2005. Characterization of Pulley Ridge coral and fish fauna. Technical Report to the Gulf of Mexico Fisheries Management Council, Tampa, FL.

Halley RB, Garrison VE, Ciembronowicz KT, Edwards R, Jaap WC, Mead G, Earle S, Hine AC, Jarrett BD, Locker SD, et al. 2003. Pulley Ridge: The US deepest coral reef? Joint conference on the science and restoration of the Greater Everglades and Florida Bay ecosystem from Kissimmee to the Keys, GEER program and abstracts, Palm Harbor, FL. p. 238–240. Available from: http://conference.ifas.ufl.edu/jc/GEER.pdf via the Internet. Accessed 28 January, 2014.

Harter S, David A, Ribera M. 2008. Survey of coral and fish assemblages on Pulley Ridge, SW Florida: A report to the Gulf of Mexico Fishery Management Council. Available from: http://data.nodc.noaa.gov/coris/library/NOAA/other/survey_coral_fish_pulley_ridge_fl.pdf via the Internet. Accessed 28 January, 2014.

Hickerson EL, Schmahl GP, Johnston MA, Nuttal MF, Embesi JA, Eckert RJ. 2012. Flower Garden Banks: A refuge in the Gulf of Mexico? *In*: Proceedings of the 12th International Coral Reef Symposium, Cairns, Australia, 9–13 July 2012.

Hickerson EL, Schmahl GP, Robbart M, Precht WF, Caldow C. 2008. The state of coral reef ecosystems of Flower Garden Banks. *In:* The state of coral reef ecosystems of the Flower Garden Banks, Stetson Bank, and other banks in the northwestern Gulf of Mexico. p. 189–217. Available from: http://ccma.nos.noaa.gov/ecosystems/coralreef/coral2008/pdf/fgbnms.pdf via the Internet. Accessed 28 January, 2014.

Hine AC, Halley RB, Locker SD, Jarrett BD, Jaap WC, Mallinson DJ, Ciembronowicz KT, Ogden NB, Donahue BT, Naar DF, et al. 2008. Coral reefs, present and past, on the West Florida Shelf and platform margin. *In*: Riegl BM, Dodge RE, editors. Coral reefs of the USA. 1st ed. Dordrecht: Springer. p. 127–174.

Hood PB, Schlieder RA. 1992. Age, growth and reproduction of gag *Mycteroperca microlepis* (Pisces: Serranidae), in the eastern Gulf of Mexico. Bull Mar Sci. 51:337–352.

Hopkins TS, Blizzard DR, Brawley SA, Earle SA, Grimm DE, Gilbert DK, Johnson PG, Livingston EH, Lutz CH, Shaw JK, et al. 1977. A preliminary characterization of the biotic components of composite strip transects on the Florida Middle Grounds, northeastern Gulf of Mexico. *In*: Taylor DL, editor. Proceedings of Third International Coral Reef Symposium. Volume 1. Biology. Miami, FL: Rosenstiel School of Marine and Atmospheric Science. p. 32–37.

Jaap WC. 1984. The ecology of the south Florida coral reefs: A community profile. Washington, DC: US Department of the Interior, Fish and Wildlife Service; Metaire, LA: Minerals Management Service.

Jaap WC, Lyons WG, Dustan P, Halas, JC. 1989. Stony coral (Scleractinia and Milleporina) community structure at Bird Key Reef, Fort Jefferson National Monument, Dry Tortugas, Florida. Florida Marine Research Publications No. 46.

Jackson JBC, Alexander K, Sala E. 2012. Shifting baselines: The past and the future of ocean fisheries. Washington, DC: Island Press.

Jarrett BD, Hine AC, Halley RB, Naar DF, Locker SD, Neumann AC, Twichell D, Hu C, Donahue BT, Jaap WC, et al. 2005. Strange bedfellows: A deep-water hermatypic coral reef superimposed on a drowned barrier island; Southern Pulley Ridge, SW Florida platform margin. Mar Geol. 214:295–307.

Kenchington RA. 1990. Managing marine environments. New York: Taylor and Francis.

Kennicutt MC, Schroeder WW, Brooks JM. 1995. Temporal and spatial variations in sediment characteristics on the Mississippi-Alabama continental shelf. Cont Shelf Res. 15:1–18.

Kim CG, Lee S, Cha HK, Yang JH, Son YS. 2011. A case study of artificial reefs in fisheries management: Enhancement of sandfish, *Arctoscopus japonicus*, by artificial reefs in the eastern waters of Korea. *In*: Bortone SA, Brandini F, Fabi G, Otake S, editors. Artificial reefs in fisheries management. Boca Raton, FL: CRC Press. p. 111–124.

Koenig CK, Coleman FC, Grimes CB, Fitzhugh GR, Scanlon KM, Gledhill CT, Grace M. 2000. Protection of fish spawning habitat for the conservation of warm temperate reef-fish fisheries of shelf-edge reefs of Florida. Bull Mar Sci. 66:593–616.

Kurz RC. 1995. Predator-prey interactions between gray triggerfish (*Balistes capriscus* Gmelin) and a guild of sand dollars around artificial reefs in the northeastern Gulf of Mexico. Bull Mar Sci. 56:150–160.

Levesque JC. 2011. Commercial fisheries in the northwestern Gulf of Mexico: Possible implications for conservation management at the Flower Garden Banks National Marine Sanctuary. J Mar Sci. 68(10):2175–2190.

Levesque JC, Richardson A. 2011 Characterization of the recreational fisheries associated with the Flower Garden Banks National Marine Sanctuary (USA). Wildl Biol Pract. 7(1):90–115.

Ludwick JC. 1964. Sediments in northeastern Gulf of Mexico. *In*: Miller RL, editor. Papers in marine geology: Shepard commemorative volume. New York: MacMillan. p. 204–238.

McGovern JC, Sedberry GR, Meister HS, Westendorff TM, Wyanski DM, Harris PJ. 2005. A tag and recapture study of gag, *Mycteroperca microlepis*, off the southeastern U.S. Bull Mar Sci. 76:47–59.

Nash HL. 2013. Trinational governance to protect ecological connectivity: Support for establishing an international Gulf of Mexico marine protected area network. PhD Dissertation, Texas A&M University, Corpus Christi, TX.

Nash HL, Furiness SJ, Tunnell JW. 2013. What is known about species richness and distribution on the outer-shelf South Texas Banks? Gulf Carib Res. 25:9–18.

NOAA Fisheries. 2013. Commercial fisheries statistics. Available from: http://www.st.nmfs.noaa.gov/commercial-fisheries/commercial-landings/annual-landings/index. Accessed 28 January, 2014.

NOAA NMFS. 2013. 2013 Status of U.S. fisheries. Second quarter. Available from: http://www.nmfs.noaa.gov/sfa/statusoffisheries/SOSmain.htm via the Internet. Accessed 29 January, 2014.

Parker RO, Colby DR, Willis TD. 1983. Estimated amount of reef habitat on a portion of the U.S. South Atlantic and Gulf of Mexico continental shelf. Bull Mar Sci. 33:935–940.

Pattengill-Semmens CV. 2007. Fish assemblages of the Gulf of Mexico, including the Flower Garden Banks National Marine Sanctuary. Proc Gulf Carib Fish Inst. 59:229–238.

Pina-Amargós F, González-Sansón G. 2009. Movement patterns of goliath grouper *Epinephelus itajara* around southeast Cuba: Implications for conservation. Endang Species Res. 7:243–247.

Redman RA, Szedlmayer ST. 2009. The effects of epibenthic communities on reef fishes in the northern Gulf of Mexico. Fish Manag Ecol. 16:360–367.

Rezak R, Bright TJ, McGrail DW. 1985. Reef and bank of the northwestern Gulf of Mexico: Their geological, physical, and biological dynamics. New York: Wiley.

Rezak R, Gittings ST, Bright TJ. 1990. Biotic assemblages and ecological controls on reefs and banks of the northwestern Gulf of Mexico. Am Zool. 30:23–35.

Rezak R, Tieh TT. 1984. Basalt from Louisiana continental shelf. Geo-Mar Lett. 4:69–76.

Sadovy Y, Eklund AM. 1999. Synopsis of the biological data on the Nassau grouper, *Epinephelus striatus* (Bloch 1972) and the jewfish, *E. itajara* (Lichtenstein 1822). US Department of Commerce, NOAA Technical Report NMFS 146, and FAO Fisheries Synopsis 157.

Sammarco PW, Atchison AD, Boland GS. 2004. Expansion of coral communities within the northern Gulf of Mexico via offshore oil and gas platforms. Mar Ecol Prog Ser. 280:129–143.

Schirripa MJ, Legault CM. 1997. Status of the gag stocks of the Gulf of Mexico: Assessment 2.0. Miami, FL: National Marine Fisheries Service, Southeast Fisheries Science Center.

Schmahl GP, Hickerson EL. 2006. McGrail Bank, a deep tropical coral reef community in the northwestern Gulf of Mexico. *In*: Proceedings of 10th International Coral Reef Symposium. June 28–July 2, 2004, Okinawa, Japan. p. 1124–1130.

Schmahl GP, Hickerson EL, Precht W. 2008. Biology and ecology of coral reefs and coral communities in the Flower Garden Banks region, northwestern Gulf of Mexico. *In*: Riegl BM, Dodge RE, editors. Coral reefs of the USA. 1st ed. Dordrecht: Springer. p. 221–261.

Shipp RL, Bortone SA. 2009. A perspective of the importance of artificial habitat on the management of red snapper in the Gulf of Mexico. Rev Fish Sci. 17:41–47.

Simmons CM, Szedlmayer ST. 2011. Recruitment of age-0 gray triggerfish to benthic structured habitat in the northern Gulf of Mexico. Trans Am Fish Soc. 140:14–20.

Simmons CM, Szedlmayer ST. 2012. Territoriality, reproductive behavior, and parental care in gray triggerfish, *Balistes capriscus*, from the northern Gulf of Mexico. Bull Mar Sci. 88:197–209.

Smith GB. 1976. Ecology and distribution of eastern Gulf of Mexico reef fishes. Florida Department of Natural Resources. St. Petersburg, FL: Florida Marine Research Publications. p. 84.

Smith GB, Austin HM, Bortone SA, Hastings RW, Ogren LH. 1975. Fishes of the Florida Middle Ground with comments on ecology and zoology. Florida Department of Natural Resources Marine Research Publication No. 9.

Smith GB, Ogren LH. 1974. Comments on the nature of the Florida Middle Ground reef ichthyofauna. *In*: Proceedings of Marine Environmental Implications of Offshore Drilling in the eastern Gulf of Mexico. Conference/workshop. State Univ Sys Fla Inst Oceanogr, St. Petersburg, FL. p. 229–232.

Smith MD, Zhang J, Coleman FC. 2006. Effectiveness of marine reserves for large-scale fisheries management. Can J Fish Aquat Sci. 63:153–164.

Solis D, Perruso L, Corral J, Stoffle B, Letson D. 2013. Measuring the initial economic effects of hurricanes on commercial fish production: The US Gulf of Mexico grouper (Serranidae) fishery. Nat Hazards. 66:271–289.

Szedlmayer ST. 2011. The artificial habitat as an accessory for improving estimates of juvenile reef fish abundance in fishery management. *In*: Bortone SA, Brandini F, Fabi G, Otake S, editors. Artificial reefs in fisheries management. Boca Raton, FL: CRC Press. p. 31–44.

Tresslar RC. 1974. Corals. *In*: Bright T, Pequegnat L, editors. Biota of the West Flower Garden Bank. 1st ed. Houston, TX: Gulf Publishing Company. p. 115–139.

Weaver DC, Hickerson EL, Schmahl GP. 2006. Deep reef fish surveys by submersible on Alderdice, McGrail, and Sonnier Banks in the northwestern Gulf of Mexico. *In*: Tayler JC, editor. Emerging technologies for reef fisheries research and management. Seattle, WA: NOAA Professional Paper NMFS. p. 69–87.

Weninger Q, Waters JR. 2003. Economic benefits of management reform in the northern Gulf of Mexico reef fish fishery. J Environ Econ Manag. 46(2):207–230.

chapter four

Implementation of Coral Habitat Areas of Particular Concern (CHAPCs)

South Atlantic Fishery Management Council process

Anna Martin

Contents

Introduction

The South Atlantic Fishery Management Council (SAFMC or South Atlantic Council) is one of eight U.S. Regional Fishery Management Councils charged with conservation and management of fisheries in the exclusive economic zones. The reauthorized Magnuson-Stevens Fishery Conservation and Management Act (2007) sets forth the authority of the councils, under fishery management plans, to designate zones to protect deepwater corals from damage caused by fishing gear. The South Atlantic Council has made considerable progress toward the conservation of deepwater corals and associated habitats in the southeastern region of the United States. Through the Coral Fishery Management Plan via designation of the Oculina Bank Habitat Area of Particular Concern (HAPC) and establishing criteria for HAPC designations, Comprehensive Ecosystem-Based Amendment 1 through designation of deepwater coral HAPCs (CHAPC), and the measures under development in Coral Amendment 8 via expansion of CHAPCs, the South Atlantic Council has taken a proactive approach to deepwater coral management in the South Atlantic region while including stakeholders in the management process.

The Coral Fishery Management Plan

The Coral Fishery Management Plan, developed originally in conjunction with the Gulf of Mexico Fishery Management Council (SAFMC and GMFMC 1982), provided guidelines to the councils for establishing criteria for essential fish habitat (EFH) designations, describing as "areas of special biological significance" those areas "which are of particular concern because of a requirement in the life cycle of the stock(s); e.g., spawning grounds, nurseries, migratory routes, etc ... [and] ... those areas which are currently or potentially threatened with destruction or degradation" (SAFMC and GMFMC 1982: 6–18). In January 1998, the ruling implementing the EFH provisions of the Magnuson-Stevens Act became effective and included specific criteria for identifying areas of habitat within EFH areas as HAPC. For an area to be considered an HAPC, one or more of the following criteria must be met: (1) the ecological function provided by the habitat is important; (2) the habitat is sensitive to human-induced environmental degradation; (3) development activities are stressing or will stress the habitat type; and (4) the habitat type is rare (SAFMC and GMFMC 1982).

Oculina Bank HAPC

The Coral Fishery Management Plan also identified the significance of the area now known as the Oculina Bank HAPC. Through the fishery management plan, a 238 km^2 (92 mi^2) area of the *Oculina* reef tract was designated an HAPC. The South Atlantic Council prohibited the use of bottom trawls, bottom longlines, dredges, fish traps, and fish pots within the HAPC in order to minimize impacts of bottom-tending fishing gear on the *Oculina varicosa* coral occurring in the area (SAFMC and GMFMC 1982). Furthering protections in this area, the South Atlantic Council prohibited fishing for and retention of snapper and grouper species within the CHAPC and prohibited anchoring by vessels fishing for snapper grouper species through measures included in Snapper Grouper Amendment 6 (SAFMC 1993). The area to which these snapper grouper gear restrictions applied became known as the Oculina Experimental Closed Area. The purpose of these restrictions was to "enhance stock stability and increase recruitment by providing an area where deep water species can grow and reproduce without being subjected to fishing mortality" (SAFMC 1993: 36). The management measures in Coral Amendment 3 (SAFMC 1995) prohibited all fishing vessels from anchoring within the Oculina Bank HAPC. Also in 1996, in an effort to minimize the impacts of the rock shrimp fishery on EFH, including the fragile coral species occurring in the Oculina Bank, the South Atlantic Council prohibited trawling for rock shrimp in an area known as the Rock Shrimp Closed Area (SAFMC 1996). In 1998, the South Atlantic Council expanded the Oculina Bank HAPC to include the Rock Shrimp Closed Area and added two satellite sites to the Oculina Bank HAPC (SAFMC 1998). Gear restrictions within the expanded HAPC include fishing with a bottom longline, bottom trawl, dredge, fish pot, fish trap or use of an anchor by a fishing vessel (SAFMC 1998). Currently, the South Atlantic Council is considering actions to expand the northern and western boundaries of the Oculina Bank HAPC based on research deepwater coral scientists have brought forward (see Coral Amendment 8 section below, SAFMC 2013 under review).

Comprehensive Ecosystem-Based Amendment 1

As noted above, management actions in the Comprehensive Ecosystem-Based Amendment 1 included the identification and designation of five deepwater CHAPCs and implementation of bottom-tending gear restrictions to preserve what is thought to be the largest known continuous distribution (58,570 km^2 or 23,000 mi^2) of deepwater coral ecosystems (SAFMC 2010). Comprehensive Ecosystem-Based Amendment 1 set forth protections for deepwater corals through designation of the following CHAPC areas: Cape Lookout, Cape Fear, Blake Ridge Diapir, Stetson-Miami Terrace, and Pourtalés Terrace CHAPCs. Currently, the only commercial fisheries operating in the areas are the wreckfish, golden crab, and royal red shrimp fisheries. Through incorporation of Fishery Access Areas into several of the CHAPC designations, measures in Comprehensive Ecosystem-Based Amendment 1 have allowed these fisheries to continue traditional practices with minimal impacts on deepwater coral habitat. A description of the South Atlantic Council–designated CHAPCs through Comprehensive Ecosystem-Based Amendment 1 follows (SAFMC 2010) (see Figure 4.1).

Cape Lookout CHAPC

The Cape Lookout CHAPC was identified based on research described by Dr. S. W. Ross in "State of the U.S. deep coral ecosystems in the Southeastern United States Region: Cape Hatteras to the Florida Straits" (Ross and Nizinski 2007). The northernmost range of the Cape Lookout CHAPC contains the most extensive *Lophelia pertusa* coral mounds to be discovered off the coast of North Carolina. The second area of coral mounds located within the CHAPC is described to be of the same general construction as the northern region, built of coral rubble matrix. Extensive fields of coral rubble surround the area, and both living and dead corals occur throughout the bank. The Cape Lookout CHAPC also supports a diverse invertebrate fauna community (SAFMC 2010). The CHAPC designation in this area protects 316 km^2 (122 mi^2) of deepwater coral and associated habitats.

Cape Fear Lophelia CHAPC

The Cape Fear *Lophelia* CHAPC is approximately 134 km^2 (52 mi^2) and demonstrates some of the most rugged habitat and vertical deviations of any area sampled. The mounds appear to be of the same general construction as those in the Cape Lookout Banks, built of coral rubble with enclosed sediments. Extensive fields of coral rubble surround the area and both living and dead corals occur on this bank (SAFMC 2010).

Stetson-Miami Terrace CHAPC

The Stetson-Miami Terrace CHAPC is by far the largest (60,937 km^2 or 23,528 mi^2) of the five deepwater CHAPCs designated through Comprehensive Ecosystem-Based Amendment 1 (SAFMC 2010) and further expansions are being considered through the development of Coral Amendment 8. The Stetson Reef is characterized by hundreds of pinnacles along the eastern Blake Plateau off the coast of South Carolina and over 200 coral mounds including live bushes of *Lophelia* coral, sponges, gorgonians, and black coral bushes (Reed et al. 2006).

The area within the CHAPC known as the Miami Terrace and Escarpment is a Miocene-age terrace off southeast Florida that supports high-relief hardbottom habitats and rich benthic communities in 600 m (1969 ft) depths (Reed et al. 2006). *Lophelia* coral mounds are

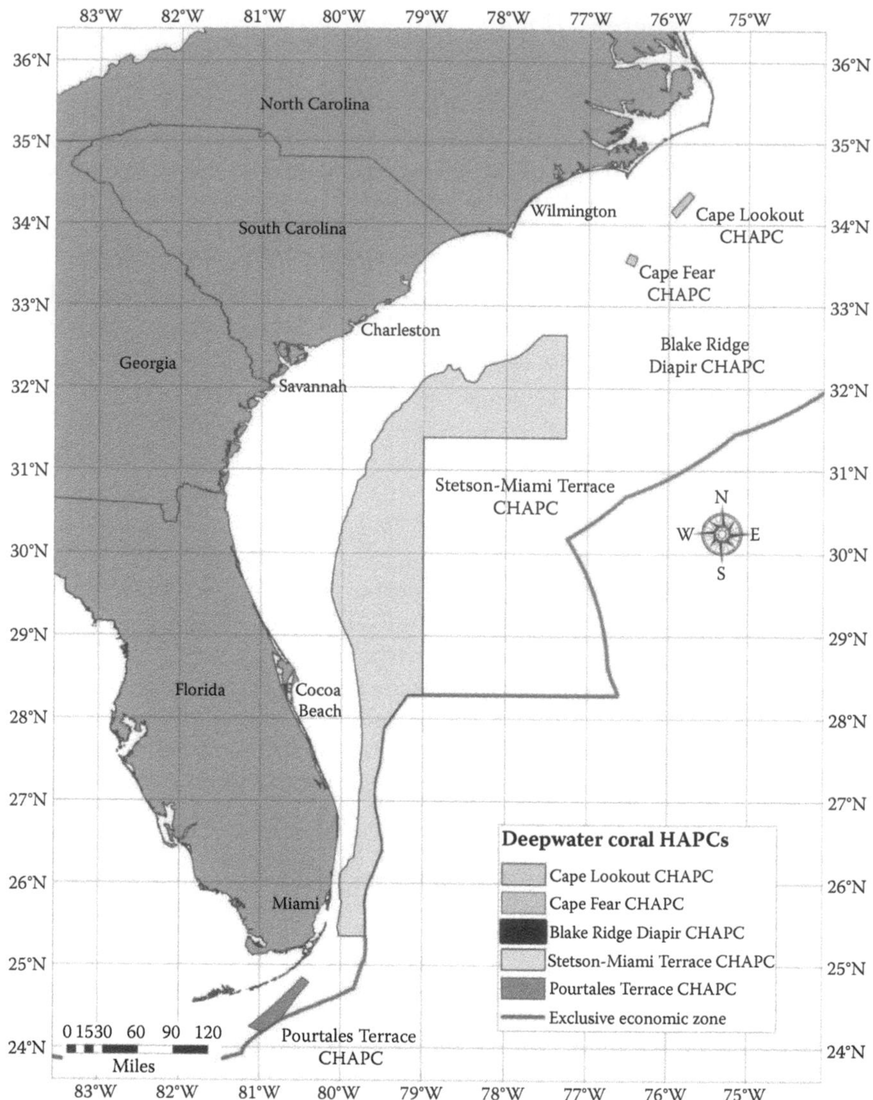

***Figure 4.1* (See color insert)** The deepwater Coral Habitat Areas of Particular Concern (CHAPCs) implemented by SAFMC through Comprehensive Ecosystem-Based Amendment 1 in 2009.

present at the base of the escarpment, but scientists know little about the abundance, distribution, or associated fauna. The escarpments, especially near the top of the ridges, are abundant with corals, octocorals, and sponges (Reed et al. 2006, SAFMC 2010).

Pourtalés Terrace CHAPC

The Pourtalés Terrace CHAPC is also a Miocene-age terrace and is located off the Florida reef tract. The CHAPC includes high-relief hardbottom habitats and rich benthic communities. Sinkholes are present on the outer edge of the terrace. The Pourtalés Terrace CHAPC designation conserves approximately 819 km^2 (509 mi^2) of deepwater habitats (SAFMC 2010).

Blake Ridge Diapir CHAPC

A 2001 National Oceanic and Atmospheric Administration (NOAA)-sponsored deepwater coral research cruise marked the first discovery of a methane gas hydrate seep on the sea floor of the Blake Ridge (SAFMC 2010). The Blake Ridge Diapir Methane Seep CHAPC protects approximately 13.4 km^2 (4 mi^2) of deepwater habitats. This area incorporates benthic habitats that do not occur elsewhere in the region (SAFMC 2010).

South Atlantic Coral Amendment 8

Recommendations for boundary modifications to several of the CHAPCs originally designated through the Coral Fishery Management Plan and Comprehensive Ecosystem-Based Amendment 1 (Oculina Bank, Stetson-Miami Terrace, Cape Lookout) were brought forward by the South Atlantic Coral Advisory Panel in October 2011. Deepwater coral scientists serving on the Coral Advisory Panel presented research that utilized NOAA regional bathymetric charts to pinpoint locations of high-relief features off the coast of Florida (Daytona, Florida and Titusville, Florida) for further mapping with multibeam sonar and groundtruth using submersibles and remotely operated vehicle technology (Reed and Farrington 2011). In addition, scientists presented research to the Coral Advisory Panel detailing observations of a shallow-water *L. pertusa* ecosystem outside the western boundary of the existing Stetson-Miami Terrace CHAPC (Ross et al. 2012a). Additionally, deepwater coral scientists reported on observations of a *L. pertusa* mound occurring north of the Cape Lookout CHAPC. As a result, the Coral Advisory Panel developed a suite of recommendations for the South Atlantic Council's consideration to expand the CHAPCs.

The South Atlantic Council has worked cooperatively with their Coral, Habitat, Deepwater Shrimp, Law Enforcement, and Snapper Grouper Advisory Panels to refine the modifications of the CHAPC areas in Coral Amendment 8 to ensure protection of sensitive deepwater coral resources while minimizing fishery activity on traditional grounds (see Figures 4.2 through 4.5). The South Atlantic Council has held numerous Advisory Panel meetings and gathered public scoping and public hearing input on Coral Amendment 8 during 2012–2013. The South Atlantic Council approved Coral Amendment 8 for review by the Secretary of Commerce at its September 2013 meeting (SAFMC 2013 underreview).

Background information on deepwater coral research that resulted in the management measures under development in Coral Amendment 8 follows.

Oculina Bank HAPC

Scientists affiliated with the South Atlantic Council Coral Advisory Panel presented research observations at their meeting in October 2011 on two areas of high-relief *O. varicosa* coral mounds and hardbottom habitats discovered outside the current northern and western boundaries of the Oculina Bank HAPC (Figures 4.2 and 4.3). The locations of the sites had been previously identified from NOAA regional bathymetric charts as possible locations of high-relief habitat and were later groundtruthed utilizing multibeam sonar, a remotely operated vehicle, and submersible video surveys. One area extends from the northern boundary of the Oculina Bank HAPC up to St. Augustine, Florida. The second area is to the west of the current boundary, primarily between the two existing Oculina Bank HAPC satellite sites (Reed and Farrington 2011).

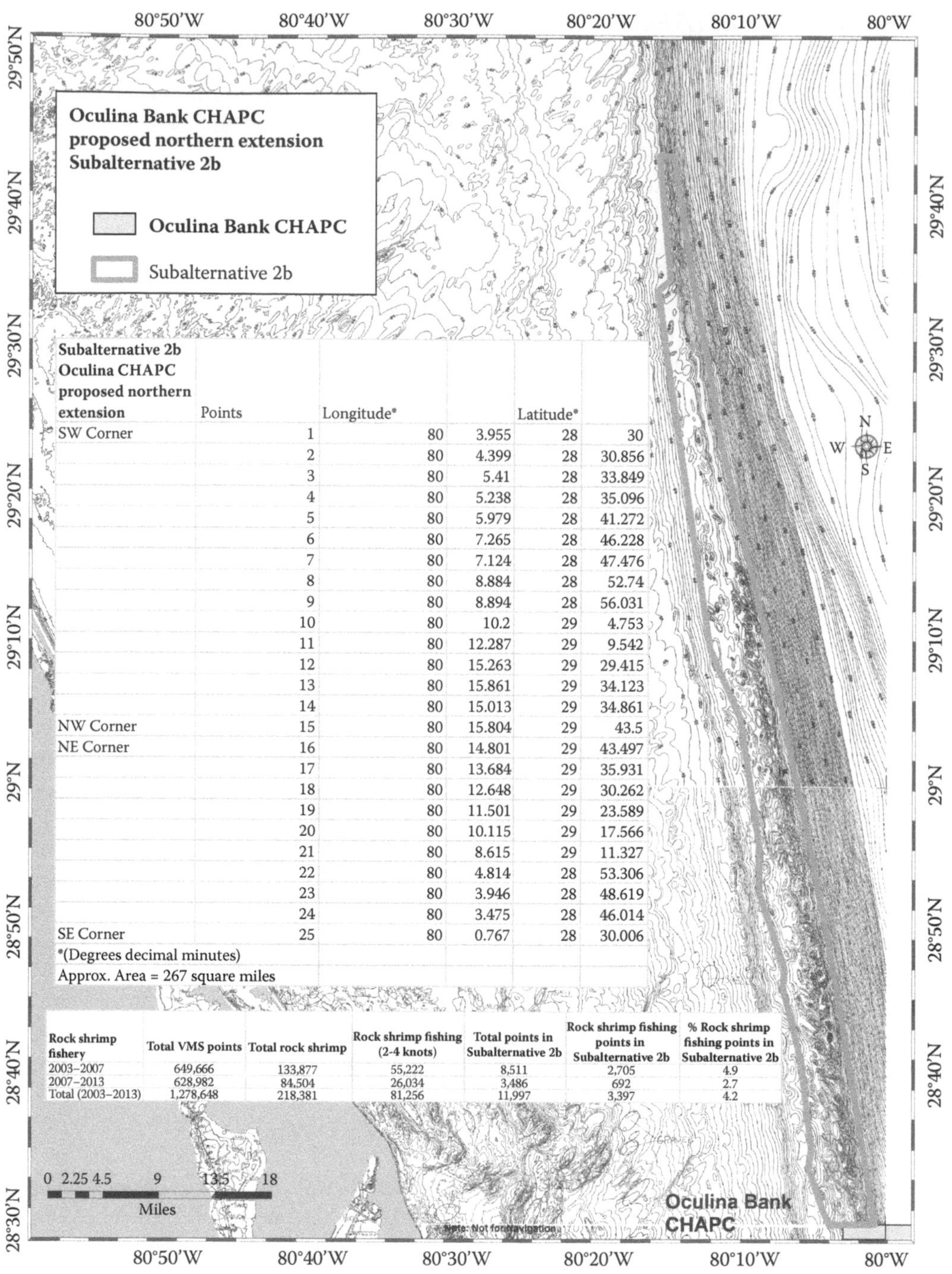

Subalternative 2b Oculina CHAPC proposed northern extension	Points	Longitude*		Latitude*	
SW Corner	1	80	3.955	28	30
	2	80	4.399	28	30.856
	3	80	5.41	28	33.849
	4	80	5.238	28	35.096
	5	80	5.979	28	41.272
	6	80	7.265	28	46.228
	7	80	7.124	28	47.476
	8	80	8.884	28	52.74
	9	80	8.894	28	56.031
	10	80	10.2	29	4.753
	11	80	12.287	29	9.542
	12	80	15.263	29	29.415
	13	80	15.861	29	34.123
	14	80	15.013	29	34.861
NW Corner	15	80	15.804	29	43.5
NE Corner	16	80	14.801	29	43.497
	17	80	13.684	29	35.931
	18	80	12.648	29	30.262
	19	80	11.501	29	23.589
	20	80	10.115	29	17.566
	21	80	8.615	29	11.327
	22	80	4.814	28	53.306
	23	80	3.946	28	48.619
	24	80	3.475	28	46.014
SE Corner	25	80	0.767	28	30.006
*(Degrees decimal minutes)					
Approx. Area = 267 square miles					

Rock shrimp fishery	Total VMS points	Total rock shrimp	Rock shrimp fishing (2-4 knots)	Total points in Subalternative 2b	Rock shrimp fishing points in Subalternative 2b	% Rock shrimp fishing points in Subalternative 2b
2003–2007	649,666	133,877	55,222	8,511	2,705	4.9
2007–2013	628,982	84,504	26,034	3,486	692	2.7
Total (2003–2013)	1,278,648	218,381	81,256	11,997	3,397	4.2

Figure 4.2 **(See color insert)** The SAFMC Preferred Alternative for a northern expansion of the Oculina Bank HAPC included in Coral Amendment 8 (under development). The chart incorporates VMS data from the rock shrimp fishery, 2003–2013.

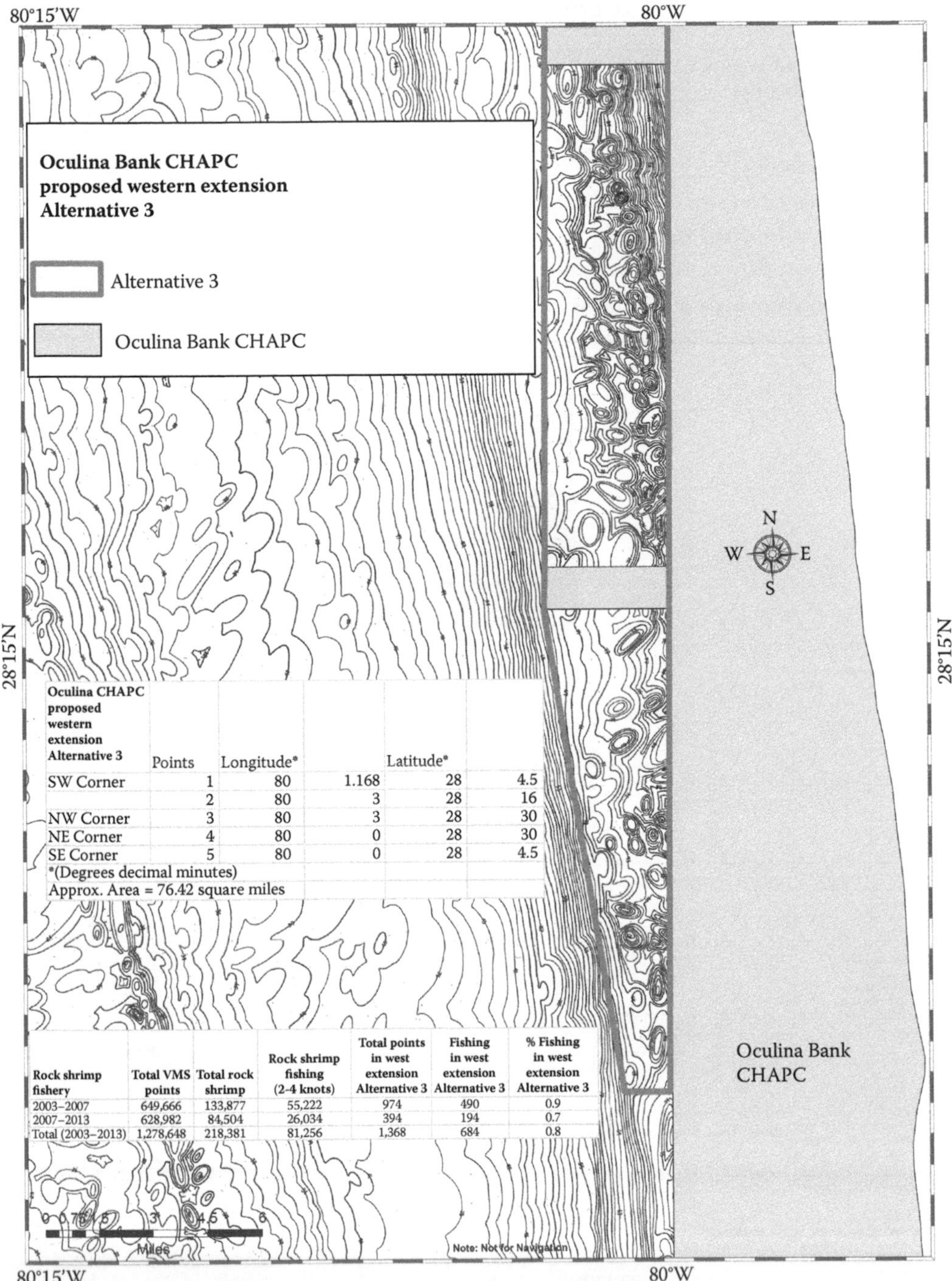

Oculina CHAPC proposed western extension Alternative 3	Points	Longitude*		Latitude*	
SW Corner	1	80	1.168	28	4.5
	2	80	3	28	16
NW Corner	3	80	3	28	30
NE Corner	4	80	0	28	30
SE Corner	5	80	0	28	4.5
*(Degrees decimal minutes)					
Approx. Area = 76.42 square miles					

Rock shrimp fishery	Total VMS points	Total rock shrimp	Rock shrimp fishing (2-4 knots)	Total points in west extension Alternative 3	Fishing in west extension Alternative 3	% Fishing in west extension Alternative 3
2003–2007	649,666	133,877	55,222	974	490	0.9
2007–2013	628,982	84,504	26,034	394	194	0.7
Total (2003–2013)	1,278,648	218,381	81,256	1,368	684	0.8

Figure 4.3 **(See color insert)** The SAFMC Preferred Alternative for a western expansion of the Oculina Bank HAPC included in Coral Amendment 8 (under development). The chart incorporates VMS data from the rock shrimp fishery, 2003–2013.

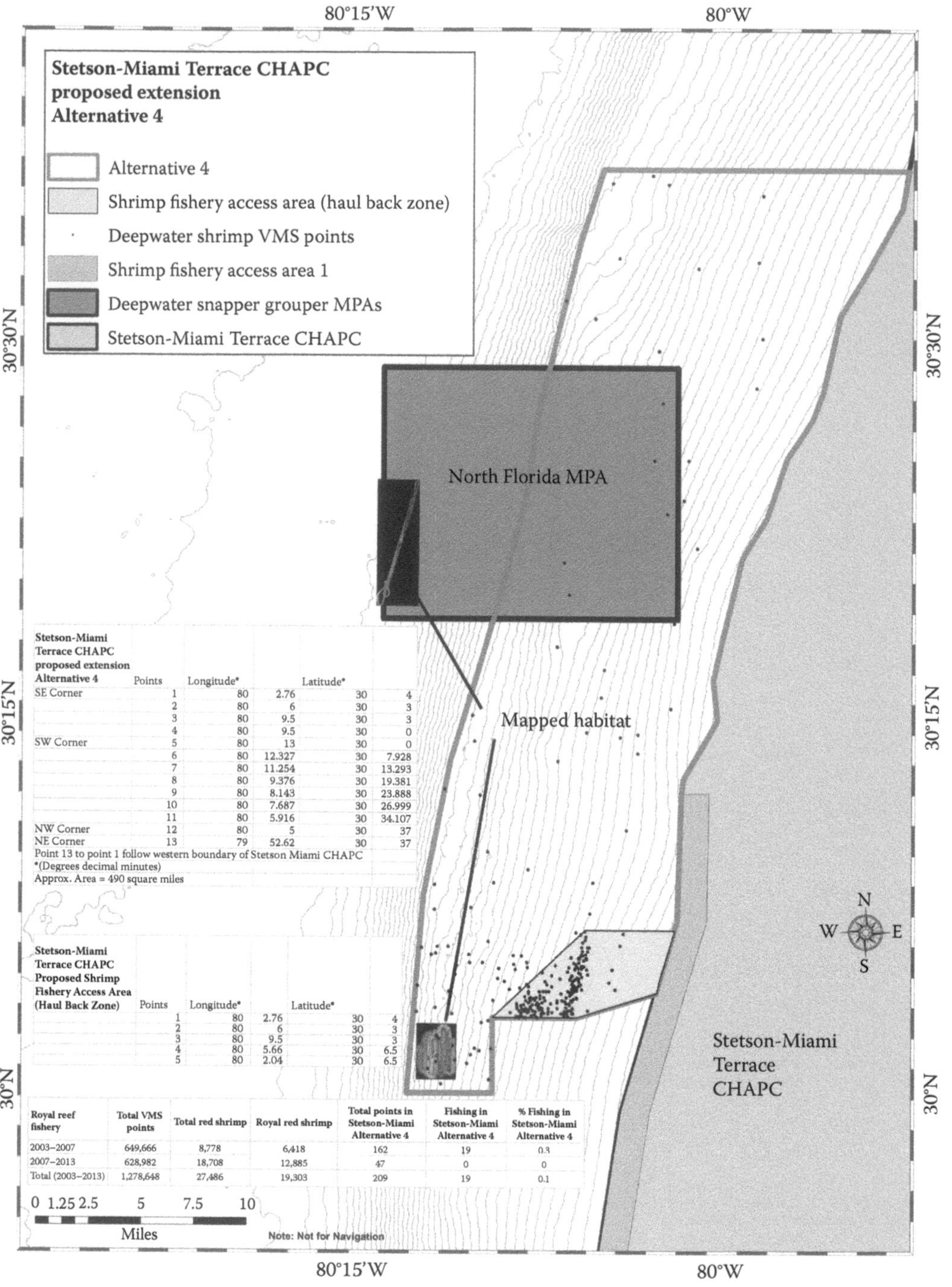

Stetson-Miami Terrace CHAPC proposed extension Alternative 4	Points	Longitude*		Latitude*	
SE Corner	1	80	2.76	30	4
	2	80	6	30	3
	3	80	9.5	30	3
	4	80	9.5	30	0
SW Corner	5	80	13	30	0
	6	80	12.327	30	7.928
	7	80	11.254	30	13.293
	8	80	9.376	30	19.381
	9	80	8.143	30	23.888
	10	80	7.687	30	26.999
	11	80	5.916	30	34.107
NW Corner	12	80	5	30	37
NE Corner	13	79	52.62	30	37
Point 13 to point 1 follow western boundary of Stetson Miami CHAPC					
*(Degrees decimal minutes)					
Approx. Area = 490 square miles					

Stetson-Miami Terrace CHAPC Proposed Shrimp Fishery Access Area (Haul Back Zone)	Points	Longitude*		Latitude*	
	1	80	2.76	30	4
	2	80	6	30	3
	3	80	9.5	30	3
	4	80	5.66	30	6.5
	5	80	2.04	30	6.5

Royal reef fishery	Total VMS points	Total red shrimp	Royal red shrimp	Total points in Stetson-Miami Alternative 4	Fishing in Stetson-Miami Alternative 4	% Fishing in Stetson-Miami Alternative 4
2003–2007	649,666	8,778	6,418	162	19	0.3
2007–2013	628,982	18,708	12,885	47	0	0
Total (2003–2013)	1,278,648	27,486	19,303	209	19	0.1

Figure 4.4 (See color insert) The SAFMC Preferred Alternative for a western expansion of the Stetson-Miami Terrace CHAPC included in Coral Amendment 8 (under development). The chart incorporates VMS data from the rock shrimp fishery 2003–2013 indicating royal red fishery activity. MPA, marine protected area.

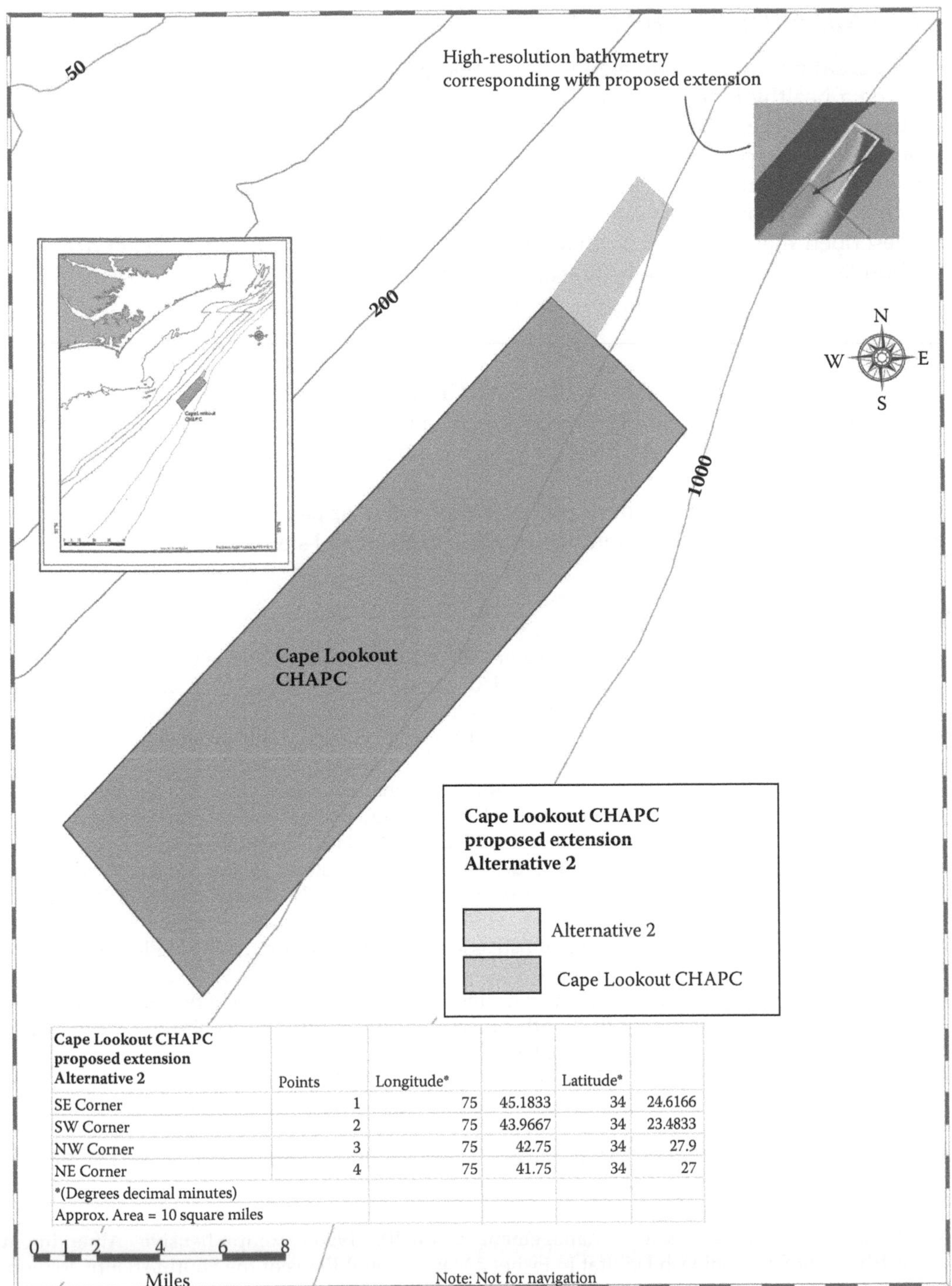

Cape Lookout CHAPC proposed extension Alternative 2	Points	Longitude*		Latitude*	
SE Corner	1	75	45.1833	34	24.6166
SW Corner	2	75	43.9667	34	23.4833
NW Corner	3	75	42.75	34	27.9
NE Corner	4	75	41.75	34	27
*(Degrees decimal minutes)					
Approx. Area = 10 square miles					

Figure 4.5 **(See color insert)** The SAFMC Preferred Alternative for a northern expansion of the Cape Lookout CHAPC included in Coral Amendment 8 (under development).

Stetson Miami Terrace CHAPC

In 2010, coral researchers affiliated with the South Atlantic Council Coral Advisory Panel observed a healthy *L. pertusa* ecosystem thriving in unusually shallow waters (180–250 m depths) west of the existing boundary of the Stetson-Miami Terrace CHAPC (Figure 4.4). Before this notable observation, coral scientists assumed *L. pertusa* occurred in water deeper than 300 m and bounded by temperatures occurring in these depths. Scientists associated the finding with colder than normal temperatures at the depths of the *Lophelia* discovery, and described the area as one with an unusual upwelling oceanographic event driven by the abundance of hardbottom habitat and proximity to the Gulf Stream (Ross et al. 2012a).

Cape Lookout CHAPC

The expansion proposed for the Cape Lookout CHAPC (waters off the coast of Wilmington, North Carolina) was mapped with multibeam sonar during a research cruise that transited through the area. The resulting multibeam data identified several low-relief *L. pertusa* mounds north of the Cape Lookout CHAPC (Ross and Quattrini 2009) (Figure 4.5). The CHAPC and proposed northern extension area appear to be the northernmost deepwater coral habitat on the southeastern U.S. continental slope.

References

Reed J, Farrington S. 2011. A proposal for extension of the boundaries of the *Oculina* coral habitat area of particular concern (Oculina Bank HAPC). Submitted to the SAFMC. 2 December, 2010.

Reed JK, Weaver DC, Pomponi SA. 2006. Habitat and fauna of deep-water *Lophelia pertusa* coral reefs off the Southeastern U.S.: Blake Plateau, Straits of Florida, and Gulf of Mexico. Bull Mar Sci. 78:343–375.

Ross SW, Brooke SD, Quattrini AM. 2012. An unusually shallow and productive deep-water coral community discovered off the Southeastern United States. Poster Presentation, 5th International Symposium on Deep-Sea Corals, 1–6 April, 2012, Amsterdam, The Netherlands.

Ross SW, Nizinski MS. 2007. State of the U.S. deep coral ecosystems in the Southeastern United States Region: Cape Hatteras to the Florida Straits. NOAA Tech. Memo. NMFS-OPR-29. Silver Spring, MD.

Ross SW, Quattrini AM. 2009. Deep-sea reef fish assemblage patterns on the Blake Plateau (Western North Atlantic Ocean). Mar Ecol. 30:74–92.

SAFMC (South Atlantic Fishery Management Council). 1993. Amendment 6 to the Snapper Grouper Fishery Management Plan for the South Atlantic Region. South Atlantic Fishery Management Council, 1 Southpark Cir., Suite 306, Charleston, SC 29407-4699.

SAFMC (South Atlantic Fishery Management Council). 1995. Amendment 3 to the Fishery Management Plan for Coral and Coral Reefs of the South Atlantic Region. South Atlantic Fishery Management Council, 1 Southpark Cir., Suite 306, Charleston, SC 29407-4699.

SAFMC (South Atlantic Fishery Management Council). 1996. Amendment 1 to the Fishery Management Plan for the Shrimp Fishery of the South Atlantic Region (Rock Shrimp). South Atlantic Fishery Management Council, 1 Southpark Cir., Suite 306, Charleston, SC 29407-4699.

SAFMC (South Atlantic Fishery Management Council). 1998. Comprehensive Amendment Addressing Essential Fish Habitat in Fishery Management Plans of the South Atlantic Region. South Atlantic Fishery Management Council, 1 Southpark Cir., Suite 306, Charleston, SC 29407-4699.

SAFMC (South Atlantic Fishery Management Council). 2010. Comprehensive Ecosystem-Based Amendment 1 for the South Atlantic Region. (Amendment 6 to the Coral Fishery Management Plan). South Atlantic Fishery Management Council, 4055 Faber Place Drive, Suite 201, North Charleston, SC 29405.

SAFMC (South Atlantic Fishery Management Council). (2013, under review). Amendment 8 to the Coral Fishery Management Plan for the South Atlantic Region. South Atlantic Fishery Management Council, 4055 Faber Place Drive, Suite 201, North Charleston, SC 29405.

SAFMC (South Atlantic Fishery Management Council) and GMFMC (Gulf of Mexico Fishery Management Council). 1982. Fishery Management Plan for Coral and Coral Reefs of the Gulf of Mexico and South Atlantic. South Atlantic Fishery Management Council, 1 Southpark Cir., Suite 306, Charleston, South Carolina and Gulf of Mexico Fishery Management Council, 2203 N. Lois Avenue Suite 1100, Tampa, FL.

chapter five

Characterization and interrelationships of deepwater coral/sponge habitats and fish communities off Florida

John K. Reed, Stacey Harter, Stephanie Farrington, and Andrew David

Contents

Introduction

In 2009 and 2010, the South Atlantic Fishery Management Council (SAFMC) and the U.S. National Oceanographic and Atmospheric Administration (NOAA) through the Magnuson-Stevens Fishery Management Act established eight deepwater, shelf-edge marine protected areas (MPAs) and five deepwater coral habitat areas of particular concern (CHAPCs) along the outer continental shelf and upper slope off the southeastern United States (NOAA 2010). This network of protected areas was established to sustain and restore reef fish populations and to protect deep-sea coral and sponge ecosystems (DSCEs) from destructive fishing practices, such as bottom trawling and the use of longlines of crab pots, which may extend several kilometers with dozens of traps on each one. This study focused on the deepwater protected managed areas on Pourtalès Terrace, south of the Florida Keys, which includes the East Hump MPA and the Pourtalès Terrace CHAPC, and documented the benthic habitats, fish communities, and benthic macrofaunal communities associated with deepwater, high-relief geological features.

In September 2011, a 3-week research cruise was conducted by the Cooperative Institute for Ocean Exploration, Research, and Technology (CIOERT) at Harbor Branch Oceanographic Institute, Florida Atlantic University (HBOI-FAU) in collaboration with NOAA. The NOAA ship R/V *Nancy Foster* and the University of Connecticut's (UCONN) *Kraken 2* remotely operated vehicle (ROV) were used to survey 14 sites inside and outside the MPA and CHAPC on Pourtalès Terrace (Reed et al. 2013a). This is the first detailed quantitative characterization of the deepwater reef habitats and fish communities within these managed areas in the southern Straits of Florida, and the first extensive multibeam sonar mapping of the terrace.

The primary goal of this research was to document and characterize the deepwater benthic habitats and associated fish communities within these newly established managed areas. These data were analyzed specifically to better understand the interrelationships of the fish communities, including commercially and recreationally important species, relative to the DSCE habitats. These data may then be used as a relative baseline to document changes in these areas due to the implementation of fishing restrictions and to monitor the efficacy and health of these newly designated managed areas. These data will be of value to the SAFMC, NOAA Fisheries Service, and NOAA Office of National Sanctuaries for management decisions on these habitats and managed key species.

Methods

Site description: Pourtalès Terrace

Pourtalès Terrace lies in the southern Straits of Florida, south of the Florida Keys, and consists of extensive, high-relief, hardbottom habitat, and essential fish habitat (EFH) covering 3429 km^2 at depths of 200–450 m (Figure 5.1). The terrace parallels the Florida Keys for 213 km and has a maximum width of 32 km (Land and Paull 2000). The complex karstlike topography of the terrace surface consists of Tertiary limestone of highly phosphatized, biocalcarenite Eocene and Miocene bedrock. High-relief, hardbottom, topographic features consist of a chain of sinkholes extending for ~100 km along the southwest terrace margin, and numerous high-relief knolls and ridges with elevations up to 90 m (Jordan et al. 1964, Malloy and Hurley 1970, Reed et al. 2005).

ROV survey protocol

ROV video and photographic surveys were conducted at each site to ground-truth new multibeam sonar maps, and to quantify and characterize the benthic habitats, fish

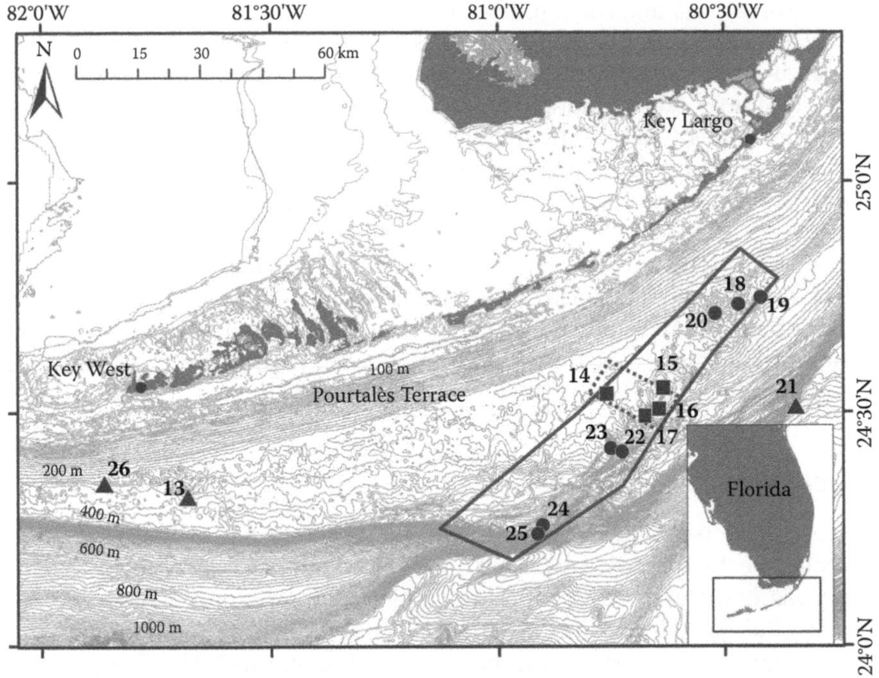

Figure 5.1 Map of Pourtalès Terrace showing *Kraken* ROV dive sites from the 2011 NOAA ship R/V *Nancy Foster* cruise. The NOAA bathymetric contour map shows high-relief topography on Pourtalès Terrace. The East Hump MPA site is indicated with a dotted-line polygon; the deepwater CHAPC is indicated with a black polygon; circles indicate dive sites within the CHAPC, squares indicate MPA dive sites, and triangles indicate dive sites outside management areas.

communities, benthic macrobiota, and coral/sponge cover. Shipboard multibeam echo sounder surveys were conducted at dive sites for which there were no previous multibeam sonar maps. The new sonar maps were used to select dive sites, and surveys were conducted with the UCONN *Kraken 2* ROV, which was equipped with digital video and still cameras mounted with parallel lasers for scale, a CTD (conductivity, temperature, and depth recorder), and a manipulator. Each ROV dive was ~1–2 km in length for a duration of 3–4 h and was documented with continuously recording digital video and digital photographs. The ROV used an integrated navigation system that provided real-time tracking of the ROV every 2 s. Georeferenced multibeam TIFF files, obtained from the sonar surveys, were provided as background files to display target sites and geological features of interest to aid in ROV navigation. All data documentation (digital still images, video, and dive annotations) was georeferenced to the ROV position by matching the time and date to the ROV navigation files.

Fish surveys

An on-screen display video overlay recorded time, date, ROV heading, and ROV depth. The video was digitally recorded continuously throughout each dive from surface to surface. The video camera was typically angled downward ~30° from the horizontal to record fish aggregations and habitat, both near and far to the horizon. The protocol for the fish analyses was to divide the continuous video into 5 min segments, or whenever there was a change in habitat type, whichever came first. Consequently, each video segment consisted of only one habitat type. These habitat designations are described below in the benthic analyses (see Benthic habitat characterization section). During each ROV dive, all fish were identified to the lowest taxonomic level possible and counted. The total distance (km) of each dive was used to calculate the linear density (number of individuals per kilometer) of each fish species. The video camera angle precludes an accurate calculation of areal density of the fish (i.e., number per square kilometer); however, we estimate that the field of view width was generally about 10 m, and most fish were identified within a 5 m distance. So the densities listed in Table 5.1 could be multiplied by 0.1 to estimate the number of fish per square kilometer (based on an average 10 m wide field of view).

Benthic macrobiota characterization

Photographic transects were conducted throughout each ROV dive using the digital still camera directed vertically downward (or as perpendicular as possible to the substrate). The camera was equipped with two parallel lasers (10 cm apart) for scale. In general, two to four digital images were recorded per minute. Each image filename was coded with the corresponding UTC time and date code (using Stamp 2.8 by Tempest Solutions), which was imported into Microsoft Access 2010 and linked to the ROV navigation data using the date/time field of each image. Poor and unusable photos (e.g., blurred, black, off-bottom) or overlapping photos were not included in the analyses. The benthic macrobiota were quantified by analyzing the images for each dive using three measures: (1) species occurrence (presence/absence), (2) percentage of cover of benthic biota, and (3) density of benthic biota.

Percentage of cover of benthic macrobiota was determined by analyzing the quantitative transect images with Coral Point Count with Excel extensions (CPCe 4.1; Kohler and Gill 2006) and following protocols established in part by Vinick et al. (2012) for offshore, deepwater surveys in this region. Fifty random points overlaid on each image with CPCe were identified as to substrate type and associated benthic biota, and then percentage of cover was calculated. The density of the benthic biota was determined by using the parallel lasers for scale and CPCe to calculate the area of each image. All benthic macrobiota (usually >3 cm total length) were then identified to the lowest taxon level possible, counted, and density calculated (number of organisms per square meter). For this report we used the term "coral" as defined by the NOAA Deep-Sea Coral Program (Partyka et al. 2007) as including hard or stony corals (Scleractinia), other taxa with solid calcareous skeletons (hydrozoan lace corals [Stylasteridae]), and nonaccreting taxa such as gorgonians (Octocorallia) and black corals (Antipatharia).

Benthic habitat characterization

Each ROV dive was divided into transects based on several habitat descriptors that were used as factors to characterize and define the benthic habitats. These factors were used to plot percentage of cover and density of benthic macrobiota and density of fish, and to plot

Table 5.1 Fish Densities Counted at Each *Kraken* 2 ROV Dive Site on Pourtalès Terrace

		Dive site													
Scientific name	**Common name**	**13**	**14**	**15**	**16**	**17**	**18**	**19**	**20**	**21**	**22**	**23**	**24**	**25**	**26**
Acropomatidae	Lanternbelly											8.6			
Anthias nicholsi Firth, 1933	Yellowfin bass	344.4	171.9	92.6	9.7	5.2	12.7	2.4	66.1			10.0			743.2
Anthias woodsi Anderson and Heemstra, 1980	Swallowtail bass														1.1
Anthiinae	Anthiid	100.0	338.8	203.7	228.4	47.0	246.0	27.6	373.9		15.8	12.9			475.8
Antigonia capros Lowe, 1843	Deepbody boarfish	75.9	21.5	167.9	178.7	1.5	75.4	3.3	8.7		0.9	8.6			5.3
Bathypterois grallator (Goode and Bean, 1886)	Tripod fish									1.3					
Beryx decadactylus Cuvier, 1829	Red bream												3.8		
Brotula barbata (Bloch and Schneider, 1801)	Bearded brotula								0.9						
Caulolatilus microps Goode and Bean, 1878	Blueline tilefish	3.7	3.3	2.5			7.1	3.3	11.3						2.1
Chaetodon sedentarius Poey, 1860	Reef butterflyfish				0.6										
Chaunax sp. Lowe, 1846	Gaper														
Chaunax stigmaeus Fowler, 1946	Redeye gaper	1.9													
Chlorophthalmus agassizi Bonaparte, 1840	Shortnose greeneye			7.4		0.7					14.0	42.1	3.8		

(continued)

Table 5.1 (Continued) Fish Densities Counted from ROV Video Surveys on Pourtalès Terrace

		Dive site													
Scientific name	**Common name**	13	14	15	16	17	18	19	20	21	22	23	24	25	26
Congridae	Conger eel	1.9								5.0				2.3	
Cyttopsis rosea (Lowe, 1843)	Rosy dory											0.7			
Decodon puellaris (Poey, 1860)	Red hogfish		0.8			1.5	2.4	1.6	0.9			0.7			
Diodontidae	Pufferfish				0.6										
Emmelichthyidae	Rover			14.8	3.2	5.2									
Epigonus sp. Rafinesque, 1810	Deepwater cardinalfish											8.6			
Etelis oculatus (Valenciennes, 1828)	Queen snapper				0.6	2.2					0.9				
Gephyroberyx darwinii (Johnson, 1866)	Big roughy	13.0	4.1	9.9		91.0	5.6	26.8	24.3		0.9	25.7			6.3
Gymnothorax funebris Ranzani, 1839	Green moray				0.6										
Gymnothorax polygonius Poey, 1875	Polygon moray								0.9						
Gymnothorax sp. Bloch, 1795	Moray eel								0.9						
Helicolenus dactylopterus (Delaroche, 1809)	Blackbelly rosefish	57.4	7.4	8.6	2.6	2.2	7.1	5.7	7.8		19.3	11.4	6.8	4.6	54.7
Hemanthias vivanus (Jordan and Swain, 1885)	Red barbier	5.6	0.8				7.9		4.3			2.1			2.1

Hyperoglyphe perciformis (Mitchill, 1818)	Barrelfish					20.1			0.9			0.7			
Hyporthodus niveatus (Valenciennes, 1828)	Snowy grouper	25.9	8.3	2.5	10.3	10.4	6.3	8.1	5.2			0.7			12.6
Jeboehlkia gladifer Robins, 1967	Bladefin bass	1.9	0.8			0.7			1.7		1.8				
Laemonema barbatulum Goode and Bean, 1883	Shortbeard codling	38.9	3.3	4.9	0.6	2.2	19.8	0.8	3.5	1.3	0.9	6.4	14.3	10.7	26.3
Laemonema melanurum Goode and Bean, 1896	Codling													9.2	
Laemonema sp. Günther, 1862	Morid cod	11.1	4.9		1.3		1.6	10.6	3.5		0.9	21.4	34.6	20.6	62.1
Leucoraja lentiginosa (Bigelow and Schroeder, 1951)	Speckled skate									1.3					
Lophiodes beroe Caruso, 1981	Goosefish													0.8	1.1
Lutjanus vivanus (Cuvier, 1828)	Silk snapper				0.6										
Macroramphosus scolopax (Linnaeus, 1758)	Longspine snipefish	3.7	1.7			1.5	8.7	4.3				7.9			8.4
Merluccius albidus (Mitchill, 1818)	Offshore hake												0.8		
Muraenidae	Moray				0.6				0.9		0.9				
Nezumia sp. Jordan and Starks, 1904	Grenadier									2.5			65.4	42.7	

(continued)

Table 5.1 (Continued) Fish Densities Counted from ROV Video Surveys on Pourtalès Terrace

		Dive site													
Scientific name	Common name	13	14	15	16	17	18	19	20	21	22	23	24	25	26
Ophidiidae	Cusk-eel												3.0	3.1	
Oxynotus centrina (Linnaeus, 1758)	Angular roughshark														1.1
Pagrus pagrus (Linnaeus, 1758)	Red porgy								0.9						
Parahollardia lineata (Longley, 1935)	Jambeau						0.8								
Peristedion miniatum Goode, 1880	Armored searobin									1.3					
Plectranthias garrupellus Robins and Starck, 1961	Apricot bass		3.3	2.5	18.1	6.0	5.6		20.0		1.8				
Polymixia sp. Lowe, 1836	Beardfish												45.1		
Priacanthus arenatus Cuvier, 1829	Bigeye				1.9										
Prognathodes aya (Jordan, 1886)	Bank butterflyfish				1.3										
Prognathodes guyanensis (Durand, 1960)	French butterflyfish				1.3										
Pronotogrammus martinicensis (Guichenot, 1868)	Roughtongue bass		52.1	98.8	587.1	22.4	119.8		23.5		67.5				

Raja sp. Linnaeus, 1758	Skate											0.7			
Rajidae	Skates													0.8	
Scyliorhinidae	Catshark							0.8					0.8		1.1
Scyliorhinus meadi Springer, 1966	Blotched catshark													3.8	
Scyliorhinus retifer (Garman, 1881)	Chain catshark	1.9													
Seriola dumerili (Risso, 1810)	Greater amberjack				0.6						4.4				
Seriola rivoliana Valenciennes, 1833	Almaco jack				0.6						0.9				
Seriola sp. Cuvier, 1816	Amberjack	9.3			7.1						4.4				
Serranidae	Sea bass														3.2
Synchiropus sp. Gill, 1859	Dragonet				0.6										
Torpedo nobiliana Bonaparte, 1835	Atlantic torpedo							0.8							
Urophycis sp. Gill, 1863	Phycid hake						0.8								
Total density		696.5	623.0	616.1	1057.0	219.8	527.6	96.1	560.1	12.7	135.3	169.2	178.4	98.6	1406.5
Total number of species		16	15	12	23	16	16	13	20	6	15	17	10	10	16

Note: All fish were identified from each ROV dive to the lowest taxonomic level possible and counted. The total distance (km) of each dive was used to calculate the linear density (number of individuals per kilometer) of each fish species.

transects on the multibeam sonar maps in ArcGIS 10.0. The factors included the following habitat descriptors:

1. Geomorphology: The geological feature generally was defined from the multibeam sonar images, for example, mound-top (peak of rock mound), mound-slope (flank of rock mound), mound-wall (steep, near vertical upper slope of rock mound), *Lophelia* mound (deepwater reef, composed chiefly of the branching azooxanthellate scleractinian coral *Lophelia pertusa*), sinkhole (deepwater sinkhole), valley (flat, low slope areas off reef, typically at the base of the mounds), pavement (flat, low-relief hardbottom areas), and mound-deep (isolated mound separate and at the base of the primary rock mound).
2. Substrate: This was a subset of the Southeastern United States Deep-Sea Corals (SEADESC) habitat categories that were developed by the NOAA Deep-Sea Coral Program for use in analysis of deep-sea coral dive surveys (Partyka et al. 2007). On Pourtalès Terrace, the substrate descriptors included soft bottom (unconsolidated sand/mud), and the following hardbottom types: rock pavement, pavement with ledges, pavement with sediment veneer, rock wall, and *L. pertusa* coral. These are illustrated with ROV images in Figure 5.2.
3. Depth: The depth range of the transect or dive.
4. Slope: Slope was estimated from the ROV video: flat = 0°–5°, low = 5°–30°, moderate = 30°–60°, high (wall) = 60°–90°.

Management status was also used to characterize a dive site, that is, whether it was within the protected managed areas (MPA or CHAPC) or outside the protected areas (no protection).

Statistical analyses

Multivariate analyses were used to assess differences in benthic faunal assemblages and fish communities among the dive sites and habitat factors. All analyses were conducted with PRIMER 6.1.13 analytical software based on guidelines outlined in Clarke and Warwick (2001) and Clarke and Gorley (2006). The ROV transects were characterized by the habitat factors described above (geomorphology, substrate, depth, and slope). In addition, the dive sites were compared on the basis of management status (i.e., protected [within the MPA or CHAPC] or unprotected). For the benthic analyses, the number of individuals for each species of benthic macrobiota was counted in each image, then summed by transect, and divided by the total area of the digital still images examined within that transect. This resulted in the density of each benthic species (number per square meter) by transect. The densities were then averaged in PRIMER by site and habitat factors and square-root transformed to reduce the effect of area on the similarity coefficients.

For analyses of the fish communities, the number of individuals of each species was counted within each transect, summed for the entire transect, and then divided by the total distance of that transect. This resulted in the linear density of each species by transect (number per kilometer). The counts were then averaged in PRIMER by site and habitat factors and presence/absence transformed to reduce the dominating influences of abundant species to the similarity matrix. Similarity among dive sites and habitat factors for both fish and benthic biota were then calculated separately using the S17 Bray-Curtis similarity coefficient. A nonmetric multidimensional scaling ordination (NMDS) plot and a dendrogram with group-average linking were created to depict the results of a

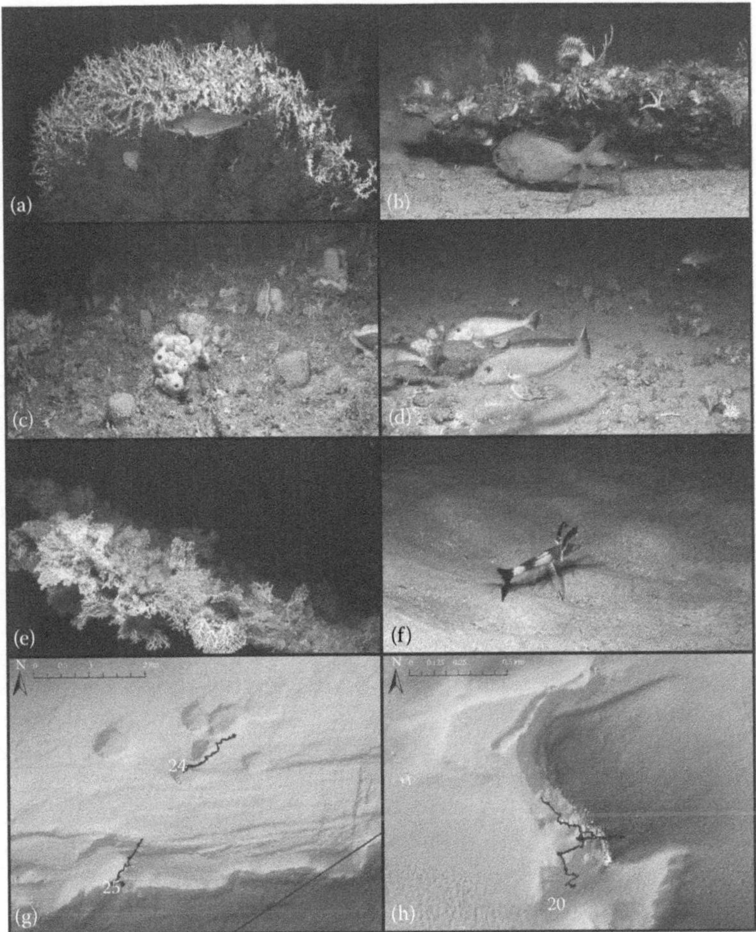

Figure 5.2 **(See color insert)** Representative habitats found on Pourtalès Terrace from ROV dive images. (a) Habitat: deepwater *L. pertusa* coral mound; live coral colony (~1 m diameter), black coral (*Leiopathes* sp.), and mora codling (*Laemonema melanurum*); CHAPC, Site 25, 509 m. (b) Habitat: flat rock pavement with low-relief ledges; big roughy (*G. darwinii*), anemones (*Actinoscyphia* sp.), various gorgonians, hydroids, and crinoids; CHAPC, Mound no. 311 Wall, Site 23, 302 m. (c) Habitat: mound-slope, rock pavement; dense cover of Stylasteridae corals and numerous demosponge species; HAPC, Mound no. 311, Site 22, 233 m. (d) Habitat: flat pavement with sediment; blueline tilefish (*C. microps*), stylasterid corals; CHAPC, Alligator Bioherm no. 4, Site 18, 183 m. (e) Habitat: mound-wall, upper slope and edge of mound; dense cover of Stylasteridae coral; MPA, NW Mound, Site 14, 191 m. (f) Habitat: soft bottom; tripod fish (*B. grallator*); unprotected, Pourtalès Escarpment, Site 21, 836 m. (g) Multibeam sonar of CHAPC, Sinkhole site (Site 24) and *Lophelia* mound site (Site 25); dive tracks are shown by solid red lines. (h) Multibeam sonar of CHAPC, Alligator Bioherm no. 3 (Site 20).

concurrent similarities profile (SIMPROF). Similarity percentages (SIMPER) was utilized to determine which species contributed to the dissimilarities between the group pairs. In addition, one-way ANOVA was used to test for differences in the fish densities of the four most abundant commercially and recreationally important species among the management areas.

Results

In September 2011, the 3-week CIOERT Florida Shelf-Edge Exploration II Cruise with the NOAA ship R/V *Nancy Foster* resulted in 14 ROV dives on Pourtalès Terrace for a total bottom time of 52 h, covering 16.02 km at depths ranging from 154 to 838 m (Figure 5.1). A total of 2866 *in situ* digital images were recorded, including 2253 transect photographs that were used for the quantitative benthic data analyses. Ten sites were surveyed on Pourtalès Terrace with multibeam sonar by the survey team of the R/V *Nancy Foster* and covered a total of 397.1 km^2. Except for one site, these sites had not been previously surveyed with high-resolution, multibeam sonar. Three sonar maps were also made for the first time of areas inside the recently designated East Hump MPA site. These new sonar maps enabled the discovery of many new features and EFH. In particular, four deepwater sinkholes were discovered where only one had been previously known (Marathon Sinkhole, Site 24). We also discovered a large deepwater *Lophelia* coral mound, which is the southernmost *Lophelia* mound known in U.S. waters (Site 25). The ROV dives documented and characterized 14 sites on Pourtalès Terrace including 11 within the CHAPC, of which four were within the East Hump MPA; three dives were outside the protected areas (Table 5.2).

Habitat characterization

A SEADESC Level II Report describing each dive site from this cruise was submitted to NOAA as an unpublished technical report (Reed et al. 2013a). In general, many of the sites were high-relief, flat-topped mounds consisting of eroded Tertiary rock. These were somewhat similar in shape to buttes on land but not as large as seamounts or guyots. Each ROV dive was approximately 1–2 km in length, crossing and surveying a mound or other geological feature. The ROV dives over these topographical features were divided into transects based on various habitat factors. These included geomorphology descriptors, which consisted of mound-slope (the lower slope of a mound was typically inclined at 10°–30° to 30°–60°), mound-wall (typically the upper slope was nearly vertical [60°–90°] and undercut with a series of ledges, almost like a pagoda), and mound-top (flat top or rounded peak). A zone of deepwater sinkholes was discovered at one site (Site 24) that was quite distinct from the mound habitats. Another dive discovered a deepwater *Lophelia* coral mound (Site 25) that also had distinct habitat, benthic fauna, and fish communities. The Pourtalès Escarpment site (Site 21), which was the deepest site and entirely soft bottom, was also distinct in benthic fauna and fish. Examples of the various types of geomorphology and substrate are shown in Figure 5.2.

Site descriptions: Sites within the East Hump MPA (also within CHAPC)

The majority of the newly designated East Hump MPA lies within the Pourtalès Terrace CHAPC. Four ROV dives made within the MPA were the first visual surveys (ROV or submersible) to characterize this MPA. These are described below.

East Hump MPA, Northwest Peak (Site 14). The multibeam sonar showed a large 25 m tall, flat-topped mound (180 m depth at top, 205 m at base), oriented N–S and nearly 1.8 km long. The ROV traveled up the southeast slope, then north across the top, covering about half of the top plateau. The lower slope was 30° rock and sediment, and the upper slope was nearly vertical with ledges of up to 2 m vertical relief. The top edge and ledges had dense cover of stylasterid coral and diverse demosponges.

Table 5.2 Summary of ROV Video/Photographic Survey Sites on Pourtalès Terrace

ROV site number	Site name	Managed area	Depth (m)	% Hardbottom	Fish, species (no.)/density (number/km)	Benthic biota, density (number/m²)	% Cover, benthic biota	% Cover, coral	% Cover, gorgonians	% Cover, Antipatharia	% Cover, Porifera
13	Jordan Site F	None	183–284	100	16/696	4.29	4.4	1.7	0.2	1.1	0.4
14	NW Mound	MPA	182–213	85.1	15/623	8.38	5.8	2.5	0.07	0	1.9
15	SE Mound	MPA	207–281	100	12/616	10.62	10.7	0.8	0.1	0	7.7
16	Tennessee Bioherm 1	MPA	182–269	100	23/1057	25.1	20.4	4.5	1	0	14.2
17	Tennessee Bioherm 1, East Wall	MPA	203–269	100	16/220	17.79	18.1	1.03	4.08	0	11.9
18	Alligator Bioherm 4	HAPC	176–229	87.7	16/528	11.98	7.2	1.6	0.02	0	4.6
19	Alligator Bioherm 4, East Wall	HAPC	203–249	87.4	13/96	11.94	10.6	1.7	4.1	0.1	2.8
20	Alligator Bioherm 3	HAPC	163–233	98.3	20/560	18.23	16.6	5.7	2.1	0.03	5.2
21	Pourtalès Escarpment	None	817–838	1.5	6/13	0.07	0.2	0	0	0	0
22	Mound 311	HAPC	154–320	100	15/135	29.24	22.2	9	0.02	0	11.8
23	Mound 311, SE Wall	HAPC	280–315	97.1	17/169	2.96	3.9	0.1	0.9	0.08	1
24	Sinkhole	HAPC	473–494	100	10/178	30.44	13.8	0.8	8.6	0.7	2.4
25	*Lophelia* Mound	HAPC	467–574	99.5	10/99	13.09	23	15	3.3	0.8	1.6
26	Key West Bioherm	None	195–255	96.4	16/1406	11.69	6.5	1.4	0.8	0.3	0.6

Note: Managed areas are areas inside the protected management areas (MPA and CHAPC) or outside (None); fish density (number of individuals per kilometer) and number of species from video analysis; density and percentage of cover of benthic biota from CPCe Point Count analysis of quantitative photo transects.

East Hump MPA, Southeast Peak (Site 15). The multibeam sonar showed a mound near the border of the MPA. The NW–SE oriented mound was 941 m long and 600 m wide, with a depth of 197 m at the top, and a trench scoured to 292 m at the northeast base. The south slope was paved with sediment, some boulders, and ledges; the north face was a 60°–90° rock slope with numerous ledges. The benthic cover was dense stylasterid coral, diverse demosponges, hexactinellid sponges, and black coral. Fishing line was observed on the coral, slopes, and peak of the mound.

East Hump MPA, Tennessee Bioherm no. 1 (Site 16). The multibeam survey showed Tennessee Bioherm no. 1 as an elongated mound (170 m depth at top), oriented N–S and 1450 m long, with a depth of 256 m at the south base and 322 m in a deep scoured trench at the NE base. The south face of the mound was a 10°–30° rock pavement slope and ledges; the east slope was 45° with 1–2 m ledges; the top was a rounded rock peak covered with stylasterid coral and rubble. The dominant fauna were dense and diverse demosponges, hexactinellids, stylasterids, and plexaurid gorgonians.

East Hump MPA, escarpment SE of Tennessee Bioherm no. 1 (Site 17). The multibeam sonar showed a 3 km long escarpment oriented NE–SW; the top was 190 m deep and at the base of the wall was a trench at a depth of 270 m. The east face was a 30°–60° rock slope, with steep facies, ledges, and boulders. Dense biota included demosponges, stylasterid coral, and primnoid gorgonians.

Site descriptions: Sites inside the Pourtalès Terrace CHAPC (but outside the East Hump MPA)

In addition to the four MPA sites, seven additional sites were surveyed within the Pourtalès CHAPC (Sites 18–20, 22–25). These included a variety of high-relief mounds, escarpments, and a newly discovered deepwater *Lophelia* coral reef and deepwater sinkholes. Some of these sites were qualitatively described in Reed et al. (2005) and were described in more detail in the NOAA technical report (Reed et al. 2013a). Newly surveyed sites are described below.

Escarpment east of Alligator Bioherm no. 4 (Site 19). The multibeam sonar showed a long escarpment at the easternmost end of the humps region of Pourtalès Terrace (outside the MPA). At the base of the escarpment was a scoured depression, 246 m maximum depth, and the top of the wall was 186 m deep. The slope was 45°–60° rock with vertical escarpments and ledges. Biota consisted of areas with nearly 100% cover of living stylasterid coral and stylasterid rubble, dense primnoid gorgonians, 1.5 m tall black coral, and dense sponges. Large amounts of fishing line were observed on the bottom.

Mound no. 311 (Site 22). This site was first discovered and dived on by one of us (JKR) with the *Johnson-Sea-Link* submersible in 2006. The multibeam sonar deployed from the R/V *Nancy Foster* showed an elongate mound (minimum depth 150 m) oriented NW–SE, 920 m long, with a maximum depth of 374 m at the SE base. The east face was very steep from 40°–60° rock slope to 80°–90° rock escarpments and ledges. The cover was dense in areas, with up to 100% living stylasterid coral and stylasterid rubble. Other hard corals included *Madrepora* sp., *Madracis* sp., and *Dendrophyllia* sp. This site appeared the most heavily impacted by fishing as few large fish were present and the mound was crisscrossed with fishing line and tackle.

Marathon Sinkhole (Site 24). The Marathon Sinkhole was first discovered in the 1950s and surveyed by Jordan (1954) and then later described by Land and Paull (2000) using the Navy NR-1 submarine to fly above it. Reed et al. (2005) entered the sinkhole with *Johnson-Sea-Link* submersible dives, but only one sinkhole was known at this site at that time. The multibeam sonar deployed from the R/V *Nancy Foster* in 2011 revealed three additional sinkholes within 900 m of the Marathon Sinkhole. Sinkhole 1 (the actual Marathon Sinkhole) is a double sinkhole, 900 m in diameter E–W, with a maximum depth of 527 m at the base, and 455 m at the top north rim. Sinkholes 2, 3, and 4 range in diameter from 400 to 525 m, with a maximum depth of 529 m (Figure 5.2g). The upper rims of the sinkholes consisted of very rugged, karstlike limestone with undercut ledges and vertical escarpments. The substrate between the sinkholes was flat rock pavement with low-relief ledges and small boulders. This area was dominated by demosponges, dense gorgonians, and black coral. Scleractinian hard corals occurred around the rims and included *L. pertusa*, *Enallopsammia profunda*, *Madrepora* sp., and *Dendrophyllia* sp. Fish included beardfish (*Polymyxia* sp.), sharks, and red bream (*Beryx decadactylus*). The commercially fished golden crab (*Chaceon fenneri*) was also observed.

Lophelia coral mound (Site 25). The multibeam sonar showed an extensive escarpment south of the sinkholes (Figure 5.2g). At the base of this escarpment are several mounds. One surveyed mound was a 48 m tall *Lophelia* coral bioherm, with a depth of 548 m at the south base, and 500 m at the peak. This is the southernmost *Lophelia* coral reef yet discovered in U.S. waters. It appears from the multibeam sonar that other coral mounds may be present and even common along this outer edge of Pourtalès Terrace. The peak and upper slope of the coral mound were covered with thickets of live *L. pertusa*, in 5–8 m long hedge-like rows and up to 50 cm tall, and coral rubble. Hexactinellid sponges and black coral were common, but few fish other than mora codling (*Laemonema* sp.) were observed. The escarpment to the north was smooth rock pavement, with rock slabs, cobble, and outcrops. Sponges and gorgonians were common.

Site descriptions: Sites outside the CHAPC

Three dives were made on Pourtalès Terrace outside the CHAPC boundaries. Extensive areas of high-relief topography are apparent in NOAA regional bathymetric charts extending all along the terrace to the west of the CHAPC boundary (Figure 5.1). Two dives south of Key West (Sites 13 and 26) were in this area of high relief and may be characteristic of the other unprotected areas on the terrace; they appear to provide coral and sponge habitat as well as EFH. The Magnuson-Stevens Act defines EFH as waters and substrates necessary to fish for spawning, breeding, feeding, or growth to maturity. Site 21, which was the deepest (838 m) of the dive sites, was on the outer edge of Pourtalès Terrace Escarpment near the boundary of the exclusive economic zone where the NOAA bathymetric contour chart shows a steep slope from 700 to 850 m. Because this site lies within the axis of the Florida Current, which may exceed 2 m s^{-1}, tracking and maneuvering of the ROV was difficult. Consequently, we were unable to reach the intended location. The seafloor at the dive site was flat sandy-mud bottom where we recorded video of royal red shrimp (*Pleoticus robustus*), tripod fish (*Bathypterois grallator*), mora codling, armored sea robin (*Peristedion miniatum*), and red deep-sea crab (*Chaceon quinquedens*).

Benthic macrobiota

Some common taxa were identified to genus or species level from the visual recordings, but many could only be identified to a higher taxonomic level such as family, class, order, or even phylum. Sponges, gorgonians, and black coral in this region are especially difficult to identify without a specimen in hand. Many deepwater species in this region appear nearly identical, such as fan sponges that may be in different orders or even classes (i.e., Demospongiae or Hexactinellida). A total of 146 taxa were identified from the quantitative photograph transects and were used for CPCe percentage of cover and density analyses (see Reed et al. 2013a for a complete list, percentage of cover, and densities). These included 58 taxa of Porifera and 47 Cnidaria, which included the following corals: Scleractinia (*L. pertusa, Madracis myriaster, Madrepora oculata, Solenosmilia variabilis,* and Dendrophylliidae); Stylasteridae (five species); Octocorallia (16 gorgonian taxa including, *Ellisella* spp., Plexauridae, Isididae, *Muricea* spp., *Paramuricea* spp., *Plumarella* spp., and Primnoidae); and Antipatharia (*Bathypathes alternata, Leiopathes* spp., *Stichopathes lutkeni,* and *Tanacetipathes* spp.). Noncoral Cnidaria included Actiniaria, Zoanthidea, Cerianthidae, Alcyonacea (Nephtheidae, *Anthomastus* sp., *Capnella* sp.), and Hydroidolina.

The dominant sponges included the following Demospongiae (recently revised by Redmond et al. 2013): order Tetractinellida (*Geodia* spp., including three new species [Paco Cardenas, Evolutionary Biology Center, Uppsala University, Sweden, personal communication], *Pachastrella* sp., and the lithistids Corallistidae, *Theonella* spp., *Leiodermatium* sp.); Hadromerida (encrusting Spirastrellidae); Poecilosclerida (Raspailliidae, encrusting *Hymedesmia* sp.); Halichondrida (*Auletta* sp., *Topsentia* sp., *Phakellia* sp.); and Homoscleromorpha (*Plakortis* spp.). Glass sponges, class Hexactinellida, included *Aphrocallistes beatrix beatrix, Hexactinella* sp., *Iphiteon* sp., and a new genus *Nodastrella* gen. nov. (Dohrmann et al. 2012).

Other benthic macrobiota included Annelida, Mollusca, Arthropoda, Bryozoa, Echinodermata, Hemichordata, and Ascidiacea. Some of the shallower mound tops also included algae: Cyanobacteria, Chlorophyta, and Rhodophyta (primarily crustose coralline algae). Alligator Bioherm no. 3 (Site 20) had algae on the top plateau of the mound: Cyanophyta (depth 168 m), thin encrusting Chlorophyta (164 m), and Corallinales (encrusting Rhodophyta, 164–172 m). The Key West Bioherm (Site 26) had encrusting green algae at 195 m, and some unidentified coralline red algae at 193–222 m. The deepest algae recorded in the Atlantic were at 268 m for an unidentified crustose coralline, 210 m for *Ostreobium* sp. (Chlorophyta), and 189 m for *Peyssonnelia* sp. (Rhodophyta), all from San Salvador seamount in the Bahamas (Littler et al. 1985).

Density of benthic macrobiota

Table 5.2 lists, for each site, the percentage of cover of hard substrate, density and numbers of fish species, density of benthic macrobiota, and percentage of cover for other dominant benthic taxa (corals, sponges, and gorgonians). The Pourtalès Escarpment site (Site 21) was the only site that was nearly 100% soft-mud bottom; as a result, it had no corals, sponges, or other sessile benthic fauna. All the other sites were predominantly hard bottom (85%–100% cover) with a relatively high percentage of cover of corals, sponges, and numerous motile invertebrates. Percentage of cover of macrobiota on hardbottom sites ranged from 3.9% (Mound no. 311 SE Wall, Site 23) to 23% (*Lophelia* mound, Site 25). Mean percentage of cover for the four MPA dive sites was 13.7%; Tennessee Bioherm no. 1 (Site 16) had the highest cover of the MPA sites (20.4%). Within the seven CHAPC sites (outside the MPA), the greatest percentages of cover were at the *Lophelia* mound (23% cover), Mound no. 311 (22.2%), and Alligator Bioherm no. 3 (16.6%). The lowest percentage of cover on any hardbottom site was at Mound no. 311, SE Wall (3.9% cover).

The greatest cover by scleractinian framework coral was at the *Lophelia* mound site (15% mean cover); however, the dive track also included a pavement valley and a rock mound slope north of the coral mound. The actual mean cover of standing *L. pertusa* on the coral mound itself was 17.8% on top and 19.7% on the slope. The maximum *L. pertusa* cover in a single photographic image was 44%. *L. pertusa* was also found along the edge of the sinkholes (Site 24, 0.13%), along with *M. oculata* (0.06%). Stylasterid coral had relatively high coverage at most sites: 0.7%–4.1% within the MPA sites, 5.1% at Alligator Bioherm no. 3, and 6.2% on Mound no. 311. The maximum mean stylasterid coral cover was on top of Mound no. 311 (44%). Many of the mound tops were covered with thick layers of stylasterid coral rubble as well as standing live Stylasteridae. Nonscleractinian corals included gorgonians, which were densest at the sinkhole (8.6% cover), the wall east of Tennessee Bioherm no. 1 (4.1%), and Alligator Bioherm no. 4 (4.1%). Antipatharians were present at most sites but in low numbers where present (0.03%–1.1%). Sponges were common and diverse at most sites (0.4%–14.2% cover) and were most abundant at the MPA sites.

Benthic macrobiota: Site relationships

Dive sites within and outside the managed areas (MPA and CHAPC) were compared using a nonmetric multidimensional scaling plot of Bray-Curtis similarity for the benthic macrobiota (species densities averaged by site and then square-root transformed) (Figure 5.3). The muddy Pourtalès Escarpment (Site 21), which did not have hardbottom habitat, was removed from the plot because it was such a strong outlier. The remaining sites could be categorized into five statistically different faunal groups; these are shown by letter designations in the plots (SIMPROF, $p < .05$). All the MPA sites (Sites 14–17) and some of the CHAPC sites (Sites 18–20, 22) clustered together at 40% similarity (Groups B, C, D). The CHAPC sites (Sites 24 and 25), which were considerably deeper (467–574 m), formed a statistically separate group (Group A). The unprotected sites off Key West (Sites 13, 26)

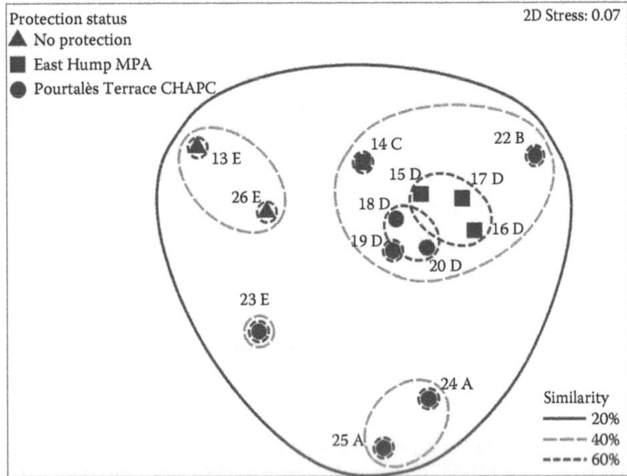

Figure 5.3 Similarity of sites within and outside the protected management areas (MPA and CHAPC) based on the benthic macrobiota densities (nonmetric, multidimensional scaling plot based on the Bray-Curtis similarity matrix calculated from benthic biota densities averaged by site with square-root transformation). Assemblage similarities at 20%, 40%, and 60% are shown. Sites are indicated by numbers; statistically similar groups (SIMPROF, $p < .05$) are indicated by the same letters (A through E).

were also distinct and comprised a third group (E). The wall off Mound no. 311 (Site 23) was distinct from the other CHAPC sites; it also had the lowest benthic faunal density and percentage of cover.

Pairwise tests using SIMPER showed which species contributed most to the differences between these groups. The group of CHAPC sites showed 70.2 average dissimilarity from the unprotected sites (outside the CHAPC). This dissimilarity was due primarily to the occurrence of the following taxa: sagartiid anemones (contributing 6.3% of the dissimilarity), stylasterid corals 5.8%, unidentified white gorgonians 4.7%, and primnoids 4.7%. The East Hump MPA group and the unprotected group showed 66.2% average dissimilarity, with demosponges contributing 9.8% of the dissimilarity, sagartiid anemones 7.0%, *Hymedesmia* sp. 3.6%, and primnoids 3.5%.

Fish communities: Site relationships

All fish were identified for each ROV dive site to the lowest taxon practicable and counted. The 14 ROV dives covered a 16.02 km distance and video analysis recorded a total of 7273 individual fish consisting of 62 taxa in 38 families (Table 5.1). The number of species per site ranged from six on the mud slope of the Pourtalès Escarpment (Site 21) to 23 at Tennessee Bioherm no. 1 within the MPA (Site 16) (Table 5.2). Excluding the escarpment mud site, the mean density ranged from 96.1 fish per kilometer at Alligator Bioherm no. 4, East Wall (Site 19) to 1406 per kilometer at Key West Bioherm (Site 26). The four MPA sites had 12–23 species per site and densities of 220–1057 fish per kilometer, with Tennessee Bioherm no. 1 supporting the most diverse and dense populations of the four. The seven CHAPC sites had 10–20 species per site and densities of 96–560, with Alligator Bioherm no. 3 the most diverse and dense. It is important to note that the two unprotected, hardbottom sites, both off Key West (Sites 13, 26), had relatively high diversities and densities of fish. The Key West Bioherm, in particular, had the greatest density of fish. Overall, the fish taxa with the greatest mean densities were unidentified anthiins (138 per kilometer), yellowfin bass (97; *Anthias nicholsi*), roughtongue bass (65; *Pronotogrammus martinicensis*), deepbody boarfish (36; *Antigonia capros*), big roughy (14; *Gephyroberyx darwinii*), and blackbelly rosefish (13; *Helicolenus dactylopterus*).

Sites within and outside the managed areas (MPA and CHAPC) were compared using a nonmetric multidimensional scaling plot of the Bray-Curtis similarity coefficient using the presence/absence transformation of fish species (Figure 5.4a). Six statistically different groups resulted from the SIMPROF test ($p < .05$). The letters by the site numbers in the figure indicate the statistically significant groups. The Pourtalès Escarpment site (Site 21) was distinct from the other sites; this site was the deepest and entirely soft bottom. The Sinkhole and *Lophelia* mound sites (Sites 24 and 25, respectively) formed a distinct group (F) and were also considerably deeper than the other CHAPC sites. All the MPA sites (dives 14–17) formed a statistically significant distinct group (D) with CHAPC Sites 18 and 20. The unprotected sites off Key West (Sites 13 and 26) formed a distinct group (A) with the Alligator Bioherm no. 4 wall site (Site 19).

Pairwise tests using SIMPER showed which species contributed the most to the differences between these groups. The East Hump MPA sites showed 61.2 average dissimilarity from the unprotected sites. Roughtongue bass contributed 12.4% to the dissimilarity, yellowfin bass 9.5%, blackbelly rosefish 9.1%, and deepbody boarfish 8.8%. The CHAPC sites and the unprotected sites showed 63.9 average dissimilarity, with yellowfin bass contributing 11.7% to the dissimilarity, blackbelly rosefish 9.9%, and mora codling 8.7%.

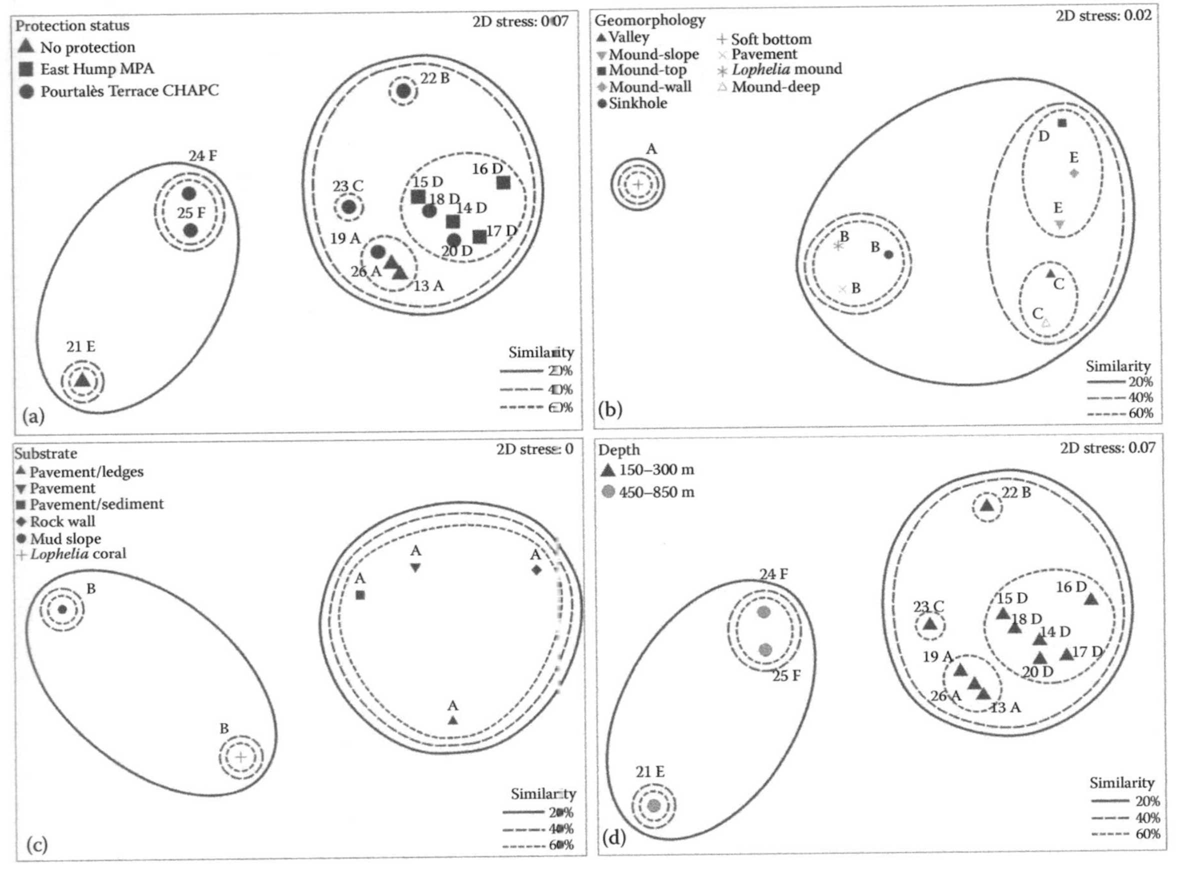

Figure 5.4 Relationship of fish populations with various habitat factors (nonmetric, multidimensional scaling plots based on the Bray-Curtis similarity matrix calculated from fish species data with average presence/absence transformation). Factors included: (a) protection status, (b) geomorphology, (c) substrate, and (d) depth. (a) and (d) are averaged by site; (b) and (c) are averaged by habitat factor among all sites. Assemblage similarities at 20%, 40%, and 60% are shown. Sites are indicated by numbers; statistically similar groups (SIMPROF, $p < .05$) are indicated by the same letters.

Interrelationships of fish communities and habitat

The interrelationships of the fish communities with habitat factors were analyzed with MDS plots of similarity (Figure 5.4b–d). Depth was the most influential factor contributing to fish species composition. The MDS plot of depth (Figure 5.4d) shows two distinct groupings with all the deep dives (Sites 24, 25, 21; depths of 450–850 m) clustering together at 20% similarity. The shallower sites (depths of 150–300 m) also clustered together at 20%. The two depth categories had 87.6 average dissimilarity between them; grenadier (*Nezumia* sp.) contributed to 15.7% of the dissimilarity, mora codling 13%, anthiins 9.7%, and blackbelly rosefish 9.7%.

Geomorphology was the second most influential factor in determining fish species composition (Figure 5.4b). Five statistically different groups resulted from the SIMPROF test ($p < .05$). Again, Site 21, the soft-bottom site on Pourtalès Escarpment, was distinct from the other sites. The geomorphology classes of *Lophelia* mound, sinkhole, and pavement formed a statistically distinct group (Group B), clustering together at 60% similarity. Valley and mound-deep sites formed another distinct group (C). Mound-slope, mound-wall, and mound-top grouped together at 60% similarity. The mound-top habitat was slightly different in fish composition from the mound-slope and mound-wall (SIMPER). The species responsible for this dissimilarity were higher abundances of deepbody boarfish, roughtongue bass, and yellowfin bass in the mound-top zone.

Substrate was the third most influential factor in fish community composition. The MDS plot (Figure 5.4c) shows two statistically different groups (SIMPROF, $p < .05$), with pavement/ledges, pavement, pavement/sediment, and rock wall habitats all forming one group (Group A), and the *Lophelia* coral and mud slope habitats forming the other (B). The average dissimilarities between mud slope and the other habitat zones ranged from 92.3 to 98.2 (SIMPER). The species responsible for these differences were conger eel, which was more abundant on the mud slope, and blackbelly rosefish and mora codling, which occurred in greater densities on the pavement and rock wall substrates. Average dissimilarities between the *Lophelia* coral and other habitat zones ranged from 76.0 to 91.4 (SIMPER). The species responsible for these differences were grenadier and mora codling, which were more abundant in the *Lophelia* coral habitat, and blackbelly rosefish, which were more abundant on the pavement and rock wall substrates.

In addition, all commercially and recreationally important species, including managed species, were analyzed separately. Although they had relatively lower densities than many of the smaller-sized species, these species are important to the SAFMC and NOAA fisheries for management purposes. Eleven commercially and recreationally important fish species were observed, and their densities for each site are listed in Table 5.3. Currently, species that are targeted and managed by the fishery include: blueline tilefish (*Caulolatilus microps*), snowy grouper (*Hyporthodus niveatus*), queen snapper (*Etelis oculatus*), red porgy (*Pagrus pagrus*), greater amberjack (*Seriola dumerili*), almaco jack (*Seriola rivoliana*), and silk snapper (*Lutjanus vivanus*). All of the species in Table 5.3 are represented in the National Marine Fisheries Service landings statistics, with the exception of big roughy. A one-way ANOVA was used to determine significant differences in fish densities among the management areas for the four most abundant targeted species (blueline tilefish, big roughy, snowy grouper, and blackbelly rosefish) (Figure 5.5). The three deeper sites (21, 24, and 25; 450–850 m depths) were omitted from this analysis as no commercial or recreational fish species were observed. The remaining 11 dives compared were between depths of 150 and 300 m. There was no significant difference ($p = .39$) in mean densities of blueline tilefish among management areas; however, in general, the CHAPC sites had the greatest

Table 5.3 Densities (number of individuals per kilometer) of Commercially and Recreationally Important Fish Species for Each ROV Dive Site

ROV site number	Managed area	Blueline tilefish	Big roughy	Snowy grouper	Queen snapper	Blackbelly rosefish	Barrelfish	Red porgy	Greater amberjack	Almaco jack	Red bream	Silk snapper
13	None	3.7	13	25.9		57.4						
14	MPA	3.3	4.1	8.3		7.4						
15	MPA	2.5	9.9	2.5		8.6						
16	MPA			10.3	0.6	2.6			0.6	0.6		0.6
17	MPA		91	10.4	2.2	2.2	20.1					
18	CHAPC	7.1	5.6	6.3		7.1						
19	CHAPC	3.3	26.8	8.1		5.7						
20	CHAPC	11.3	24.3	5.2		7.8	0.9	0.9				
21	None											
22	CHAPC		0.9		0.9	19.3			4.4	0.9		
23	CHAPC		25.7	0.7		11.4	0.7					
24	CHAPC					6.8					3.8	
25	CHAPC					4.6						
26	None	2.1	6.3	12.6		54.7						

Note: Managed areas are inside the protected management areas (MPA and CHAPC) or outside (None). See Table 5.1 for scientific names.

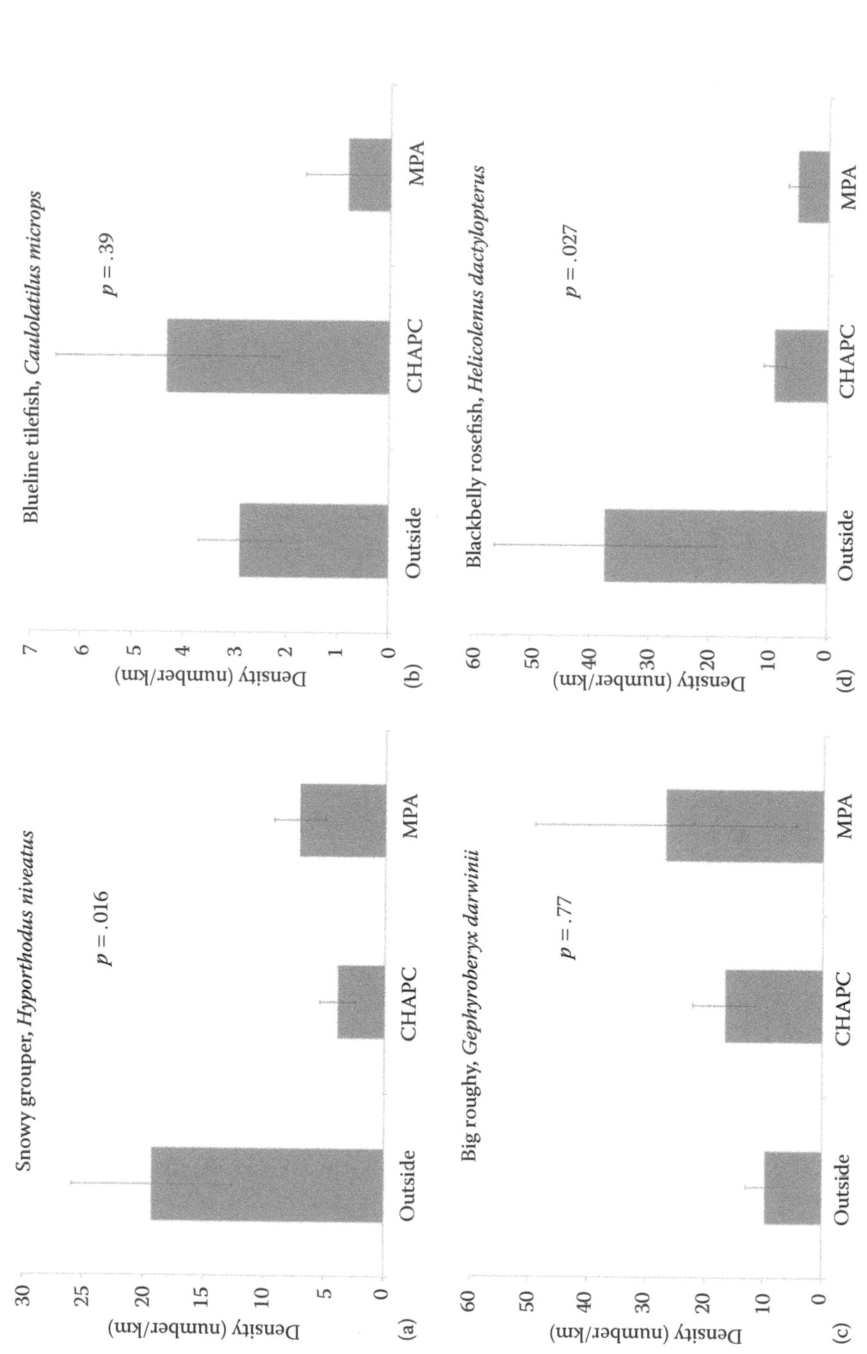

Figure 5.5 (a–d) Mean density (standard error is indicated by the vertical line on each bar) of commercially and recreationally important fish species from ROV dive sites within and outside the East Hump MPA, CHAPC, and nonprotected areas on Pourtalès Terrace. *p* values ($p < .05$) indicate statistically significant differences between the groups.

densities and the MPA sites had the lowest densities. Mean densities of big roughy were highest in the protected areas, but variances were too great to make it significant ($p = .77$). Mean densities of snowy grouper and blackbelly rosefish, however, were significantly greater ($p = .016$ and $p = .027$, respectively) in the nonprotected sites than in the protected areas. It is important to note that the MPA has only been in effect since 2009 and the CHAPC since 2010. These data will be helpful as relative baseline data for future surveys by SAFMC and NOAA Fisheries Service.

Discussion

Deepwater reef habitat and fish surveys off southeastern United States

Various habitat classifications have been used in surveys of aquatic ecosystems, that is, the Coastal and Marine Ecological Classification Standard (FGDC 2012), the EUNIS system in Europe (Connor et al. 1997), the Integrated Marine and Coastal Regionalisation for Australia (IMCRA 1998), and the Classification of Wetlands and Deepwater Habitats of the United States (Cowardin et al. 1979). No one classification is sufficient for all deepwater reef habitats. The Cowardin classification system was adopted by the U.S. Fish and Wildlife Service, and although the title implies use for deepwater habitats, "deepwater" is defined as areas where surface water is permanent and often deep. More recently in the United States, NOAA (Partyka et al. 2007) developed a classification system for deepwater coral reefs termed the Southeastern United States Deep-Sea Corals (SEADESC) Initiative to characterize areas of deep-sea corals. Auster et al. (2005) developed a system for seamounts. Regardless of the habitat classification used, it is important to include both geological and biological components. We have applied pertinent SEADESC codes as well as descriptors from other classifications for habitat and geological descriptions that we also have used on other monitoring surveys of shelf-edge MPA and CHAPC sites off the southeastern United States. In this study, we used four factors of habitat classification for analyzing the interrelationships of the fish communities; these included geomorphology (e.g., sinkhole, *Lophelia* coral mound, rock mound-top, mound-slope, mound-wall), substrate (e.g., soft bottom, rock pavement, rock escarpment, coral), depth, and slope. In addition, we considered the management status of the site as another factor, that is, whether it was within the protected managed areas (MPA and CHAPC) or outside the protected areas (no protection).

Several studies have characterized the deepwater reef habitats and associated fish communities off the southeastern United States. Each has used various habitat classifications and most have been concentrated in the South Atlantic Bight of the Blake Plateau from north Florida to North Carolina (Sedberry 2001, Ross and Quattrini 2007, Fraser and Sedberry 2008, Schobernd and Sedberry 2009). In submersible surveys of deepwater *Lophelia* reefs from North Carolina to north Florida (356–910 m depth), Ross and Quattrini (2007) quantified three habitat types to compare fish distributions: prime reef (*Lophelia* coral), transition zones, and off reef. They documented 99 fish species; the prime reef habitat was characterized by mora codling (*Laemonema melanurum*) (21% of total), roughtip grenadier (*Nezumia sclerorhynchus*) (17%), red bream (14%), and blackbelly rosefish (10%). Several species (e.g., swallowtail bass [*Anthias woodsi*], red bream, American conger [*Conger oceanicus*], and cutthroat eel [*Dysommina rugosa*]) demonstrated specificity to deep-reef habitats, while others (e.g., shortnose greeneye [*Chlorophthalmus agassizi*], electric stingray [*Benthobatis marcida*], Pluto skate [*Fenestraja plutonia*], and longfin hake [*Phycis chesteri*]) were always more common away from reefs. The following 13 species were common to both the deepwater *Lophelia* reefs (Ross and Quattrini 2007) and Pourtalès Terrace (this study, Reed et al. 2005, 2006): red bream, blackbelly rosefish, swordfish (*Xiphias gladius*), shortnose greeneye, mora codling (*L. melanurum, L. barbatulum*), grenadier, catshark (*Scyliorhinus meadi, S. retifer*), offshore hake (*Merluccius albidus*), goosefish (*Lophiodes beroe*), redeye gaper (*Chaunax stigmaeus*), and swallowtail bass.

Off central eastern Florida, Harter et al. (2009) analyzed fish assemblages and benthic habitats inside and outside the deepwater *Oculina* coral MPA by using ROV video and photographic transects to determine whether *Oculina varicosa* forms an essential habitat compared to other structure-forming habitats and to examine the effectiveness of the MPA. Comparison of five habitat types (rock pavement, rubble, rock outcrops, standing dead *O. varicosa*, live *O. varicosa*) by multivariate analyses of the 62 fish species indicated no differences in fish communities or diversity among the hardbottom habitat types. However, grouper densities were significantly higher on the most structurally complex habitats (i.e., live *O. varicosa*, standing dead *O. varicosa*, and rock outcrops) compared to the less complex habitats (pavement and rubble).

Only two studies have characterized the deepwater reef habitat and fish assemblages in the southern Straits of Florida and in particular on the Pourtalès Terrace (Reed et al. 2005, 2006). Using submersible-based qualitative photographic and video surveys, the studies documented 30 species (presence/absence only) of fish at two sinkhole sites and several high-relief rocky mounds on the terrace. Six species were found in the 2006 study that were not found in this present survey: silky shark (*Carcharhinus falciformis*), Warsaw grouper (*Hyporthodus nigritus*), marbled catshark (*Galeus area*), ocean sunfish (*Mola mola*), and several swordfish that often attacked the submersible while it was inside the sinkholes. Schools of squid were also common in the sinkholes, which undoubtedly attracted the swordfish. Also, a large spawning aggregation of beardfish (*Polymixia lowei*) in densities up to 117 m^{-2} was found in the bottom of one of the sinkholes, and ovarian histology indicated recent spawning activity (Baumberger et al. 2010).

Popenoe and Manheim (2001) described the habitat relationships of wreckfish (*Polyprion americanus*) populations on Charleston Bump, a deepwater limestone feature at depths of 250–1000 m on the Blake Plateau, which is somewhat similar to the hard-ground, limestone habitat found on Pourtalès Terrace. Both features are of Tertiary, phosphoritic limestone that has formed highly resistant pavements and complex erosional features including rock piles, scarps, and undercut ledges. These provide habitat and hiding places for larger fish, which at the Charleston Bump included wreckfish, red bream, and roughy (Trachichthyidae). We found the latter two taxa but did not observe wreckfish on Pourtalès Terrace. However at Miami Terrace, which is just to the north of Pourtalès and at similar depths, we have documented a large permanent population of wreckfish on an isolated pinnacle that may be the southernmost known breeding population off the southeastern United States.

Interrelationships of fish communities and benthic habitats

Sites within and outside the managed areas (MPA and CHAPC) were compared for similarity to determine if the newly designated managed areas were similar to one another or were different from unprotected sites. The MDS plots for both the benthic macrobiota and fish communities show that the sites within the managed areas (MPA and CHAPC sites) were relatively similar, demonstrating a close relationship between benthos and fish at these managed sites (Figures 5.3 and 5.4a). The sinkhole and *Lophelia* mound sites were a distinct group in the plots for both the benthic biota and fish; both sites are within the

CHAPC but deeper than the remaining hardbottom sites. The two Mound no. 311 sites also formed distinct groups for both the benthic biota and fish. Mound no. 311 had the greatest percentage of cover of macrobiota of the CHAPC sites (22.2%) and a high cover of Stylasteridae corals; however, it had few large fish. This may be due to fishing. This site was also one of the most impacted by human debris (described later in this section). The southeast wall of Mound no. 311 was a steep escarpment that had few ledges or other microhabitat; as a result, this site only supported a low percentage of macrobiota (3.9%), and, as expected, was one of the sites with a lower density of fish as well (Table 5.2). Other relationships between benthic macrobiota and fish were apparent at Tennessee Bioherm no. 1 within the MPA. This site had the highest number of fish species (23) as well as the second highest cover of macrobiota (20.4%). In addition, Alligator Bioherm no. 3 within the CHAPC had the third highest cover of macrobiota (16.6%) with very dense populations of stylasterid corals and the second highest species richness (20 species) of fish.

The unprotected sites outside the managed areas (Jordan Site F and Key West Bioherm) formed a distinct group in the MDS plots. Both sites had relatively low cover of benthic macrobiota, but Key West Bioherm had the highest density of fish among all sites (Table 5.2) due mostly to a large number of anthiin fish. However, few large fish were observed here, partly because it is the closest to heavy fishing pressure from Key West. This site was also highly impacted by human debris and fishing gear.

The interrelationships of the fish communities and habitat factors were also analyzed with MDS plots of similarity (Figure 5.4b–d). Rather than comparing by sites, fish community data were compiled from the ROV transects based on four habitat factors: geomorphology, substrate, depth, and slope. Depth was the most influential factor contributing to fish species composition, and the similarity plot shows two similar groups: the deep (450–850 m) and the shallower sites (150–300 m). Geomorphology was the second most influential factor and divided the sites into five statistically different groups. One group consisted of the rock mound sites (mound-slope, mound-wall, and mound-top), a second group consisted of the *Lophelia* coral mound and sinkhole sites, and the third group consisted of the deeper sites in the valleys and deep-mounds. Substrate was the third most influential factor consisting of two groups. One group consisted of the rock substrate categories: pavement, pavement and ledges, pavement and sediment, and rock wall. The second group was the *Lophelia* coral habitat and the deep mud habitat on Pourtalès Escarpment, although these two habitats only showed low similarity to one another (<20%). Slope showed little relationship to the fish species composition and any influence of slope was likely overshadowed by the other factors.

While it is well known that deep coral habitats support relatively high diversity and densities of fish species (Costello et al. 2005, Koenig et al. 2005, Ross and Quattrini 2007), it is unclear whether the fish are attracted to live coral or just structure made by the coral habitat. Some research shows a link between deepwater coral habitat and fish nursery grounds (Etnoyer and Warrenchuk 2007, Buhl-Mortensen et al. 2010). In addition to deepwater hard coral habitat, distributional data have shown that deepwater soft corals, gorgonians, sponges, and even sea pens (Pennatulacea) can be important as potential fish habitat (Edinger et al. 2007, Baillon et al. 2012). However, Auster (2007) noted that cooccurrence does not imply a mechanistic relationship between particular habitat types and fish populations and the cooccurrence of fishes with corals does not necessarily mean there is a functional link to population processes. Also, most studies seem to imply that shelf and slope deepwater species exhibit facultative versus obligate habitat use patterns (Auster et al. 1995), although the linkages between coral-associated fishes and their more widely distributed populations remain undefined (Auster 2007). However, in all cases the presence of high-density aggregations of coral-associated fishes suggests that deepwater coral is important habitat (Auster 2007). Certainly, areas of complex high-relief topography in the South Atlantic Bight and throughout the Straits of Florida, including the deepwater *Lophelia* reefs and Pourtalès Terrace, provide habitat for dense and diverse populations of fish. Combined with the complex bottom topography, these areas have high productivity, strong currents, and upstream larval sources from the Florida Current, which support many ecologically and economically important reef fish species. Many of these species live and spawn on rocky reefs on the edge of the continental shelf and upper continental slope (Koenig et al. 2000, Sedberry 2001, Quattrini et al. 2004). These deepwater reefs, whether rocky or coral, may provide increased food availability, and often are found in areas of upwelling that can concentrate zooplankton. Although food supply generally decreases with depth in the ocean, it can be concentrated by topographic features such as seamounts and pinnacles (Koslow 1997). The deepwater reef habitat may also provide a refuge from predators, and habitat from which to ambush prey (Costello et al. 2005).

Deepwater protected areas

Since 2009, eight deepwater, shelf-edge MPAs and five deepwater CHAPCs have been established by NOAA along the outer continental shelf of the southeastern United States from North Carolina to south Florida (NOAA 2010). This network of protected areas was established to sustain and restore reef fish populations and to protect deep-sea coral/sponge habitat from destructive fishing practices, such as bottom trawling for royal red shrimp and the use of longlines of crab pots that may extend several kilometers with dozens of traps on each (NOAA 2010). The eight MPAs are classified as Type II by the SAFMC and were designated in particular to protect species of the deepwater snapper–grouper complex. Regulations applicable inside the MPAs prohibit fishing for, or possession of, any snapper/grouper species or use of shark bottom longline gear. Transiting through the MPAs and trolling for pelagic species, however, is allowed. The closures will provide ecosystem-level benefits to the entire complex as well as protect the shelf-edge reef habitat utilized by the protected species. These consist of five species of grouper: snowy grouper, yellowedge grouper (*Hyporthodus flavolimbatus*), Warsaw grouper, speckled hind (*Epinephelus drummondhayi*), and misty grouper (*Hyporthodus mystacinus*), and two species of tilefish: tilefish (*Lopholatilus chamaeleonticeps*) and blueline tilefish. The deepwater, shelf-edge MPAs are known to contain reef habitat utilized by these species of grouper as well as deep mud banks used by the two tilefish species.

The primary goal of this research was to gather additional data on benthic habitat and fish assemblages in the South Atlantic MPAs and HAPCs. Our long-term sampling program was designed to document changes in these assemblages after fishing restrictions are implemented. Efficacy testing of this management option will aid fishery managers in future use of area restrictions for the protection of valuable habitat and fishery resources. Area closures constitute a politically charged issue that is unlikely to retain support without evidence indicating increases in the target species. As such, decisions to create future area closures will be based upon the efficacy of these areas and the lessons learned during their implementation.

Additionally, although no lionfish (*Pterois volitans, P. miles*) were observed in this survey of Pourtalès Terrace, our recent surveys of the shelf-edge MPAs from North Florida to Carolinas and the Pulley Ridge HAPC in the Gulf of Mexico found areas infested with the invasive lionfish to depths of 100 m. In the Atlantic, lionfish have been recorded to depths of 300 m and their population continues to expand rapidly throughout the western

Atlantic, Caribbean and Gulf of Mexico (Schofield et al. 2013). Their presence on Pourtalès Terrace reefs in the future is probable. Future monitoring will assist in evaluating the effects of this invasion on these ecosystems. The monitoring program for the deepwater MPAs and HAPCs will ensure that SAFMC and NOAA Fisheries remain well informed of changes among reef fish populations and coral habitats associated with these protected areas.

The seafloor in the Straits of Florida has a variety of extensive DSCE habitats including: deepwater coral mounds; various hardbottom habitats off Florida including the Miami Terrace, Pourtalès Terrace, and deepwater canyons (Agassiz and Tortugas Valleys); and deep island slopes off western Bahamas and northern Cuba. In U.S. waters, much of this is now protected as deepwater CHAPCs. NOAA bathymetric contour maps and digital elevation models were used to identify and delineate the areal extent of potential DSCE habitat within and outside the CHAPCs off Florida (Reed et al. 2013b). The total area of the CHAPCs off the southeastern United States, from North Carolina to south Florida, is 62,714 km^2. Reed et al. (2013b) calculated that 22,057 km^2 of potential DSCE habitat occurs in U.S. waters off eastern and southern Florida, and that 15,503 km^2 (70.3%) is within the Florida CHAPCs. This leaves approximately 6554 km^2 of DSCE habitat that remains unprotected (29.7%) and outside the boundaries of the CHAPCs in U.S. waters off Florida. This includes large portions of hardbottom habitat on the Pourtalès Terrace and the escarpment walls of the Tortugas and Agassiz Valleys, which are just west of Pourtalès Terrace. These could remain vulnerable to bottom-tending fishing gear.

Impacts from fishing and trawling

Bottom-tending fishing gear, which has deleterious effects upon deepwater coral habitat, is now prohibited in the MPAs and HAPCs except in allowable fishing areas. In 1984, a portion of the deepwater *O. varicosa* coral reef ecosystem off eastern Florida at depths of 80–100 m was protected as the *Oculina* Habitat Area of Particular Concern (OHAPC), prohibiting bottom trawls, longlines, dredges, and anchors. Unfortunately, the northern two-thirds of the known reef system remained open to such gear until 2000 when the OHAPC boundaries were expanded to 1029 km^2. In the 1970s, the *Oculina* reefs were teeming with large spawning aggregations of grouper and snapper (Gilmore and Jones 1992). By the early 1990s, commercial and recreational fishing had decimated the fish populations, and the coral had been severely impacted by bottom trawling for rock shrimp. Quantitative analyses of photographic images by point count have revealed drastic losses of live coral cover between 1975 and 2001 (Reed et al. 2007). Six coral reef sites had nearly 100% loss of live coral, whereas only two reefs within the boundaries of the original OHAPC since 1984 survived and were not impacted by trawling. The decline in fish populations, primarily the grouper gag (*Mycteroperca microlepis*) and scamp (*Mycteroperca phenax*), on the *Oculina* reefs over that past 20 years is well documented (Gilmore and Jones 1992, Koenig et al. 2000, 2005) and may be attributed to both habitat loss and overfishing. Population densities of the dominant basses (roughtongue bass and red barbier [*Hemanthias vivanus*]), dominant grouper (scamp, gag, and speckled hind), and pelagic species (greater amberjack and almaco jack) all showed a positive association with intact coral habitat (either sparse or dense live coral) compared to unconsolidated coral rubble habitat (Koenig et al. 2005). Scamp density in intact coral habitat was significantly greater ($p = .05$) than in other habitats (sparse live coral or rubble). Only one commercially important taxon (snapper *Lutjanus* spp.) was primarily associated with the coral rubble habitat.

At Pourtalès Terrace we found evidence of lost fishing gear (lines and tackle) on the deepwater reefs. In the point count analysis of the digital images from the photographic transects, categories were included for "human debris" and subcategories for "fishing gear" (fishing line, longline, fish/crab traps, and nets). Sites 13, 22, 24, and 26 recorded the largest impact of human debris (fishing gear and other debris). In general, the most fishing gear and other debris were at the sites directly off Key West (Sites 13 and 26), which is understandable as they are the closest sites to heavy fishing and boating traffic. Site 16, which is within the newly designated MPA, also had considerable lost fishing gear on the reef and coral habitats. Many other sites also had fishing gear; for example, at Site 14, within the MPA and closest to shore off Marathon in the Florida Keys, we observed considerable amounts of fishing line, leaders, and tackle on the peak of the mound. We also observed stylasterid corals that apparently had been recently knocked down and still retained their living color. Lost fishing gear was also common on top of the Southeast Mound within the MPA (Site 15) and Mound no. 311 just west of the MPA (Site 22). Site 24 had a lot of debris but is far offshore, and this may be due to shipping traffic following the Florida Current. It also is near the allowable crab fishing area within the CHAPC, where there is a considerable amount of lost trap gear. Deepwater crab fishers estimate that there may be 1000–2000 lost or discarded crab traps within the region of Pourtalès Terrace and several hundred traps within the Pourtalès CHAPC (T. Matthews, Florida Wildlife Research Institute, personal communication). Not all of these traps had biodegradable panels; some crab fishers have indicated that they and their colleagues sealed the decay panel using nonbiodegradable materials to prevent the panel from opening during trap pulling in the strong Florida Current. Such ghost traps could continue fishing for years, indiscriminately killing crabs and fish.

Worldwide, bottom trawling has severely impacted deep-sea coral reef habitats and continues to be a major concern and threat (Rogers 1999, Butler and Gass 2001, Morgan et al. 2005, Mortensen et al. 2005). Bottom trawling causes severe mechanical damage as evident on deepwater *Lophelia* reefs in the northeast Atlantic (Rogers 1999, Fosså et al. 2002), the deepwater *Oculina* reefs (Reed et al. 2007), hardbottom habitats off the southeastern United States (Van Dolah et al. 1987), and deepwater seamounts off New Zealand and Tasmania (Koslow et al. 2001).

Future work and conclusions

This research cruise has resulted in a rich set of new data documenting and characterizing the deepwater benthic habitats and associated fish communities within the newly established MPAs on Pourtalès Terrace. New multibeam sonar maps, ground-truthed by ROV dives, enabled the discovery of the southernmost known deepwater *Lophelia* coral reef in U.S. waters as well as previously unknown deepwater sinkholes. It has provided baseline data for characterizing the newly designated East Hump MPA site, and eight additional sites within the newly designated deepwater CHAPC on Pourtalès Terrace.

These data were analyzed specifically to better understand the interrelationships of the deepwater fish communities, including commercially and recreationally important species, relative to the DSCE habitats. Eleven commercially and recreationally important fish species were observed and are important to the SAFMC and NOAA Fisheries for management purposes. Species that are targeted and managed by the fishery included: blueline tilefish (*Caulolatilus microps*), snowy grouper (*Hyporthodus niveatus*), queen snapper (*Etelis oculatus*), red porgy (*P. pagrus*), greater amberjack (*Seriola dumerili*), almaco jack (*Seriola rivoliana*), and silk snapper (*L. vivanus*). Statistical analyses showed clear relationships of habitat types (geomorphology, substrate, and depth) with the deepwater fish communities. While depth was the most influential factor contributing to fish species composition among the sites, the geomorphology factor was second most important, and substrate type was the third most influential factor. It is also interesting to note that the statistical plots for both the benthic macrobiota and fish communities show that the sites within the managed areas (MPA and CHAPC sites) were quite similar compared to the non-protected sites outside the CHAPC. This cruise also provided baseline documentation of the populations of commercially and recreationally important species within and outside the MPAs and their interrelationships with habitat and benthic biota such as deepwater corals.

New information was collected on several high-relief features outside the MPAs, which showed that extensive coral/sponge habitat and potential EFH exist outside the protected CHAPC boundaries. These unprotected areas should be of priority for future research and for possible inclusion in the managed areas. These data provided in this study will be important for managers and scientists with NOAA Fisheries, the SAFMC, Florida Keys National Marine Sanctuary, NOAA Deep-Sea Coral Research and Technology Program, NOAA Coral Reef Conservation Program, and NOAA Mesophotic Reef Ecosystem Program. These data may then be compared with the results of future research cruises to document changes in these areas due to the implementation of fishing restrictions and to monitor the efficacy and health of these newly designated managed areas.

Acknowledgments

We gratefully acknowledge the NOAA Cooperative Institute for Ocean Exploration, Research, and Technology (CIOERT) at Harbor Branch Oceanographic Institute, Florida Atlantic University (HBOI-FAU), and thank the Robertson Coral Reef Research and Conservation Program at HBOI. The crews of the NOAA ship R/V *Nancy Foster* and the University of Connecticut's *Kraken 2* ROV are thanked for their support. CIOERT gratefully acknowledges funding provided by the NOAA Office of Ocean Exploration and Research (OER Award #: NA09OAR4320073), NOAA Deep Sea Coral Research and Technology Program, and NOAA Office of Marine and Aviation Operations (OMAO) in support of the research, ship operations, and ROV operations. Funding for data analysis was also provided to CIOERT by the NOAA Fisheries Coral Reef Conservation Program through the South Atlantic Fishery Management Council (Award #: NA11NMF4410061). This is HBOI-FAU Contribution Number 1915.

References

Auster PJ. 2007. Linking deep-water corals and fish populations. Bull Mar Sci. 81(Suppl 1):93–99.

Auster PJ, Malatesta RJ, LaRosa SC. 1995. Patterns of microhabitat utilization by mobile megafauna on the southern New England (USA) continental shelf and slope. Mar Ecol Prog Ser. 127:77–85.

Auster P, Moore J, Heinonen K, Watling L. 2005. A habitat classification scheme for seamount landscapes: Assessing the functional role of deep-water corals as fish habitat. *In*: Freiwald A, Roberts J, editors. Cold-water corals and ecosystems. New York: Springer-Verlag.

Baillon S, Hamel J-F, Wareham VE, Mercier A. 2012. Deep cold-water corals as nurseries for fish larvae. Front Ecol Environ. 10(7):351–356.

Baumberger RE, Brown-Peterson NJ, Reed JK, Gilmore RG. 2010. Spawning aggregation of beardfish, *Polymixia lowei*, in a deep-water sinkhole off the Florida Keys. Copeia. 2010(1):41–46.

Buhl-Mortensen L, Vanreusel A, Gooday AJ, Levin LA, Priede IG, Buhl-Mortensen P, Gheerardyn H, King NJ, Raes M. 2010. Biological structures as a source of habitat heterogeneity and biodiversity on the deep ocean margins. Mar Ecol. 31(1):21–50.

Butler M, Gass S. 2001. How to protect corals in Atlantic Canada. *In*: Willison JHM, Hall J, Gass SE, Kenchington ELR, Butler M, Doherty P, editors. Proceedings of the 1st International Symposium on Deep-sea Corals. Halifax, Nova Scotia: Ecology Action Centre, Nova Scotia Museum. p. 156–165.

Clarke K, Gorley R. 2006. PRIMER v6: User manual/tutorial. Plymouth, UK: PRIMER-E.

Clarke K, Warwick R. 2001. Changes in marine communities: An approach to statistical analysis and interpretation. 2nd ed. Plymouth, UK: PRIMER-E.

Connor D, Brazier D, Hill T, Northen K. 1997. Marine Nature Conservation Review: Marine biotope classification for Britain and Ireland. Volume 1. Littoral biotopes. JNCC Report(229). Peterborough, UK: Joint Nature Conservation Committee.

Costello MJ, McCrea M, Freiwald A, Lundälv T, Jonsson L, Bett BJ, van Weering TC, de Haas H, Roberts JM, Allen D, et al. 2005. Role of cold-water *Lophelia pertusa* coral reefs as fish habitat in the NE Atlantic. *In*: Freiwald A, Roberts J, editors. Cold-water corals and ecosystems. New York: Springer-Verlag.

Cowardin LM, Carter V, Golet FC, LaRoe ET. 1979. Classification of wetlands and deepwater habitats of the United States. FWS/OBS-79/31, GPO 024-010-00524-6. Washington, DC: U.S. Fish and Wildlife Service.

Dohrmann M, Gocke C, Reed J, Janussen D. 2012. Integrative taxonomy justifies a new genus, *Nodastrella* gen. nov. for North Atlantic "*Rosella*" species (Porifera: Hexactinellida: Rosellidae). Zootaxa. 3383:1–13.

Edinger EN, Wareham VE, Haedrich RL. 2007. Patterns of groundfish diversity and abundance in relation to deep-sea coral distributions in Newfoundland and Labrador waters. Bull Mar Sci. 81(Suppl 1):101–122.

Etnoyer P, Warrenchuk J. 2007. A catshark nursery in a deep gorgonian field in the Mississippi Canyon, Gulf of Mexico. Bull Mar Sci. 81(3):553–559.

FGDC (Federal Geographic Data Committee). 2012. Coastal and marine ecological classification standard. FGDC-STD-018-2012. Reston, VA: Federal Geographic Data Committee.

Fosså JH, Mortensen PB, Furevik DM. 2002. The deep-water coral *Lophelia pertusa* in Norwegian waters: Distribution and fishery impacts. Hydrobiologia. 471(1–3):1–12.

Fraser SB, Sedberry GR. 2008. Reef morphology and invertebrate distribution at continental shelf edge reefs in the South Atlantic Bight. Southeastern Nat. 7(2):191–206.

Gilmore GR, Jones RS. 1992. Color variation and associated behavior in the epinepheline groupers, *Mycteroperca microlepis* (Goode and Bean) and *M. phenax* Jordan and Swain. Bull Mar Sci. 51(1):83–103.

Harter SL, Ribera MM, Shepard AN, Reed JK. 2009. Assessment of fish populations and habitat on *Oculina* Bank, a deep-sea coral marine protected area off Eastern Florida. Fish B-NOAA. 107(2):195–206.

IMCRA (Interim Marine Coastal Regionalisation for Australia). 1998. Interim Marine Coastal Regionalisation for Australia Technical Group. An ecosystem-based classification for marine and coastal environments Version 3.3. Canberra: Environment Australia, Commonwealth Department of the Environment.

Jordan G. 1954. Large sink holes in Straits of Florida. Bull Am Ass Petrol Geol. 38:1810–1817.

Jordan G, Malloy R, Kofoed J. 1964. Bathymetry and geology of Pourtalès Terrace. Mar Geol. 1:259–287.

Koenig CC, Coleman FC, Grimes C, Fitzhugh G, Scanlon K, Gledhill C, Grace M. 2000. Protection of fish spawning habitat for the conservation of warm-temperate reef-fish fisheries of shelf-edge reefs of Florida. Bull Mar Sci. 66:593–616.

Koenig CC, Shepard AN, Reed JK, Coleman FC, Brooke SD, Brusher J, Scanlon KM. 2005. Habitat and fish populations in the deep-sea *Oculina* coral ecosystem of the Western Atlantic. Amer Fish Soc Symp. 41:795–805.

Kohler KE, Gill SM. 2006. Coral Point Count with Excel extensions (CPCe): A visual basic program for the determination of coral and substrate cover using random point count methodology. Comput Geosci. 32:1259–1269.

Koslow JA. 1997. Seamounts and the ecology of deep-sea fisheries: The firm-bodied fishes that feed around seamounts are biologically distinct from their deepwater neighbors—and may be especially vulnerable to overfishing. Am Sci. 85(2):168–176.

Koslow JA, Gowlett-Holmes K, Lowry J, O'Hara T, Poore G, Williams A. 2001. Seamount benthic macrofauna off southern Tasmania: Community structure and impacts of trawling. Mar Ecol Prog Ser. 21:111–125.

Land L, Paull C. 2000. Submarine karst belt rimming the continental slope in the Straits of Florida. Geo-Mar Lett. 20(2):123–132.

Littler MM, Littler DS, Blair SM, Norris JN. 1985. Deepest known plant life discovered on an uncharted seamount. Science. 227:57–59.

Malloy RJ, Hurley R. 1970. Geomorphology and geologic structure: Straits of Florida. Geol Soc Am Bull. 81:1947–1972.

Morgan LE, Etnoyer P, Scholz AJ, Mertens M, Powell M. 2005. Conservation and management implications of deep-sea coral and fishing effort distributions in the northeast Pacific Ocean. *In*: Freiwald A, Roberts JM, editors. Cold-water corals and ecosystems. New York: Springer-Verlag.

Mortensen PB, Buhl-Mortensen L, Gordon DC, Fader GBJ, McKeown DL, Fenton DG. 2005. Effects of fisheries on deepwater gorgonian corals in the Northeast Channel, Nova Scotia. Am Fish Soc Symp. 41:369–382.

NOAA (National Oceanic and Atmospheric Administration). 2010. Fisheries of the Caribbean, Gulf of Mexico, and South Atlantic; comprehensive ecosystem-based amendment for the South Atlantic region. Fed Reg. 75(119):50 CFR Part 633, 35330–35335.

Partyka ML, Ross SW, Quattrini AM, Sedberry GR, Birdsong TW, Potter J, Gottfried S. 2007. Southeastern United States deep-sea corals (SEADESC) initiative: A collaborative effort to characterize areas of habitat-forming deep-sea corals. NOAA Technical Memorandum OAR OER 1, Silver Spring, MD.

Popenoe P, Manheim FT. 2001. Origin and history of the Charleston Bump: Geological formations, currents, bottom conditions, and their relationship to wreckfish habitats on the Blake Plateau. *In*: Sedberry GR, editor. Island in the stream: Oceanography and fisheries of the Charleston Bump. American Fisheries Society. Bethesda, MD, p. 43–94.

Quattrini AM, Ross SW, Sulak KJ, Necaise AM, Casazza TL, Dennis GD. 2004. Marine fishes new to continental United States waters, North Carolina, and the Gulf of Mexico. Southeast Nat. 3(1):155–172.

Redmond NE, Morrow CC, Thacker RW, Diaz MC, Boury-Esnault N, Cárdenas P, Hajdu E, Lôbo-Hadju G, Picton BE, Pomponi SA, et al. 2013. Phylogeny and systematics of Demospongiae in light of new small-subunit ribosomal DNA (18S) sequences. Integr Comp Biol. 53(3):388–415.

Reed JK, Farrington S, David A, Harter S, Murfin D, Stierhoff K. 2013a. CIOERT SEADESC II Report: Extreme Corals 2011: South Atlantic Deep Coral Survey. NOAA Ship *Pisces* Cruise PC-11-03, May 31–June 11, 2011. Report To: NOAA Office of Ocean Exploration and Research and NOAA Deep Sea Coral Research and Technology Program. HBOI Miscellaneous Contribution Number 863.

Reed JK, Koenig CC, Shepard AN. 2007. Impacts of bottom trawling on a deep-water *Oculina* coral ecosystem off Florida. Bull Mar Sci. 81(3):481–496.

Reed JK, Messing C, Walker BK, Brooke S, Correa TBS, Brouwer M, Udouj T, Farrington S. 2013b. Habitat characterization, distribution, and areal extent of deep-sea coral ecosystems off Florida, southeastern U.S.A. Caribb J Sci. 47(1):13–30.

Reed JK, Pomponi SA, Weaver D, Paull CK, Wright AE. 2005. Deep-water sinkholes and bioherms of South Florida and the Pourtalès Terrace: Habitat and fauna. Bull Mar Sci. 77(2):267–296.

Reed JK, Weaver DC, Pomponi SA. 2006. Habitat and fauna of deep-water *Lophelia pertusa* coral reefs off the southeastern US: Blake Plateau, Straits of Florida, and Gulf of Mexico. Bull Mar Sci. 78(2):343–375.

Rogers AD. 1999. The biology of *Lophelia pertusa* (Linnaeus 1758) and other deep-water reef-forming corals and impacts from human activities. Int Rev Ges Hydrobiol. 84:315–406.

Ross SW, Quattrini AM. 2007. The fish fauna associated with deep coral banks off the southeastern United States. Deep Sea Res Pt I: Oceanogr Res Pap. 54(6):975–1007.

Schobernd CM, Sedberry GR. 2009. Shelf-edge and upper-slope reef fish assemblages in the South Atlantic Bight: Habitat characteristics, spatial variation, and reproductive behavior. Bull Mar Sci. 84(1):67–92.

Schofield PJ, Morris Jr. JA, Langston JN, Fuller PL. 2013. *Pterois volitans/miles*. USGS nonindigenous aquatic species database, Gainesville, FL. http://nas.er.usgs.gov/queries/FactSheet.aspx?speciesID=963. Revision date: 18 September, 2012.

Sedberry GR. 2001. Island in the stream: Oceanography and fisheries of the Charleston Bump. Volume 25. American Fisheries Society Symposium 25. Bethesda, MD: American Fisheries Society.

Van Dolah RF, Wendt PH, Nicholson N. 1987. Effects of a research trawl on a hardbottom assemblage of sponges and corals. Fish Res. 5:39–64.

Vinick C, Riccobono A, Messing CG, Walker BK, Reed JK, Farrington S. 2012. Siting study for a hydrokinetic energy project located offshore southeastern Florida: Protocols for survey methodology for offshore marine hydrokinetic energy projects, U.S. Department of Energy. www.osti.gov/servlets/purl/1035555/.

chapter six

Factors affecting coral reef fisheries in the eastern Gulf of Mexico

Walter C. Jaap, Steve W. Ross, Sandra Brooke, and William S. Arnold

Contents

Introduction

The Gulf of Mexico is a marginal sea, 1.5 million km^2 in surface area with no major islands, open to the south but surrounded by continental land masses (North and Central America) to the west, north, and east. Barrier islands parallel much of the coast of the eastern Gulf of Mexico. North of Tarpon Springs, Florida (the mouth of the Anclote River, 25°10′ N), salt marshes and coastal swamps replace mangroves along the coastline. Specific to the area under consideration here (Figure 6.1), the continental shelf off the west coast of Florida is 800 km long and 25–250 km wide; the depth gradient on the shelf is 0.2–4 m km^{-1}, but at the upper slope it steepens to 6–9 m km^{-1}. Shelf and slope sediments off northwestern Florida are siliceous, but are mostly carbonate off the west coast of the peninsula (Brooks and Holmes 2011, Hine and Locker 2011).

The principle current system in the eastern Gulf of Mexico is the Loop Current, which "forms" in the Straits of Yucatan (Fairbridge 1966) and traverses northward into the northern gulf before looping southward and exiting and merging with the Florida Current in the Straits of Florida. Transport volume and the extent of penetration into the northern gulf varies (Lee et al. 1992, 1994, 2002), which has important implications for smaller-scale current patterns and associated biological parameters such as larval transport and population connectivity; nutrient supply and primary production; the initiation, maintenance,

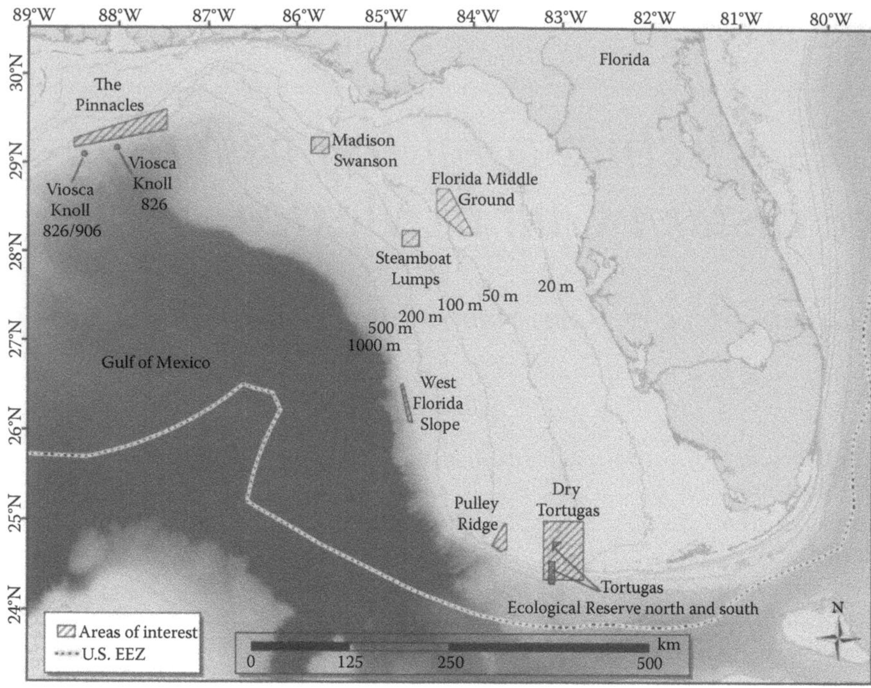

Figure 6.1 **(See color insert)** Eastern Gulf of Mexico areas of interest: the Dry Tortugas, Pulley Ridge, Florida Middle Grounds, Madison Swanson, Steamboat Lumps, the Pinnacles, and Viosca Knolls. EEZ, exclusive economic zone. (From Cartography by Mark Mueller, Gulf of Mexico Fishery Management Council.)

and transport of harmful algal blooms; and general water clarity (Collard et al. 2000, Muller-Karger 2000, Barth et al. 2008). The Loop Current is generally situated west of the 100 m depth contour; however, filaments from the current may at times intrude into shallower water (Cooper et al. 1990) and the position of the axis of the current also varies. This wavering of the current axis influences nearshore and potentially even deepwater patterns of upwelling and downwelling, again with important biological implications. Near the Florida Keys, the Florida Current creates rotating water masses (gyres) that persist for 60–100 days (Lee et al. 1994, 2002).

In general, zooxanthellate coral species live in shallower depths, while azooxanthellate coral species are found in depths where solar illumination does not support photosynthesis. Scleractinia (Cnidaria, Anthozoa) are important foundations for biodiversity in the eastern Gulf of Mexico (Cairns et al. 2009). These corals construct limestone reefs that provide habitat for numerous species, such as snappers, groupers, and lobster, which support important commercial and recreational fisheries. This habitat consists not only of the obvious benthic structure, but also of the overlying water column due to the influence of the topographic structure on local hydrodynamics. This important effect of the coral reef on current patterns above the reef, while poorly understood, is a significant driver of reef fish aggregations observed to form over many shallow- and deepwater coral reefs (Carleton et al. 2001, Wolanski 2001). Thus, negative impacts on coral reef structure result in both direct effects on the organisms inhabiting the reef and indirect effects on organisms occupying habitats surrounding the reef, and potentially on organisms at considerable distances from the reef. The teleplanic nature of these organisms results in biological connectivity patterns among biota from otherwise isolated reefs, a primary concern in this era of worldwide coral reef demise.

Coral reef habitat takes many forms over a broad area and depth range (Jaap and Hallock 1990, Brooke and Schroeder 2007, Hine et al. 2008, Jaap et al. 2008, Schroeder and Brooke 2011). The shallow reef communities are Caribbean-like (reef structures constructed by coral reef organisms) in the Dry Tortugas but then transition into temperate low-relief, sponge–octocoral–stony coral epibenthic communities in central and northern areas of the Gulf of Mexico (Giammona 1978, Cairns et al. 1993, Hine et al. 2008, Cairns and Bayer 2009, Etnoyer 2009, Quattrini et al. 2014). The species richness of the Scleractinia follows this trend, with the greater number of species concentrated in the Dry Tortugas (Table 6.1). Deep-sea coral habitats (>300 m) harbor unique communities of fishes and invertebrates, unlike those of the surrounding soft substrata (Henry and Roberts 2007, Ross and Quattrini 2007). The depth range included in this discussion includes the shelf as well as the upper continental slope to at least 1000 m.

Reef-related fisheries are an important economic asset for commercial and recreational fishers who target the Palinuridae (lobster), Lutjanidae (snapper), Serranidae (grouper and sea bass), and Sparidae (porgy) populations associated with reef habitat. Essential habitat for reef and coastal migratory pelagic fishes in the Gulf of Mexico is defined as that area from the U.S./Mexico border to the Dry Tortugas and from estuarine waters out to depths of 183 m (Gulf of Mexico Fishery Management Council 2004). A unique fishery for *Mennipe mercenaria* (stone crab), where a claw is removed and the crab is returned to the sea, is associated with the epibenthic habitat in the eastern Gulf of Mexico. Periodically, there is also a deepwater fishery for *Chaceon fenneri* (golden crab) near the deep-sea reefs of the West Florida Slope, and the north-central gulf fishery for *Hymenopenaeus robustus* (royal red shrimp) appears to harvest near deep-sea reefs. Recently, intense fish harvesting at shelf depths has resulted in repeated closures for some Serranidae and Lutjanidae species to allow populations to recover. Deep-sea coral reefs at present appear to support few, if any, significant commercial or recreational fisheries; however, deep-drop recreational fishing is becoming more popular. Large demersal fisheries below 300 m are uncommon in the Gulf of Mexico (Stiles et al. 2007).

Reef fish are landed in all U.S. states bordering the Gulf of Mexico, although most landings occur in Florida (Table 6.2). Landings range from just a few thousand pounds per year for *Balistes capriscus* (gray triggerfish) to several million pounds per year for species such as *Rhomboplites aurorubens* (vermilion snapper), *Epinephelus morio* (red grouper), and *Panuliris argus* (spiny lobster). In many cases, landings described in Table 6.2 correspond to species distribution patterns. For example, the distribution of *P. argus* (spiny lobster) in the Gulf of Mexico reflects the abundance and distribution of that species along the coast of southwest Florida and in the Florida Keys archipelago (Figure 6.2). In other cases, market conditions and species preference strongly influence landing patterns. *B. capriscus* (gray triggerfish) are found in offshore waters bordering each of the Gulf of Mexico coastal states, but are not landed in Florida, apparently due to a lack of market interest. In general, though, reef fish species, particularly snappers and groupers, are extremely popular market fish, supporting a major industry and eliciting substantial political and special-interest attention. This attention focuses almost exclusively on allowable harvest levels and the interplay between commercial and recreational harvest allocations, with considerably less attention paid to the potential direct and indirect impacts of such harvest activities on the coral reef ecosystems that support these fisheries.

Table 6.1 Millepora and Scleractinia Distribution

Species	DT	PR	EGHB 0–50 m	EGHB > 50 m	FMG
Millepora alcicornis	Z		Z		Z
Millepora complanata	Z				
Stephanocoenia intersepta	Z	Z	Z		Z
Madracis asperula				A	
Madracis decactis	Z		Z		Z
Madracis formosa	Z				
Madracis mirabilis	Z				
Madracis pharensis luciphila	Z				
Madracis pharensis pharensis	Z	Z			Z
Acropora cervicornis	Z				
Acropora palmata	Z				
Acropora prolifera	Z				
Agaricia agaricites	Z				
Agaricia fragilis	Z	Z	Z		Z
Agaricia humilis	Z				
Agaricia lamarcki	Z	Z			
Agaricia undata	Z	Z			
Leptoseris cucullata	Z	Z			
Siderastrea radians	Z		Z		Z
Siderastrea siderea	Z				
Porites astreoides	Z				
Porites branneri	Z				Z
Porites porites f. *divaricata*	Z		Z		Z
Porites porites f. *furcata*	Z				
Porites porites f. *porites*	Z				
Colpophyllia natans	Z				
Diploria clivosa	Z				
Diploria labyrinythiformis	Z				
Diploria strigosa	Z				
Favia fragum	Z				
Manicina areolata	Z		Z		Z
Montastraea annularis	Z				
Montastraea cavernosa	Z	Z			
Montastraea faveolata	Z				
Montastraea franksi	Z				
Solenastrea bournoni	Z				
Solenastrea hyades	Z		A		Z
Astrangia poculata	Z		A		Z
Astrangia solitaria	Z				
Oculina diffusa	Z		Z		Z
Oculina robusta	Z		Z		Z
Oculina tenella			A	A	
Dendrogyra cylindrus	Z				

Table 6.1 (Continued) Millepora and Scleractinia Distribution

Species	DT	PR	EGHB 0–50 m	EGHB > 50 m	FMG
Dichocoenia stokesi	Z				Z
Meandrina meandrites	Z				Z
Isophyllastraea rigida	Z				
Isophyllia sinuosa	Z		Z		Z
Mussa angulosa	Z				
Mycetophyllia aliciae	Z				
Mycetophyllia ferox	Z				
Mycetophyllia lamarckiana	Z				
Scolymia cubensis	Z				
Scolymia lacera	Z		Z		Z
Anomocora fecunda				A	
Anomocora marchadi				A	
Anomocora prolifera				A	
Caryophyllia ambrosia					
Caryophyllia berteriana					
Caryophyllia horologium					
Caryophyllia polygona					
Cladocora arbuscula	Z		Z		Z
Cladocora debilis				A	
Coenocyathus parvulus				A	
Coenosmilia arbuscula				A	
Concentrotheca laevigata				A	
Dasmosmilia lymani				A	
Dasmosmilia variegata				A	
Deltocyathus calcar				A	
Deltocyathus eccentricus				A	
Deltocyathus italicus				A	
Desmophyllum dianthus				A	
Eusmilia fastigiata	Z				
Lophelia pertusa				A	
Oxysmilia rotundifolia				A	
Paracyathus pulchellus				A	
Phacelocyathus flos				A	
Phyllangia americana americana	A		A		A
Phyllangia pequegnatae				A	
Polycyathus senegalensis				A	
Pourtalosmilia conferta				A	
Premocyathus cornuformis				A	
Rhizosmilia maculata				A	
Stephanocyathus (Odontocyathus) coronatus				A	
Stephanocyathus (Stephanocyathus) diadema				A	

(continued)

Table 6.1 (Continued) Millepora and Scleractinia Distribution

Species	DT	PR	EGHB 0–50 m	EGHB > 50 m	FMG
Stephanocyathus (Stephanocyathus) paliferus				A	
Trochocyathus rawsonii				A	
Sphenotrochus andrewianus moorei				A	
Stenocyathus vermiformis				A	
Flabellum atlanticum				A	
Flabellum floridanum				A	
Flabellum moseleyi				A	
Javania cailleti				A	
Polymyces fragilis				A	
Guynia annulata				A	
Schizocyathus fissilis				A	
Balanophyllia floridana			A	A	
Bathypsammia tintinnabulum				A	
Cladopsammia manuelensis				A	
Eguchipsammia cornucopia				A	
Enallopsammia profunda				A	
Rhizopsammia goesi				A	

Source: Cairns, S.D., Stony corals: I. Caryophylliina and Dendrophylliina (Anthozoa: Scleractinia), Memoirs of the Hourglass Cruises, Volume 3, pt. 4. St. Petersburg: FMRI, p. 27, 1977.

Notes: A, azooxanthellate species; DT, Dry Tortugas; EGHB, eastern gulf hard bottom; FMG, Florida Middle Ground; PR, Pulley Ridge; Z, zooxanthellate species.

Examination of historic landing patterns may provide insights into trends in availability of the associated species, and perhaps some guidance regarding the health of the ecosystems that support those fisheries. However, patterns of commercial landing trends in Gulf of Mexico reef fish fisheries are mixed. For species such as *Lutjanus griseus* (gray snapper), *Lutjanus synagris* (lane snapper), *Mycteroperca microlepis* (gag), *Mycteroperca bonaci* (black grouper), *Seriola dumerili* and *S. fasciata* (greater and lesser amberjack), and *B. capriscus* (gray triggerfish), commercial landings during 1993–2012 demonstrate a clear downward trend, suggesting population declines (Figure 6.3). But for other species, trends over the period are mixed or even substantially upward. Commercial landings of *R. aurorubens* (vermilion snapper), *Ocyurus chrysurus* (yellowtail snapper), *Mycteroperca phenax* (scamp), *E. morio* (red grouper), *Malacanthus plumieri* (sand tilefish), and, to some degree, *Panulirus argus* (spiny lobster) are essentially similar between 1993 and 2012, although, for many of these species, there is substantial variability from year to year (Figure 6.3). In the unique case of red snapper, one of the most popular and politically controversial species pursued by commercial (and recreational) fishers in Gulf of Mexico waters, reported commercial landings have increased threefold since 1993 (Figure 6.3). These somewhat confused commercial catch patterns reflect the complex interactions of commercial fishing activities, changing abundance patterns of reef fish within a dynamic community context, and the poorly understood relationships between relative reef fish abundance and the ever-changing reef landscape (including predator–prey relationships and significant disturbances such as red tides), which influence species abundance (Tolimieri 1998, Ault et al. 2006).

Table 6.2 2012 Commercial Landings for U.S. States Bordering the Gulf of Mexico

Species	Alabama	Florida	Louisiana	Mississippi	Texas	Total	Maximum (%)
Lutjanus griseus	488	227,237	32,320	891		260,936	87.1 (FL)
Gray snapper	(0.22)	(103.08)	(14.56)	(0.40)		(118.36)	
Lutjanus synagris	234	25,853	2,632		1,182	29,951	86.3 (FL)
Lane snapper	(0.11)	(11.73)	(1.22)		(0.54)	(13.59)	
Lutjanus campechanus	77,955	1,698,102	1,033,475	114,895	1,122,665	4,047,092	42.0 (FL)
Red snapper	(35.36)	(770.25)	(468.77)	(52.12)	(509.23)	(1,835.73)	
Rhomboplites aurorubens	132,500	1,508,965	291,545		511,224	2,444,234	61.7 (FL)
Vermilion snapper	(60.10)	(684.45)	(132.24)		(231.89)	(1,108.69)	
Ocyurus chrysurus		1,885,464			449	1,885,913	99.9 (FL)
Yellowtail snapper		(855.23)			(0.21)	(855.44)	
Mycteroperca microlepis	322	611,575	5,410		2,139	619,446	98.7 (FL)
Gag	(0.15)	(277.41)	(2.45)		(0.97)	(280.98)	
Mycteroperca phenax	4,103	254,664	21,170		15,053	294,990	86.3 (FL)
Scamp	(1.86)	(115.51)	(9.60)		(6.83)	(133.81)	
Mycteroperca bonaci		87,417			593	88,010	99.3 (FL)
Black grouper		(39.65)			(0.27)	(39.92)	
Epinephelus morio	479	6,141,320	143			6,141,942	99.9 (FL)
Red grouper	(0.22)	(2,785.65)	(0.06)			(2,785.94)	
Seriola dumerili	12,451	602,479	84,954		33,953	733,837	82.1 (FL)
Greater amberjack	(5.65)	(273.28)	(38.53)		(15.40)	(332.86)	
Seriola fasciata		4,602	1,532			6,134	75.0 (FL)
Lesser amberjack		(2.09)	(0.69)			(2.78)	
Family Malacanthidae		297,872			105,354	403,226	73.9 (FL)
Tilefish		(135.11)			(47.79)	(182.90)	
Balistes capriscus	1,309		5,063		2,384	8,756	57.8 (LA)
Gray triggerfish	(0.59)		(2.30)		(1.08)	(3.97)	
Panulirus argus		3,633,402				3,633,402	100.0 (FL)
Caribbean spiny lobster		(1,648.08)				(1,648.08)	

Source: Data from NOAA Fisheries landings website (http://www.st.nmfs.noaa.gov/commercial-fisheries/commercial-landings/annual-landings/index).

Notes: All landings in pounds (metric tons) of whole fish. Also included are the totals for each species, the percent contribution for the state with the highest landings, and the state contributing those highest landings.

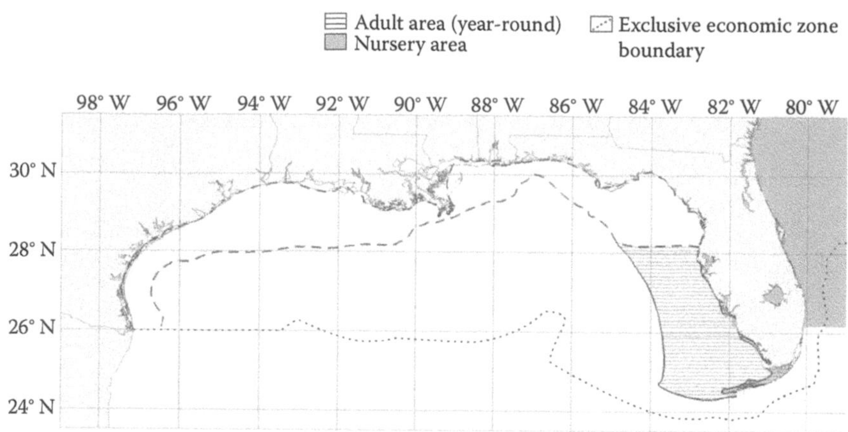

Figure 6.2 Distribution for *P. argus* (spiny lobster) in the eastern Gulf of Mexico, prepared by NOAA for the Gulf of Mexico Fishery Management Council's 1985 Gulf of Mexico Data Atlas. (http://ccma.nos.noaa.gov/products/biogeography/gom-efh/offshore.aspx#coral). Spawning, from March to July, occurs throughout adult areas.

Several marine protected areas (MPAs) have been established in the gulf to provide refuge for commercially and recreationally harvested fish populations. The largest of these is the Tortugas Ecological Reserve in the Florida Keys National Marine Sanctuary and an adjacent closed-to-fishing area inside Dry Tortugas National Park (Figure 6.1). In addition, MPA designations for the Madison Swanson and Pinnacles areas in the northeastern gulf are recent efforts to protect populations of grouper. Other areas are designated habitat areas of particular concern, including many of the deeper banks off Texas and Louisiana, the Florida Middle Grounds, and Pulley Ridge (Figure 6.1), where certain types of fishing, such as trawling, are prohibited.

Reef systems are at risk from multiple stressors that potentially could extirpate or reduce some of these communities and the multitude of populations they support (Hughes et al. 2003, Bellwood et al. 2004, Gardner et al. 2005, Hoegh-Guldberg et al. 2007, Baker et al. 2008, Pandolfi et al. 2011). One threat is a changing climate, particularly with respect to increasing water temperature. Warming is associated with greenhouse gas emissions; in 2012, those emissions grew at a rate of 1.4% annually and added 31.6 gigatonnes of CO_2 to the atmosphere (International Energy Agency 2013). As more carbon dioxide is absorbed, the ocean hydrogen ion concentration increases (Feely et al. 2004) and the seas become more acidic, resulting in impacts on organisms that are calcium carbonate producers (algae, sponges, corals, mollusks, and others). There are also concerns that the greater acidity will result in the deterioration of the reef framework (Guinotte et al. 2006) that constitutes the foundation of essential habitat for many associated species including reef fish. Additionally, the chloro-fluorocarbon component of greenhouse gases depletes atmospheric ozone, thereby increasing the exposure of organisms to destructive ultraviolet solar radiation. Increased coral bleaching and genetic problems are attributed to ozone loss. Moreover, a warmer and more acidic ocean is expected to enhance coral diseases, favor benthic algae, and increase bioerosion by sponges and other organisms. Finally, alterations in current patterns due to climate change could impact shallow and deep corals by disrupting trophic systems (e.g., by altering upwelling patterns) or interfering with larval transport and recruitment.

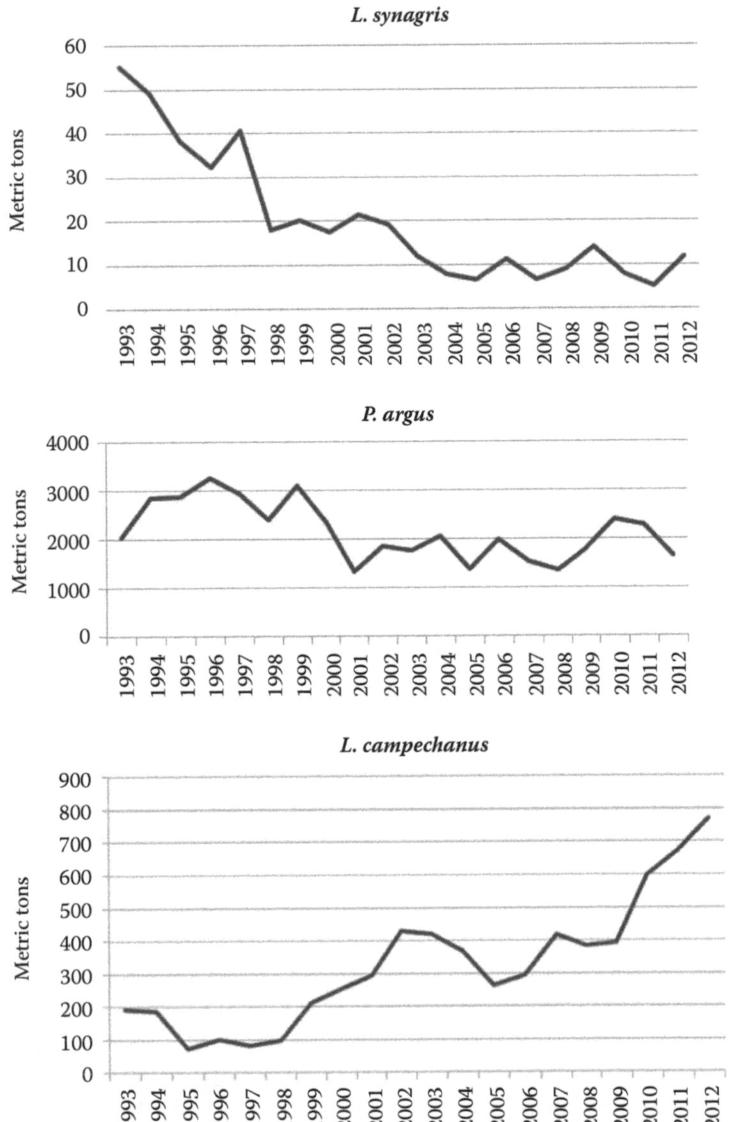

Figure 6.3 Gulf of Mexico commercial landings: *L. synagris* (lane snapper), *P. argus* (spiny lobster), and *L. campechanus* (red snapper). (From NOAA Fisheries: http://www.st.nmfs.noaa.gov/commercial-fisheries/commercial-landings/annual-landings/index.)

Another major factor potentially influencing reef fish habitat in the area of concern is the impact of human activities associated with petroleum extraction. The Deep Horizon oil-spill event in 2010 and the unprecedented use of oil dispersant chemicals to manage the resultant oil spill unquestionably impacted on biological communities in the northern Gulf of Mexico. Although much remains to be learned, preliminary evidence indicates that the oil and dispersants were transported by currents to within 145 km of the west coast of Florida and traces of the oil and dispersants were detected in organism tissues (Paul et al. 2013), although the direct and indirect impacts on coral reefs remain to be determined. Also, chronic discharge of a multiplicity of chemicals and nutrients from the Mississippi

River into the Gulf of Mexico has resulted in expansive areas of hypoxic "dead zones" (Diaz and Rosenberg 2008, Rabalais et al. 2010).

Factors associated with unchecked human population growth are a principal cause of most of these problems. Pollution, destructive farming practices, deforestation, and the overuse of fertilizers are just a few examples, and the longer we fail to pay due diligence to solving the problem of human population impact, the more difficult or impossible it will be to correct. One saying goes "No matter what your cause, it is a lost cause without human population control."

Climate change effects on reef fish fisheries potentially include reductions in species ranges, growth rates, reproduction, recruitment, and population size, and increased disease susceptibility. These effects are complex and differ depending on the species and the setting (Pinsky et al. 2013). Climate change effects may act synergistically with other coral reef stressors, such as fishing gear interactions (Zhang et al. 2011), pollution, and hypoxia events, and other insults such as invasive species, to further degrade the reef and its associated community. One of the more insidious recent events is the introduction of exotic Indo-Pacific lionfishes (*Pterois volitans* and *P. miles*), which have experienced a population explosion throughout the Gulf of Mexico, Caribbean, and eastern seaboard (Morris and Whitfield 2009). Lionfish are voracious predators of juvenile fishes and invertebrates, and are expanding into new areas at an alarming rate. No adequate solution has yet been devised to manage this problem.

The following sections examine zooxanthellate and azooxanthellate coral communities, their principal associated fisheries, and possible future trends. The reef areas covered from south to north include the Dry Tortugas, Pulley Ridge, West Florida Slope, eastern Gulf of Mexico epibenthic hardbottom communities, Florida Middle Grounds, Madison Swanson, Pinnacles, and Viosca Knolls 862, 906, and 826 (Figure 6.1).

Shelf and mesophotic reefs

Dry Tortugas area

The Dry Tortugas (Las Tortugas, or "The Turtles," so named by Ponce de Leon in 1513) is the western terminus of the Florida Reef Tract popularly referred to as the Florida Keys. It is a complex of sandy islands, seagrass meadows, open sediments, nearly emergent coral reefs, and deeper coral banks and reefs, including the Tortugas Banks (Agassiz 1885, Shinn and Jaap 2005). The area is bound within 24°21′–25°00′ N and 82°46′–83°12′ W (Figure 6.4).

In 1908, President Theodore Roosevelt designated the Dry Tortugas a wildlife refuge with the main purpose of protecting bird rookeries. The Dry Tortugas National Monument was established in 1935; bird rookeries and cultural resources (Fort Jefferson) remain the principal concerns. In 1979, United Nations Educational, Scientific and Cultural Organization (UNESCO) designated the Dry Tortugas as a World Heritage Site, and the following year, the U.S. National Park Service amended its management language to include coral reefs and resident marine life. Finally, in 1992, the area was designated Dry Tortugas National Park, occupying approximately 25,900 ha. Regulations prohibit commercial fishing (although some recreational fishing based on Florida and federal regulations is allowed) and the taking of lobster, and visits to bird rookeries and turtle nesting islands are seasonally prohibited. Tortugas Ecological Reserve, located to the west and south of the park, was established by Florida Keys National Marine Sanctuary in 2001 and is a protected area of 51,450 ha. The ecological reserve is divided into zones: some zones prohibit the taking of anything, others are open to recreational harvest, and still others

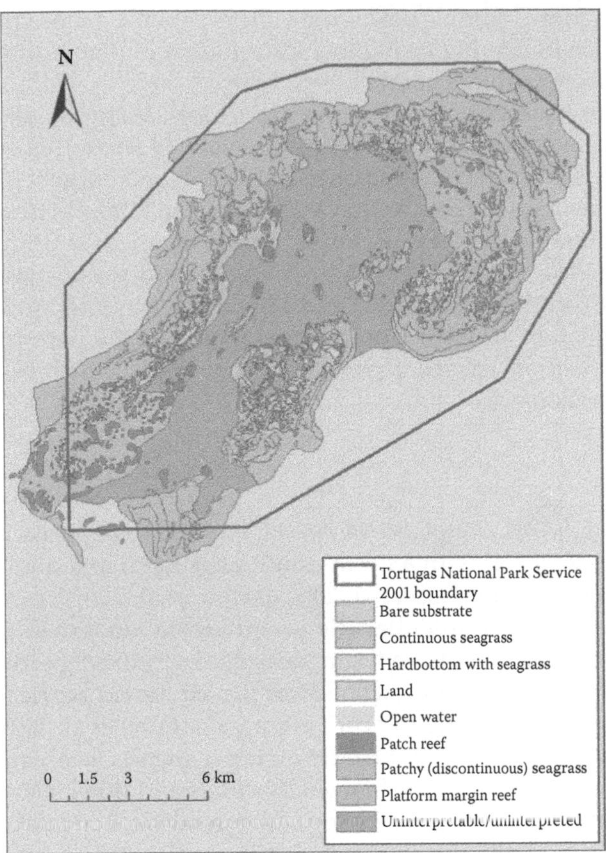

Figure 6.4 **(See color insert)** Dry Tortugas benthic habitat map. (From Florida Fish and Wildlife Conservation Commission cartography.)

are open to commercial and recreational fishers. In 2006, the park established a no-take research natural area that includes 11,900 ha with prohibitions on anchoring and fishing.

The Dry Tortugas are close to the Straits of Florida, the crossroads of the Caribbean, western Atlantic, and Gulf of Mexico. The Dry Tortugas coral fauna is rich and closely resembles the coral reef communities in the Bahamas, Cuba, and Mexico (Shinn and Jaap 2005, Jaap et al. 2008, Cairns et al. 2009). There are 47 zooxanthellate coral species at the Dry Tortugas, 38 at Looe Key, 28 in Biscayne National Park, 24 in the area north of the Miami Harbor channel in Miami-Dade County, 36 off Broward County, 24 off Palm Beach County, and 8 in Martin County on the reefs south of the St. Lucie Inlet (Jaap et al. 1989, 2008). A massive complex of reefs (the Tortugas Banks) 20–30 m deep is situated west of Loggerhead Key. The oldest Holocene coral reefs in the eastern gulf are located on the Tortugas Banks (Mallinson et al. 2003), where Holocene growth was dated to approximately 8000 years before the present (8 KYBP). The Tortugas Banks reefs failed to keep pace with rising sea levels; Mallinson et al. (2003) hypothesized that the Wisconsin glacial meltwater flowing across the Tortugas Banks lowered salinity and temperature and increased turbidity, inhibiting reef growth. A region within the Tortugas Banks, referred to as Sherwood Forest, is characterized by a low-relief foundation of Swiss cheese-like karst limestone, moderate-sized platy, and mushroom-like *Montastraea cavernosa* corals overlying the foundation. Multiple peaks and valleys provide a mosaic of horizontal, vertical, and angled surfaces, offering niches to a wide variety of plants and animals, including colonies of *Antipathes* spp. (black coral).

Major changes have occurred in Dry Tortugas reefs. Virtually all of the *Acropora palmata* (elkhorn coral) has been lost; in 1881, there were 44 ha of this species, while today there are only a few scattered colonies (Davis 1982, Jaap and Sargent 1993). Coral cover on Dry Tortugas reefs decreased from 50% in 1975 to <5% in 2010. Perturbations causing this decline include a massive die-off of *Acropora cervicornis* in 1977 from hypothermia, die-off of sea urchins (*Diadema* spp.) in 1983 from epidemic disease, major bleaching events (1985, 1987, and 1998), multiple category 3 and 4 hurricanes in 2004, and a severe hypothermia event in 2010 (Jaap et al. 2008). Coral degradation and the population collapse of the urchin, *Diadema antillarum*, caused a proliferation of algae, which resulted in less space available for coral settlement.

Pulley Ridge

Pulley Ridge (Figure 6.1) is composed of coastal carbonate deposits. It is approximately 300 km long (24°40′–26°40′ N) with a north–south alignment and a width of up to 15 km; depth ranges from 60 to 90 m (Jaap et al. 2000, Jarrett et al. 2005, Harter et al. 2008, Hine et al. 2008). Recent expeditions to Pulley Ridge employing modern seismic tools, remotely operated vehicles, submersibles, towed camera sleds, and mixed gas diving have revealed a unique reef system. The southernmost 30 km is a drowned barrier island (Figure 6.1), including dunes, spits, tidal inlets, and cat's eye ponds (Hine et al. 2008).

The island became a submerged shoal following rising sea level caused by glacial melting. After drowning, the shoal was colonized by corals and other reef organisms. Seismic data indicate that the coral structure at the top of the shoal is 1–2 m thick, formed from about 6 to 14 KYBP; vertical growth rate is estimated at 14 cm KY^{-1} (Mallinson et al. 2003). Pulley Ridge was designated an HAPC by the Gulf of Mexico Fisheries Management Council in 2005; activities involving pots, traps, longlines, nets, and anchoring are prohibited.

Autotrophic organisms (coralline and green algae and zooxanthellate corals) (Figure 6.5) are common in the southern 30 km of Pulley Ridge, which to the north is composed of rocky outcrops colonized by heterotrophic organisms (sponges, octocorals, azooxanthellate solitary and colonial Scleractinia, antipatharians, and crinoids).

The southern area is considered the deepest zooxanthellate coral reef on the continental shelf of North America. Photosynthetic active radiation reaching the reef was 5–30 μE $m^{-2}\,s^{-1}$, 1%–5% of the surface photosynthetic active radiation (Jarrett et al., 2005). The extant number of zooxanthellate coral species on south Pulley Ridge is not fully known, since identifying species based on images captured by remotely operated vehicles and submersibles is challenging. The principle corals include lettuce corals, family Agaricidae: *Leptoseris cucullata, Agarica lamarcki, A. fragilis,* and *A. undata*; branching corals (genus *Madracis*) are also present, as are pancake-like colonies of *M. cavernosa*.

Eastern Gulf of Mexico epibenthic hard bottom

This characterization is based on the "Hourglass Cruises" sampling program (Joyce and Williams 1969) dredge and trawl samples taken at 10 locations (6–73 m deep) over a 28-month period (1965–1967). Sampling stations were located off Anna Maria Island, Tampa Bay (27°37′ N) and west of Sanibel Island, Ft. Myers (26°24′ N). Hourglass Caryophyllina and Dendrophylliina were reported by Cairns (1977). Extensive information was also collected

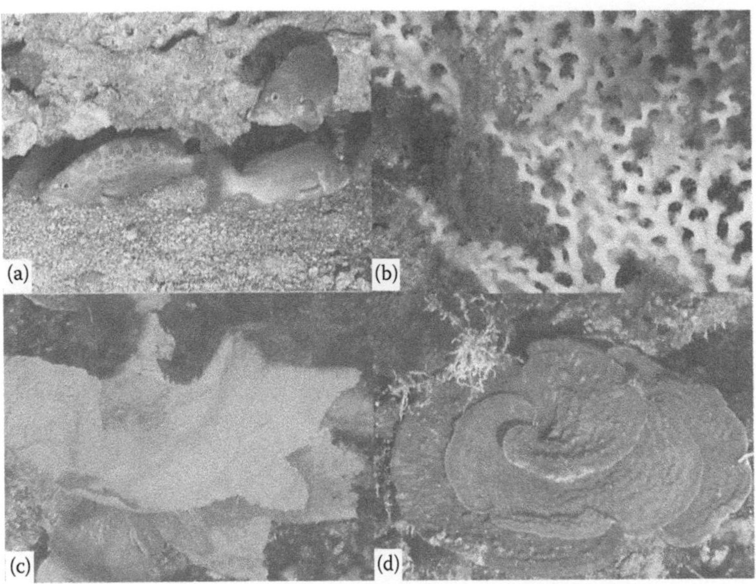

Figure 6.5 **(See color insert)** (a) Scamp grouper, *M. phenax*, under a ledge, north Pulley Ridge. (b) *O. tenella*, colonial azooxanthellate coral, north Pulley Ridge. (c) Green alga, *Anadyomene menziessi*, south Pulley Ridge. (d) Swirling plates of *L. cucullata*, zooxanthellate coral, south Pulley Ridge. (Photographs courtesy of Sustainable Seas Expedition, DeepWorker submersible.)

during the Gulfstream Natural Gas Pipeline project dating from 2001 to 2013 (ENSR 2005, Dupont et al. 2009, W. Jaap unpublished data).

Coral distribution on the eastern Gulf of Mexico hard bottom is highly related to the availability of elevated structure (i.e., ledges, rocky outcrops, etc.). Additionally, inshore areas directly influenced by large estuaries and rivers are depauperate in corals compared to areas not so influenced. In the Hourglass studies, two small species of azooxanthellate corals (*Astrangia poculata* and *Phyllangia americana*) were found in areas adjacent to Tampa Bay and San Carlos Bay; both were encrusting on exposed mollusk shells (*Arca atrina* and *Busycon* sp.). Thirteen species (Table 6.1) of zooxanthellate corals are common in eastern gulf areas with abundant hard bottom (Figure 6.6): *Cladocora arbuscula*, *Solenastrea hyades*, and *Siderastrea radians* are the principal species and are also found at depths of 15–50 m between Naples and Steinhatchee (25°00′–29°40′ N).

Beyond 50 m depth, rocky outcrops are less plentiful and two branching azooxanthellate species are uncommon (*Madracis asperula* and *Oculina tenella*). Small, solitary azooxanthellate corals are widely distributed in the eastern gulf (Table 6.1), but they are not reef building and do not provide much refuge habitat for reef-associated species. The hard-bottom habitat is a geologic structure with epibenthic communities composed of algae, sponges, octocorals, and scleractinian corals. These epibenthic communities include a wide variety of Caribbean/West Indian reef biota: the algae genera *Halimeda*, *Caulerpa*, and *Sargassum*; sponges such as *Cliona celeta*, *Spheciospongia vesperia*, and *Callyspongia vaginalis*; and octocoral genera, including *Eunicea*, *Plexaurella*, *Pseudoplexaura*, and *Muricea*; common echinoderms include black sea urchins (*Arbacia punctulata* and *Diadema antillarum*), basket starfish (*Astrophyton muricatum*), the reticulated starfish (*Oreaster reticulates*), and the five-hole sand dollar (*Mellita quinquiesperforata*).

Figure 6.6 **(See color insert)** Eastern Gulf of Mexico epibenthic (hardbottom) community, 27°59′ N, 83°14′ W, 25 m. Corals are mostly *C. arbuscula* (Lesueur); urchins are *A. punctulata* (Lamarck). (Image from Gulfstream Gas Pipeline monitoring, 2010.)

The expansive West Florida Shelf is a mosaic of carbonate ledges, outcrops, and surficial biogenic and clastic sediments (Gould and Stewart 1956, Holmes 1985). Hine et al. (2008:127) describes the central-west Florida continental shelf as a "distally-steepened, carbonate ramp platform; lacking a ridge of reefs, shoals, and islands." Limestone ledges and outcrops are a common feature of the eastern gulf seafloor. Some were created or modified by wave action during a period of lower sea level. Linear structures are typically oriented northwest to southeast, with 1 m or more of relief above the seabed. In some locations, the rock structures are covered with a few centimeters of sediments; sponges, octocorals, and stony corals protrude through the sediments, indicating a relatively recent shift in the sediments from storm activity. Ledges, caves, and outcrops provide habitat for invertebrates and fishes. Hine et al. (2008) recognized five types of hard bottom in the eastern gulf:

1. Undulating/scalloped edges, not having any discernible spatial (organized) pattern
2. Long, parallel features
3. Circular depressions with central structure (patch reefs)
4. Isolated hardbottom outcrops
5. Parallel ridges, trending NW–SE (paleo-shorelines?)

Grady (1971) and Woodward-Clyde and Continental Shelf Associates, Inc. (1983) reported that limestone outcrop exposure in depths of <40 m is more prevalent north of Florida Bay and the Ten Thousand Islands. The area between Cape Romano and Big Pine Key is the boundary region for Florida Bay and the Gulf of Mexico. Florida Bay has the most unpredictable environment of the eastern gulf due to its shallow depths and runoff from the adjacent Everglades (Schomer and Drew 1982). Florida Bay water is cooled and transported via winds and tides to the areas south and west causing hypothermic coral stress and mortalities (Davis 1982, Roberts et al. 1982); an extreme polar air mass in 2010 was long lasting and resulted in catastrophic coral deaths (Lirman et al. 2011, Colella et al. 2012). Conversely, during summer doldrums, seawater in shallow bays may become heated and result in hyperthermic and hypersaline stress and coral bleaching (Jaap 1979, 1985).

The environment of the zooxanthellate organisms in the eastern gulf is not ideal for autotrophy. Autotrophy combined with heterotrophy sustains those populations that have acclimated themselves to low light conditions (Porter 1976). Large beds of sediments supporting secondary production of interstitial invertebrates (Cahoon et al. 1990, Snelgrove 1999) support epibenthic communities in the eastern gulf. These resources are important food sources for the mixotrophic fauna.

Smith (1976) recorded at least 101 species of fishes from central West Florida Shelf hard bottom. Most of these showed a Caribbean/West Indies zoogeographic affinity. Commercially and recreationally important fish families (Serranidae [grouper], Lutjanidae [snapper], and Sparidae [porgy]) usually exhibited high species richness on these reefs.

Red tide, a harmful algal bloom caused by *Karenia brevis*, a single-cell, unarmored dinoflagellate phytoplankton (Davis 1948, Hansen et al. 2000), is a major perturbation that results in short-term catastrophic ecological change on the West Florida Shelf. Population explosions of *K. brevis* in the range of 10^5 cells per liter poison the waters and also deplete oxygen. Red tides are a well-known phenomenon in the eastern gulf, with documentation dating back to the late nineteenth century (Ingersoll 1881). Most reports describe massive fish kills and public health issues, as well as marine mammal and waterfowl deaths related to exposure to breve neurotoxin (Davis 1948, Steidinger and Ingle 1972, Landsberg 2002, Flewelling et al. 2005, Walsh et al. 2006). Anoxic conditions and the brevetoxin can extirpate nearly all the benthic organisms (Smith 1975, 1979, Dauer and Simon 1976, Simon and Dauer 1977). One recent study found that recovery takes 4–5 years (Dupont et al. 2009). Following the 2005 red tide, recovery as of May and June 2013 is complete, with 100% organic cover (algae, sponges, octocorals, and zooxanthellate corals).

The impact of red tide on fish is significant, evidenced by the tons of dead fish that wash ashore during red tide events. During the 2005 red tide, a number of charter boat captains reported that snapper and grouper had moved to deeper locations (>25 m) to avoid the toxic water (Captain Mark Roberts, Terra Ceia, Florida, USA, pers. comm.). Chronic disturbance from red tides is an important limiting factor for benthic communities, especially for nonmobile flora and fauna. It also impacts resident mobile invertebrates and fishes. Fisheries suffer because of human health concerns and localized population depressions of economically important species.

Gulfstream Natural Gas System, LLC, constructed an underwater pipeline 91 cm in diameter from Alabama–Mississippi to Port Manatee in 2001 (ENSR 2005). Trenching and burial was unsuccessful for portions of the pipeline, so Gulfstream placed limestone boulders over the pipe to comply with the permit conditions and provide protection. The rock covers 8.5 km of exposed pipe and the typical width is 11 m, forming a 9.4 ha elevated rock structure. In May 2013, the fifth inspection confirmed that the rock cover was still in place and undisturbed (Jaap and Watkins 2013).

Boulders form a bell-shaped mound across the pipe and currently are veneered with a prolific growth of algae and other sessile benthic plants and animals. The heterogeneity of the structural mass (both horizontally and vertically, creating a variety of irregular surfaces, with holes, caverns, projections, and depressions) has created diverse benthic and demersal fish communities. The most conspicuous structural species seen in pipeline videos was the brown alga (Phaeophyta), *Sargassum pteropleuron*. This alga forms dense thickets that dominate sections of the rock mound, much like a kelp forest canopy. Fishes are numerous and make use of the algal canopy for refuge. We observed *Mycetoperca phenax* (scamp), *Chaetodon sedentarius* (reef butterfly), *S. dumerili* (greater amberjack), *Archosargus probatocephalus* (sheepshead), *Haemulon plumierii* (white grunt), *H. aurolineatum* (tomtate), *Stegastes variabilis* (coco damselfish), *M. microlepis* (gag), *Lachnolaimus maximus* (hogfish), *Holacanthus* spp. (angelfishes), and *P. volitans* (lionfish, an exotic introduced species).

To compensate for pipeline construction perturbation, Gulfstream constructed multiple mitigation projects in 15–37 m depths offshore of Tampa Bay (27°34′–27°45′ N, 82°58′–83°32′ W). Each was 2.3 ha in area and received an average of 14,600 t of limestone boulders (0.5–1.5 m in length and in width) per site. Additionally, 16–17 prefabricated modules (limestone boulders set in a concrete base with internal tunnels) were deployed in nine locations at 16–17 m depths (27°33′–27°34′ N, 82°58′–83°04′ W). Structures were elevated above the seafloor. Deployments have remained in place since 2001 and have developed epibenthic community structures equivalent to naturally occurring hard bottom in similar depths. Fishes have occupied the structures, including the economically important snapper and groupers (Lutjanidae and Serranidae, respectively). In 2005, biota on mitigation sites in depths <26 m were extirpated by red tide toxins and hypoxia, but by 2007, the epibenthic and fish communities had recovered to pre–red tide status (Dupont et al. 2009). The total footprint of the boulder mitigation sites is approximately 22 ha and that of the modules is estimated at 15.5 ha. The pipeline rock cover plus the mitigation resulted in nearly 47 ha of rocky structure added to the West Florida Shelf. Monitoring in 2013 confirmed that the limestone boulders continue to support marine communities (Jaap and Watkins 2013).

Florida Middle Grounds

The Florida Middle Grounds complex parallels the shelf margin and is located between 28°10′ and 28°45′ N and 84°00′ and 84°25′ W (Figure 6.1). Reich et al. (2013) reported that the structures on the eastern Florida Middle Grounds were formed 8–9 KYBP and are constructed of fine calcareous muds capped by shell deposits of the vermetid gastropod *Petaloconchus* sp. Individual carbonate banks are 12–15 m high and crest at ~25–30 m below sea level. The Florida Middle Grounds is the northernmost reef complex in the Gulf of Mexico with numerous zooxanthellate corals (Grimm and Hopkins 1977, Jaap et al. 2000). A section of the Florida Middle Grounds was designated an HAPC in the Coral-Coral Reef Fishery Management Plan under the Magnuson-Stevens Act (Gulf of Mexico and South Atlantic Fishery Management Councils 1982) and covers 119 ha.

Environmental impact studies at the Middle Grounds for petroleum exploration and development documented 103 species of algae, 40 sponges, 75 mollusks, 56 decapod crustaceans, 41 polycheates, 23 echinoderms, and 170 species of fishes (Smith et al. 1975, Hopkins et al. 1977). The reef fish fauna (e.g., Lutjanidae, Serranidae, and Sparidae) was dominated by species with Caribbean/West Indies zoogeographic affinities (Smith 1976). Coleman et al. (2005) reported 29 species of algae, 33 species of sponges, 10 genera of octocorals, 16 species of zooxanthellate Scleractinia, and 1 *Millepora* species. Data were from a single sampling effort and from several stations reported by Grimm and Hopkins (1977).

Madison Swanson and Steamboat Lumps marine reserves

The edge of the West Florida Shelf comprises a band of drowned fossil reefs at depths of approximately 50–120 m, which provide hard substrate for colonization by sessile benthic fauna. They also provide habitat for a number of commercially important fishery species. Two of these areas, Madison Swanson and Steamboat Lumps (Figure 6.1), were established as marine reserves in 2000 by the Gulf of Mexico Fishery Management Council, primarily to protect spawning aggregations of the serranid *M. microlepis* (gag). Both reserves are approximately the same size (34.3 ha) and cover similar depth ranges (60–120 m). The Madison Swanson site has a central sandy region (80–120 m depth) that drops to 160 m to the south and west. The sandy plain is surrounded by rocky ridges along the northeastern and southern edges, with some particularly high-relief areas with pinnacles, boulders, ridges, and caves. Gag and scamp spawning sites were located on the southern rocky ridgeline of the reserve, whereas scamp and *E. morio* (red grouper) spawned on or near the high relief areas to the northeast (Coleman et al. 2011, Koenig and Coleman 2011). Steamboat Lumps marine reserve consists of a series of northeast–southwest low-relief terraces. Unlike the Madison Swanson marine reserve, there are no large rocky outcrops, but low-relief carbonate rocks provide substrate for a dense sessile invertebrate community, including sponges, octocorals, coralline algae, and occasional colonies of *Oculina* spp. No gag or scamp spawning aggregations were noted in the Steamboat Lumps marine reserve, but red grouper were abundant and were observed to spawn in the northern area of the reserve (Coleman et al. 2011, Koenig and Coleman 2011). There are other sites of similar depth and habitat along the edge of the shelf ("Twin Ridges" and "The Edges") that have been identified as spawning habitats for different grouper species (Coleman et al. 2011), but these are still open to fishing. These rocky ledge habitats not only support grouper and other reef fish but also, like other hard substrate along the West Florida Shelf, provide habitat for a diverse assemblage of corals, sponges, and other invertebrates.

Pinnacles

A series of calcareous outcrops off Alabama, known as the Pinnacles, occur in 73–101 m depths and were first described in detail by Ludwick and Walton (1957). These topographic prominences, averaging about 9 m in vertical relief, stand in stark contrast to an area of the north-central Gulf of Mexico generally dominated by riverine-derived soft sediments. Most of these spirelike, rocky features occur from about 88°18′ W to about 87°27′ W and are clustered in two bands that roughly parallel the isobaths (Sager et al. 1992). These reeflike mounds are most abundant along the 74–82 m depth zone, with another band along the 105–120 m zone. They are thought to be biogenic and to have been formed during lower sea level stands but may not be actively forming at present due to depth constraints (Sager et al. 1992). In addition to the smaller, more pinnacle-like structures noted above, this hard substratum region also contains low-relief (<2 m) ridges that run parallel to isobaths in depths of 68–76 m with lengths of up to 15 km (Sager et al. 1992). Where light is sufficient and sedimentation rate low, crustose coralline algae add to the structural complexity (Gittings et al. 1992). Continental Shelf Associates, Inc. (1992) surveyed this region, describing the hardground outcroppings and the sedimentology. The area was also mapped with multibeam sonar by the U.S. Geological Survey (Figure 6.7), and its morphology was described in detail (Gardner et al. 2001, 2002).

The hard substratum of the Pinnacles area hosts a variety of sessile invertebrates dominated by tropical and subtropical taxa, including sponges, hard corals, gorgonians, cup corals, hydroids, antipatharian corals, bryozoans, and oysters (Continental Shelf Associates, Inc. 1992). These communities add to the physical complexity of these reefs, and also contribute food resources. Gittings et al. (1992) identified at least 93 invertebrate taxa from these reefs, mostly dominated by suspension feeders. Living solitary and colonial corals found here included *Oculina* sp., *Madrepora carolina, Rhizopsammia manuelensis, Paracyathus pulchellus, Agaricia* cf. *fragilis* (Gittings et al. 1992), and *Madracis* sp. (Continental

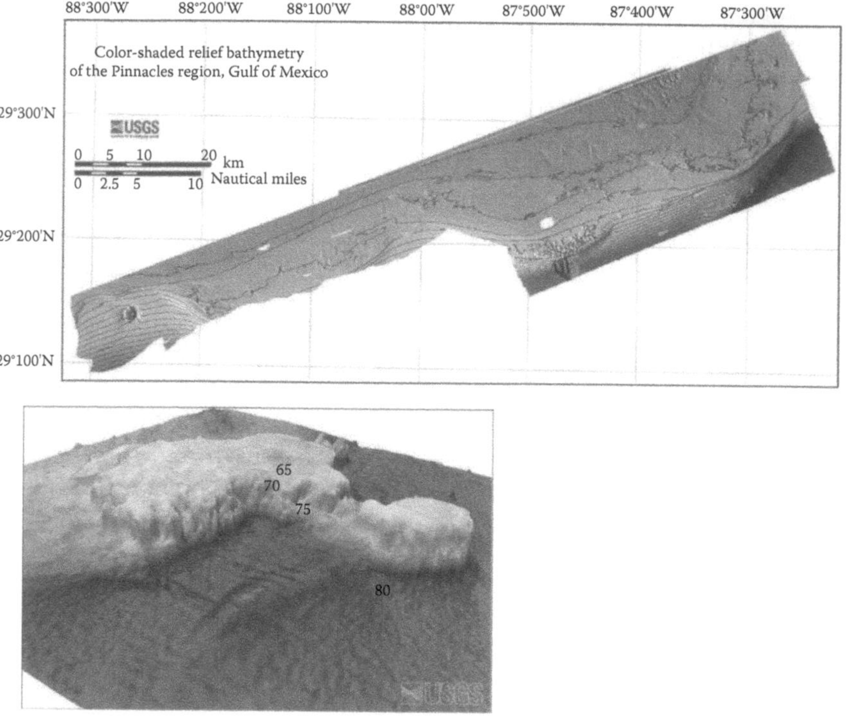

Figure 6.7 Multibeam sonar bathymetry map of the Pinnacles reef tract (top) complied by USGS (From Gardner, J.V., Dartnell, P., Sulak, K.J., Calder, B., and Hellequin, L., *Gulf Mex Sci*, 19,132–157, 2001.). Bottom image is the Roughtongue Reef study site (see figure 1.6 in Weaver, D.C., Dennis, G.D., and Sulak, K.J., Northeastern Gulf of Mexico coastal and marine ecosystem program: Community structure and trophic ecology of demersal fishes on the Pinnacles reef tract. Final synthesis report. U.S. Department of the Interior, Geological Survey USGS BSR-2001-0008; Minerals Management Service, Gulf of Mexico OCS Region, OCS Study MMS 2002-034, 2002.) showing multibeam generated topography with 9× vertical exaggeration (depth contours in meters).

Shelf Associates, Inc. 1992). *Antipathes* spp. and *Cirrhipathes* spp. were the most abundant black corals (Antipatharia) occurring on the reefs.

While the geology of the Pinnacles tract has been reasonably well studied, far less is known about the biology and ecology, particularly as regards the invertebrate communities. The Pinnacles are superficially similar in physical structure and faunal composition to temperate zone reefs elsewhere in similar depths in the Gulf of Mexico (Smith 1976) and southeastern United States (Parker and Ross 1986, Quattrini and Ross 2006, S. Ross pers. comm.), but are quite unlike the more tropical coral-based reefs of the northwestern gulf (Dennis and Bright 1988).

Deep-sea reefs

As defined here, Gulf of Mexico deep reefs occur below 300 m. They can be predominantly rocky substratum colonized by various corals and other sessile structure-forming species, or they can be formed entirely from coral growth (bioherms), as are shallow reefs. Because of the lack of sunlight, all of the deep coral species are azooxanthellate and derive their energy from heterotrophy. Most of these corals occur in areas with strong currents and are widely scattered in the Gulf of Mexico (bioherms appear more common in the north-central to eastern gulf, usually in depths of 300–800 m), and form significant habitats in certain areas (Brooke and Schroeder 2007, Schroeder and Brooke 2011). The branching scleractinian coral *Lophelia pertusa* is the dominant structure-forming deep coral species in the gulf. We briefly describe several of the best known of these deep reef sites.

West Florida Slope

The West Florida Slope, extending from the De Soto Canyon to the Straits of Florida, is a complex transition zone between the continental shelf and the deeper Gulf of Mexico (Doyle and Holmes 1985). It is a relatively smooth carbonate ramp, gradually (1°–5°) deepening from 200 to 1800–2000 m at the top of the Florida Escarpment (Doyle and Holmes 1985, Mullins et al. 1988a,b). It is punctuated in places by high relief complex topography. The southern part of the West Florida Slope is narrower and has steeper slopes and more rocky outcroppings than the northern part. At about 400–650 m depths, sediments are biogenic carbonates that change from coarse grained to fine grained with increasing depth (Mullins et al. 1988a, Gardulski et al. 1990), and sediment overlay here is often thin or lacking (Doyle and Holmes 1985). The 400–600 m depth zone is characterized by outcrops of phosphorites and carbonate hard substratum surrounded by coarse grained, winnowed sediments (Mullins et al. 1988a); however, outcrops are rare north of 26°40′ N (Doyle and Holmes 1985).

The most prominent hardground feature of the West Florida Slope is a rocky scarp about 450 m deep that traverses north–south for over 200 km, exhibiting a vertical profile ranging from about 20 to 200 m (Ross and Brooke unpublished data). To the west of this escarpment are hundreds of rocky ridges and mounds with vertical heights of a few meters to 20 m. Recent observations from submersible and remotely operated vehicle dives (led by S. Ross and S. Brooke), as well as observations by Reed et al. (2006) and Hübscher et al. (2010), documented that various corals (octocorals and black corals dominated by *Leiopathes* sp.) and sponges are attached to the rocky substratum, adding to habitat complexity. However, on many of the ridges and mounds there are well-developed stands of *L. pertusa*. Data collected by Ross and Brooke (unpublished data) indicate that *L. pertusa* has formed bioherms in many places along the ridges and mounds. Besides *L. pertusa* and *Leiopathes* sp., other corals observed here include *Madrepora occulata* and the solitary cup coral (*Thecopsammia socialis*) (Ross and Brooke unpublished data).

Aside from a partial listing of species by Reed et al. (2006), there are no published studies documenting the epibenthic fauna of the West Florida Slope. However, preliminary data on the invertebrate and fish communities of this area have been compiled by Ross and Brooke (unpublished data). Sixty-seven taxa of invertebrates were observed on the West Florida Slope reefs and nearby habitats (see Table 6.1 for a list of scleractinian corals).

Viosca Knolls 862, 906, and 826, northern Gulf of Mexico

The northern Gulf of Mexico slope is extremely complex, with seven submarine canyons, basins, hardgrounds, gullies, gas seeps, and gas hydrate deposits (Brooke and Schroeder 2007). Hard substrate in the deep northern Gulf of Mexico is primarily authigenic carbonate, produced as a byproduct of chemosynthetic activity. These carbonate rocks, slabs, and pavements can be locally abundant and provide suitable substrate for the development of sessile benthic communities. The most well-developed deep coral sites in the northern Gulf of Mexico occur on the crest and southwest flank of a mound in the southwest corner of Viosca Knoll lease block 826 (29°09.5′ N, 88°01.0′ W, 430–520 m) on the upper DeSoto slope. Viosca Knoll is a salt dome that rises from 530 m depth to approximately 450 m. *L. pertusa* is common on the hard substrate of the knoll, with individual colonies ranging in size from a few centimeters to over 1.5 m in diameter, while reefs can reach 2 m in height and 3–4 m in length. Many of these aggregated colonies appear to be in the first phase of the "thicket" building stage described by Squires (1964). Colonies <25–50 cm in diameter were usually completely alive, while large and aggregated colonies were often dead at the base and center with live outer branches. A diverse community of gorgonians, antipatharians, and sponges, together with mobile invertebrates, such as comatulid crinoids, echinoids, and bivalves, were associated with live and dead *L. pertusa* colonies. This site also had communities of chemosynthetic tubeworms often in close proximity to the corals (Brooke and Schroeder 2007). The second Viosca Knoll site is near a submarine canyon in lease block VK 862/906 (29°06.4′ N, 88°22.9′ W). This area is comprised of a topographic high on the northern edge of a rocky ridge that extends southward for over 2 km to the eastern rim of the canyon. Water depths range between 300 and 500 m. The shallower northern area consists of large carbonate slabs and boulders with individual colonies of *L. pertusa*. Several species of unidentified white anemones dominate the fauna in this shallow area and often co-occur with a small stalked hexactinellid sponge.

Dense aggregations of bamboo corals (Isididae) were observed in this relatively shallow area, as well as numerous gorgonians such as *C. americana americana* (Primnoidae) and large black corals (Antipathidiae) (Continental Shelf Associates, Inc. 2007). The southern section of this site is deeper (~500 m) and has several small mounds that are possibly coral bioherms or carbonate mounds with a thick cover of coral colonies (Quattrini et al. 2014). Coral development and associated communities are similar to those described for the Viosca Knoll 826 site. Large wreckfish (*Polyprion americanus*), barrelfish (*Hyperoglyphe perciformes*), and grouper (*Epinephelus* spp.) were observed at the shallow Viosca Knoll 862 site. Risks are similar to other sites in these depths: fishing gear impacts and potential vulnerability to oil spills.

Reef fishery risks

Using hook-and-line, nets, traps, and a hard-hat diving rig to compile the list, Longley and Hildebrandt (1941) reported 442 fish species around the Dry Tortugas off Florida, 300 of which they considered to be reef fish. Bohlke and Chaplin (1968) list 496 reef fish from the nearby Bahamas; Tilmant (1984) felt this was a reasonable estimate for Florida reefs taking into account all microniches. The ranking of fish family abundances at the Dry Tortugas from *in situ* census was, from most to fewest: (1) Pomacentridae (damselfish); (2) Scaridae (parrotfish); (3) Haemulidae (grunt); (4) Labridae (wrasse); (5) Serranidae (grouper and sea bass); and (6) Chaetodontidae (butterfly fish) (Thompson and Schmidt 1977). These relative abundances by family are similar to those at other Florida Keys reefs; however, the rankings differ from location to location (Tilmant 1984). The hydrodynamic processes in the Dry Tortugas region facilitate larval recruitment from multiple Caribbean areas. The Dry Tortugas is also a source of progeny from fish spawning aggregations that gather at Riley's Hump and other locations (Jaap et al. 2008).

Spiny lobster (*Panularis argus*) are completely protected in Dry Tortugas National Park; this has resulted in lobsters migrating outside the park where they are legally harvested. This protection has existed for decades, and the age and size of lobsters in the Dry Tortugas is greater and larger than populations found in other areas of southeast Florida (Davis 1977).

Fisheries management in the Dry Tortugas is conducted through a complex matrix of agencies and rules: the National Park Service, Florida Fish and Wildlife Commission, NOAA Sanctuaries, and two fisheries management councils all have a say in management. The fishery management plan lists 73 species found on the reef, including Serranidae, Lutjanidae, Harmulidae, Sparidae, and Labridae. An MPA initiative has placed much of the Dry Tortugas into no-take status. The Dry Tortugas shallow depths result in greater risks from coral bleaching, hurricanes, coral diseases, and spatial competition from benthic algae (Table 6.3). The reef fish species most vulnerable to habitat-related population losses are those requiring coral niches during a portion of their life cycle. There are more of these susceptible species in the Dry Tortugas because of its tropical affinities, unique crossroad location, and multitude of reef habitats. Additionally, seagrass habitat in the Dry Tortugas has experienced losses from hurricanes; thus, fish that depend upon seagrass habitats for foraging and juvenile refuge are at greater risk. The trend of coral losses over the past decade implies that this area will have reduced carrying capacity to sustain reef fish and spiny lobster populations. The concerns for resident reef fish include fewer niches due to habitat loss, and suppressed reproduction, recruitment, and growth. Careful management

Table 6.3 Coral Reef Fishery Risk Factors

Risk	DT	PR	EGHB	FMG	MS and SL	PN	VK	WFS
Coral bleaching	3		1	1	1	1	0	0
More frequent and extreme hurricanes	3	2	2	2	2	2	1	1
Ocean acidification: calcification slowdown	2	1	1	1	2	1	3	3
Ocean acidification: structural dissolution	2	3	1	1	1	1	2	2
Rising sea level	1	2	1	1	1	1	1	1
Increasing ocean temperatures	3	2	2	2	1	1	2	2
Algal competition with coral	2	1	2	2	2	1	0	0
Intensive fishing pressure	2	2	2	2	1	2	2	2
Lionfish competition	3	3	3	3	3	2	0	0
Other introduced species	2	1	1	1	1	1	1	1
Coastal pollution (e.g., Mississippi runoff)	1	1	2	2	1	2	1	1
Destructive fishing methods	1	2	2	2	1	1	3	3
Coral disease	3	2	1	1	1	1	1	1
Harmful algal blooms (red tide)	1	1	3	1	2	1	0	0
Oil spill and other anthropogenic insults	2	2	2	2	2	2	2	2
Σ Risk index value	31	25	26	24	22	20	19	19

Note: Table 6.3 is produced by the authors, based on information, experience, and best estimate.

EGHB, eastern gulf hard bottom; DT, Dry Tortugas; FMG, Florida Middle Grounds; MS, Madison Swanson; PN, Pinnacles; PR, Pulley Ridge; SL, Steamboat Lumps; VK, Viosca Knolls; WFS, West Florida Slope. Risk: 0, no risk or irrelevant; 1, low; 2, moderate; 3, high.

of the Tortugas Ecological Reserve and Research Natural Area and less fishing pressure, combined with loss of reef structure, may moderate some of the reduction in reef fishery biomass. In southeast Florida, recreational fishing effort between 2001 and 2005 was estimated to entail 6.4 million fishers averaging 27.2 million fishing trips annually. Fishing pressure resulted in the National Marine Fisheries Service reporting that 11 of the grouper and snapper stocks were overfished (National Marine Fisheries Service 2005). The coral reef fishery habitat in the Dry Tortugas is in a deteriorated condition due to the multiple stresses of global climate change; water quality issues, such as nutrient enrichment and acidification; hurricane perturbation; and a 2010 hypothermia event that extirpated many corals. Of all the areas considered in this chapter, the Dry Tortugas reef fish populations are at greatest risk from climate and oceanic chemistry perturbations.

Pulley Ridge serves largely as a commercial fishing area, with the focus on grouper, snapper, and amberjack. During multiple expeditions to Pulley Ridge, we observed commercial vessels fishing in the area. The reef structure and resident fauna provide refuge and food for higher trophic level fishes. There are approximately 90 species of fishes, and shallow and deep reef species are represented in the fauna. Harter et al. (2008) surveyed fishes on Pulley Ridge with remotely operated vehicle cameras; they reported 58 species on the south ridge, with the most commonly seen fishes on camera being *Serranus tortugarum* (chalk bass), *Chromus scotti* (purple reef fish), *C. sedentarius* (reef butterfly fish), and *Microspathodon chrysurus* (yellowtail reef fish). Several nonidentified wrasse species were also plentiful. These are not commercially important fishery species, but are foraged upon by Serranidae and Lutjanidae. They also reported that the relative species abundance of fishes on the northern portion of Pulley Ridge was remarkably different from that on the southern end in the HAPC.

Pulley Ridge is relatively pristine; remoteness and depth offer protection from anthropogenic and storm disturbances. The southern area is most susceptible to degradation due to its fragile structure. During DeepWorker submersible dives in this area, we documented layers of cemented and loose plates of *Agaricia*, *Leptoseris*, and *Montastraea* corals. On one dive, W. Jaap observed a red grouper burrow under a series of loose coral plates to find refuge. Individual plates are a few millimeters thick, somewhat round, and up to 20–30 cm in diameter. Use of anchors, traps, or trawls would result in breakage and collapse, and recovery would be very slow because of the extremely low levels of light and coincident reef growth rates. Lionfish have been reported in this area. Pulley Ridge is subject to multiple risk factors (Table 6.3) in the HAPC area, and coral is at intermediate, relative risk from disease. If sea level increases, less light will be available. Also, those corals living on the physiological margins may be unable to calcify in a more acidic ocean (Hoegh-Guldberg et al. 2007). Loss of structure will result in reduced fishery production. The northern portion of Pulley Ridge is a geological structure, so it should be less susceptible to degradation from climate and oceanic chemistry perturbations.

Commercial and recreational fishers target Serranidae, Lutjanidae, and other reef species that are found on eastern gulf epibenthic hard bottom. Lionfish are documented from many hardbottom sites (W. Jaap pers. obs.). Eastern Gulf of Mexico epibenthic communities, including algae, sponges, octocorals, hydrocorals, and scleractinian corals, colonize rocky, topographically elevated structures, such as ledges and outcroppings. Although the biological community may experience perturbation from climate and ocean chemistry changes, the refuge functions of the structures will be unaffected (Table 6.3). Corals are relatively unimportant in supporting the primary fisheries here; risks include red tide, perhaps lionfish proliferation, and the possibility of an oil spill (Paul et al. 2013).

Economically important fishery species identified during the Coleman et al. (2005) Florida Middle Grounds study included gray snapper (*L. griseus*), subadult scamp

(*M. phenax*), and subadult red grouper (*E. morio*). Gray snapper was represented by an adult population that occurred primarily along the top of the structures. *Lutjanus campechanus* (red snapper) and *M. microlepis* (gag) rarely appeared in surveys, although they are among the most economically important species in the northeastern Gulf of Mexico. The Middle Grounds has been, historically, an important commercial fishing ground; however, recently it has also become a recreational fishing destination for anglers and divers.

The Middle Grounds system is under the same threats as the eastern gulf hard bottom (Table 6.3); an additional potential problem is Mississippi River spring runoff. It is common for the Middle Grounds to be bathed in river runoff in May (Coleman et al. 2005). The risk is mostly from agricultural chemicals, since the catchment basin is huge and is an important agricultural area (Rabalais et al. 2010). However, if the reef fishery is adequately managed, reef fish production in the Middle Grounds may remain relatively stable.

At least two reports have characterized the fishes of the Pinnacles reef tract (Continental Shelf Associates, Inc. 1992, Weaver et al. 2002). The most extensive study (Weaver et al. 2002) documented 159 fish species on and near the Pinnacles reefs, with 70 of these being classified as obligate reef species and 32 species as using the reefs facultatively. These reef fishes also exhibited a Caribbean faunal affinity; however, species richness was reduced at the Pinnacles compared with other gulf reefs (Weaver et al. 2002). Planktivorous fishes, especially anthiinine serranids, were dominant, and anthiinines were suggested as keystone species (Weaver et al. 2002). Although commercially and recreationally important fishes were also important members of the reef community, and Weaver et al. (2002) alluded to fishing pressure in this area, fishing effort or the impact of fishing on these reefs does not seem to have been documented. Use of this area by commercial and recreational fisheries appears to be limited; however, aside from physical habitat destruction, overfishing probably remains the biggest threat to Pinnacles fish communities. To date we are unaware of any government initiatives to specifically protect the Pinnacles.

There do not appear to be any active bottom commercial fisheries along the deep reefs of the West Florida Slope, except for an intermittent small trap fishery for *C. fenneri* (golden crab). We observed golden crab to be abundant on both the rocky habitats and especially on *L. pertusa* colonies. Swordfish (*Xiphias gladius*) were infrequently seen along these hard grounds. Of the 50 species of bottom fishes observed in this area, most were only seen on West Florida Slope rocky substrata or on corals (Ross and Brooke unpublished data). As elsewhere (Ross and Quattrini 2007, Sulak et al. 2007), the deep reefs of the West Florida Shelf support a unique fish community that appears to be dependent on complex, structured habitats (as are shallow reef fishes).

While large demersal fisheries below 300 m are uncommon in the Gulf of Mexico (Stiles et al. 2007), several fish species could be commercially important (e.g., *Hoplostethus* spp., *Beryx* spp., *Hyperoglyphe perciformis, Helicolenus dactylopterus*), as in other parts of the world (Gordon et al. 2003, Norse et al. 2012). Aside from habitat destruction by fishing gear (especially trawls and traps), negative impacts from harvesting deep-sea fishes are due to the fact that, in general, most deep-sea fishes cannot biologically support intense fishing pressure (Roberts 2002, Norse et al. 2012). There is little information on the life histories and ecologies of potentially commercially important deep-sea fishes (see above) in this region, although such knowledge would be critical for managing emerging fisheries. Ross and Brooke (unpublished data) did not observe any obvious damage to these reefs following the Deepwater Horizon oil spill. Aside from fishing activity, environmental issues that may affect these reefs include ocean acidification and changes in currents related to climate change. To date, none of the deep reefs of the West Florida Slope have been covered by any protection measures (e.g., MPA, HAPC), but such initiatives should be considered by the Gulf of Mexico Fishery Management Council.

Conclusions

Reef habitats in the eastern Gulf of Mexico are important for multiple fisheries. The consequences of climate change and ocean acidification cannot be accurately predicted, but we believe that coral-dominated, shallow-water (<30 m) systems are most susceptible because they are at risk from warmer water, more frequent and intense storm events, coral diseases, and regime shifts from coral to non-coral-dominated benthic communities. Coral calcification may be disrupted by ocean acidification (Veron 2011). Zooxanthellate Scleractinia, such as *Pocillopora damicornis*, have shown that the effect of pCO_2 on coral recruits can be light dependent (Dufault et al. 2013).

Fish communities are expected to respond to changes in reef type and structure in several ways. First, obligate coralivorus fishes will not persist if coral loss is extreme. Second, however, herbivorous fishes may become more abundant and diverse because the standing crop of benthic algae will benefit from warmer temperatures, perhaps more nutrients, and increased living space. Because shallow, coral-dominated reefs are major destinations for recreational (hook-and-line and diver) fishers, management programs need to be proactive to adjust to changes and pressures (Mumby and Steneck 2008, Sale 2008).

Moderate-depth reefs that are mostly geological, topographic edifices with epibenthic communities including hydrocorals, octocorals, scleractinian corals, and sponges in low to moderate abundance or cover are in a more favorable situation for maintaining the *status quo*. We note the exception of Pulley Ridge, a southern HAPC and a structure built by living and dead coral plates (multiple species, mostly belonging to the family Agariicidae). Its remoteness and MPA status are beneficial to sustaining this habitat. However, climate change (including higher sea level) and ocean acidification may destabilize this area.

Deep reefs along the continental shelf margin and slope are in remote locations. An increase in seawater temperature to over 15°C at depths below 300 m would be problematic; in experiments, upper temperature lethality for *L. pertusa* was determined to be around 15°C (Brooke et al. 2013). Ocean acidification is an issue; however, several laboratory studies have shown a mixture of calcification rate responses with greater seawater acidity. Some species exhibit increased rates of calcification, while others exhibit decreased rates. If growth rates for *Lophelia* and *Madrepora* (deep-sea reef-building corals) respond to a reduced seawater pH with reduced calcification rates or dissolving of skeletal material, less refuge structure could be the result.

In sum, if reef fisheries are to maintain any kind of balance and continuing health within the Gulf of Mexico system, long-term efforts should include:

- Ongoing monitoring to gather accurate and current information on status and trends for fish and epibenthic communities
- Compliance in managing MPAs
- Education so that public support will underpin management and research

References

Agassiz A. 1885. The Tortugas and Florida reefs. Mem Am Acad Arts Sci. 11:107–134.

Ault JS, Smith SG, Bohnsack JA, Luo J, Harper DE, McClellan DB. 2006. Building sustainable fisheries in Florida's coral reef ecosystem: Positive signs in the Dry Tortugas. Bull Mar Sci. 78:633–654.

Baker AW, Glynn P, Riegl B. 2008. Climate change and coral reef bleaching: An ecological assessment of long term impacts, recovery trends and future outlook. Estuar Coast Shelf Sci. 80:435–471.

Barth A, Alvera-Azcarte A, Weisberg RH. 2008. A nested model study of the Loop Current generated variability and its impact on the West Florida Shelf. J Geophys Res. 113(C5):C05009. Available from: http://dx.doi.org/10.1029/2007JC004492 via the Internet. Accessed 31 January, 2014.

Bellwood DR, Hughes TP, Folke C, Nystrom M. 2004. Confronting coral reef crisis. Nature. 429:827–833.

Bohlke JE, Chaplin CCG. 1968. Fishes of the Bahamas and adjacent tropical waters. Wynnewood, PA: Livingston Publishing.

Brooke S, Schroeder WW. 2007. State of deep coral ecosystems in the Gulf of Mexico region: Texas to the Florida Straits. *In:* Lumsden SE, Hourigan TF, Bruckner AW, Dorr G, editors. The state of deep coral ecosystems of the United States. NOAA Technical Memorandum CRCP-3. Silver Spring: NOAA. p. 271–306.

Brooke S, Ross SW, Bane JM, Seim HE, Young CM. 2013. Temperature tolerance of the deep-sea coral *Lophelia pertusa* from the southeastern United States. Deep-Sea Res II Top Stud Oceanogr. 92:240–248.

Brooks GR, Holmes CW. 2011. West Florida continental slope. *In:* Buster NA, Holmes CW, editors. Gulf of Mexico origin, waters, and biota: Volume 3, Geology. College Station, TX: Texas A&M University. p. 129–139.

Cahoon LB, Lindquist DG, Clavijo IE. 1990. Live bottoms in the continental shelf ecosystem: A misconception. *In:* Jaap WC, editor. Proceedings of the American Academy of Underwater Sciences 10th Annual Scientific Diving Symposium. 4–7 October, 1990. St. Petersburg, FL: University of South Florida. p. 39–47.

Cairns SD. 1977. Stony corals: I. Caryophylliina and Dendrophylliina (Anthozoa: Scleractinia). Memoirs of the Hourglass Cruises. Volume 3, pt. 4. St. Petersburg: FMRI.

Cairns SD, Opresko DM, Hopkins TS, Schroeder W. 1993. New records of deepwater Cnidaria (Scleractinia and Antipatharia) from the Gulf of Mexico. NE Gulf Sci. 13:1–11.

Cairns SD, McGinley M, Podowski EL, Becker EL, Lessard-Pilon S, Viada S, Fisher CR. 2008. Coral communities of the deep Gulf of Mexico. Deep-Sea Res I Oceanogr Res Pap. 55:777–787. Available from: http://dx.doi.org/10.1016/j.dsr.2008.03.005 via the Internet. Accessed 31 January, 2014.

Cordes SD, Jaap WC, Lang JC. 2009. Scleractinia (Cnidaria) of the Gulf of Mexico. *In:* Felder DL, Camp DK, editors. Gulf of Mexico: Origin, waters, and biota: Volume 1, Biodiversity. College Station, TX: Texas A&M University Press. p. 333–341.

Cairns SD, Bayer FM. 2009. Octocorallia (Cnidaria) of the Gulf of Mexico. *In:* Felder DL, Camp DK, editors. Gulf of Mexico: Origin, waters, and biota: Volume 1, Biodiversity. College Station, TX: Texas A&M University Press. p. 321–331.

Carleton JH, Brinkman R, Doherty PJ. 2001. The effects of water flow around coral reefs on the distribution of pre-settlement fish (Great Barrier Reef, Australia). *In:* Wolanski E, editor. Oceanographic processes of coral reefs. Boca Raton: CRC Press. p. 209–230.

Colella MA, Ruzicka RR, Kidney JA, Morrison JM, Brinkhuis VB. 2012. Cold-water event of January 2010 results in catastrophic benthic mortality on patch reefs in the Florida Keys. Coral Reefs. 31:621–632.

Coleman FC, Dennis G, Jaap WC, Schmahl GP, Koenig C, Reed S, Beaver C. 2005. Status and trends of the Florida Middle Grounds. Technical Report to the Gulf of Mexico Fisheries Management Council, Tampa, FL.

Coleman FC, Scanlon KC, Koenig CC. 2011. Groupers on the edge: Shelf-edge spawning habitat in and around marine reserves of the northeastern Gulf of Mexico. Prof Geogr. 63:456–474.

Collard SB, Lugo-Fernandez A, Fitzhugh G, Brusher J, Shaffer R. 2000. A mass mortality event in the coastal waters of the central Florida Panhandle during spring and summer 1998. Gulf Mex Sci. 18:68–71.

Continental Shelf Associates, Inc. 1992. Mississippi-Alabama Shelf Pinnacle trend habitat mapping study. OCS Study/MMS 92-0026. New Orleans: US Department of the Interior, Minerals Management Service, Gulf of Mexico OCS Regional Offices.

Continental Shelf Associates, Inc. 2007. Characterization of northern Gulf of Mexico deepwater hard bottom communities with emphasis on *Lophelia* coral. OCS Study MMS 2007-044. New Orleans: U.S. Department of the Interior, Minerals Management Service, Gulf of Mexico OCS Regional Offices.

Cooper C, Forristall GZ, Joyce TM. 1990. Velocity and hydrographic structure of two Gulf of Mexico warm-core rings. J Geophys Res. 95(C2):1663–1679. Available from: http://dx.doi.org/10.1029/JC095iC02p01663 via the Internet. Accessed 31 January, 2014.

Dauer DM, Simon JL. 1976. Repopulation of the polychaete fauna of an intertidal habitat following natural defaunation: Species equilibrium. Oecologia. 22:99–117.

Davis CC. 1948. *Gymnodinium brevis* sp. nov., a cause of discolored water and animal mortalities in the Gulf of Mexico. Bot Gaz. 109:358–360.

Davis GE. 1977. Fishery harvest in an underwater park. *In*: Proceedings of the 3rd International Coral Reef Symposium. Miami: University of Miami, Volume 2. p. 605–608.

Davis GE. 1982. A century of natural change in coral distribution at the Dry Tortugas: A comparison of reef maps from 1881–1976. Bull Mar Sci. 32:608–623.

Dennis GD, Bright TJ. 1988. Reef fish assemblages on hard banks in the northwestern Gulf of Mexico. Bull Mar Sci. 43:280–307.

Diaz RJ, Rosenberg R. 2008. Spreading dead zones and consequences for marine ecosystems. Science. 321(5891):926–929.

Doyle LJ, Holmes CW. 1985. Shallow structure, stratigraphy, and carbonate sedimentary processes of West Florida upper continental slope. Am Assoc Pet Geol Bull. 69:1133–1144.

Dufault AM, Ninokawa A, Bramanti L, Cumbo VR, Fan T-Y, Edmunds PJ. 2013. The role of light in mediating the effects of ocean acidification on coral calcification. J Exp Biol. 216:1570–1577.

Dupont JM, Hallock P, Jaap WC. 2009. Ecological impacts of the 2005 red tide on artificial reef epibenthic macroinvertebrate and fish communities in the eastern Gulf of Mexico. Mar Ecol Prog Ser. 415:189–200.

ENSR. 2005. Gulfstream Natural Gas System, 2005 federal waters monitoring report. St. Petersburg: ENSR.

Etnoyer PJ. 2009. Distribution and diversity of deep water octocorals in the Gulf of Mexico. PhD dissertation, Texas A&M University, Corpus Christi, TX.

Fairbridge RW, editor. 1966. The encyclopedia of oceanography. New York: Van Nostrand-Reinhold.

Feely RA, Sabine CL, Lee K, Berelson W, Kleypas J, Fabry VJ, Millero FJ. 2004. Impact of anthropogenic CO_2 on the $CaCO_3$ system in the oceans. Science. 305:362–366.

Flewelling LJ, Naar JP, Abbott JP, Baden DG, Barros NB, Bossart GD, Bottein MD, Hammond DG, Haubold EM, Heil CA, et al. 2005. Red tides and marine mammal mortalities. Nature. 435:755–756.

Gardner JV, Dartnell P, Sulak KJ, Calder B, Hellequin L. 2001. Physiography and late Quaternary-Holocene processes of northeastern Gulf of Mexico outer continental shelf off Mississippi and Alabama. Gulf Mex Sci. 19:132–157.

Gardner JV, Dartnell P, Sulak KJ. 2002. Multibeam mapping of the Pinnacles Region, Gulf of Mexico. U.S. Geological Survey Open-File Report OF02-6. CD-ROM. Available from: http://geopubs.wr.usgs.gov/open-file/of02-006/ via the Internet. Accessed 31 January, 2014.

Gardner TA, Côte IM, Gill JA, Grant A, Watkinson AR. 2005. Hurricanes and Caribbean coral reefs: Impacts, recovery patterns, and the role of long-term decline. Ecology. 86:174–184.

Gardulski AF, Mullins HT, Weiterman S. 1990. Carbonate mineral cycles generated by foraminiferal and pteropod response to Pleistocene climate: West Florida ramp slope. Sedimentology. 37:727–743.

Giammona CP. 1978. Octocorals in the Gulf of Mexico: Their taxonomy and distribution with remarks on their paleontology. PhD thesis, Texas A&M University, College Station, TX.

Gittings SR, Bright TJ, Schroeder WW, Sager WW, Laswell JS, Rezak R. 1992. Invertebrate assemblages and ecological controls on topographic features in the northeast Gulf of Mexico. Bull Mar Sci. 50:435–455.

Gordon JDM, Bergstad OA, Figueiredo I, Menezes G. 2003. Deep-water fisheries of the northeast Atlantic: I. Description and current trends. J Northwest Atl Fish Soc. 31:137–150.

Gould HR, Stewart RH. 1956. Continental terrace sediments in the northeastern Gulf of Mexico. *In:* Hough JL, Menard HW, editors. Finding ancient shorelines. Special Publication No. 3. Tulsa, OK: Society of Economic Paleontologists Mineralogists. p. 2–19.

Grady JR. 1971. The distribution of sedimentary properties and shrimp catch on two shrimping grounds on the continental shelf of the Gulf of Mexico. *In*: Proceedings of the 23rd Annual Meeting of the Gulf and Caribbean Fisheries Institute. Coral Gables, FL, p. 139–148.

Grimm D, Hopkins T. 1977. Preliminary characterization of the octocorallian and scleractinian diversity at the Florida Middle Ground. *In*: Proceedings of the 3rd International Coral Reef Symposium. Miami, FL: University of Miami, Volume 1. p. 135–142.

Guinotte JM, Orr JC, Cairns SD, Freiwald A, Morgan L, George R. 2006. Will human-induced changes in seawater chemistry alter the distribution of deep-sea scleractinian corals? Front Ecol Environ. 4:141–146.

Gulf of Mexico Fishery Management Council. 2004. Final environmental impact statement for the generic essential fish habitat amendment to the following fishery management plans of the Gulf of Mexico (GOM): Shrimp fishery of the Gulf of Mexico, red drum fishery of the Gulf of Mexico, reef fish fishery of the Gulf of Mexico, stone crab fishery of the Gulf of Mexico and coral reef fishery of the Gulf of Mexico, spiny lobster fishery of the Gulf of Mexico and south Atlantic Fishery Management Council, and coastal migratory pelagic resources of the Gulf of Mexico and South Atlantic Fisher Management Councils. Tampa, FL: Gulf of Mexico Fishery Management Council.

Hansen G, Moestrup Ø, Roberts K. 2000. Light and electron microscopical observations on the type species of Gymnodinium, *G. fuscum* (Dinophyceae). Phycologia. 39:365–376.

Harter S, David A, Ribera M. 2008. Survey of coral and fish assemblages on Pulley Ridge, SW Florida. Report to the Gulf of Mexico Fishery Management Council.

Henry LA, Roberts JM. 2007. Biodiversity and ecological composition of macrobenthos on cold-water coral mounds and adjacent off-mound habitat in the bathyal Porcupine Seabight, NE Atlantic. Deep Sea Res I. 54:654–672.

Hine AC, Halley RB, Locker SD, Jarrett BD, Jaap WC, Mallinson DJ, Ciembronowicz KT, Ogden NB, Donahue BT, Naar DF, et al. 2008. Coral reefs, present and past on the West Florida shelf and platform margin. *In:* Riegl BM, Dodge RE, editors. Coral reefs of the USA. Berlin: Springer. p. 127–173.

Hine AC, Locker SD. 2011. The Gulf of Mexico continental shelf: Great contrasts and significant transitions. *In:* Buster NA, Holmes CW, editors. Gulf of Mexico: Volume 3, Geology. College Station, TX: Harte Research Institute for Gulf of Mexico Studies, Texas A&M University. p. 101–127.

Holmes CW. 1985. Accretion of the South Florida platform, late Quaternary development. AAPG Bull. 69:149–160.

Hopkins TS, Brawley DR, Earle SA, Grimm DE, Gilbert DK, Johnson PG, Livingston EH, Lutz CH, Shaw JK, Shaw BB, et al. 1977. A preliminary characterization of the biotic components of composite strip transects on the Florida Middle Ground, northeastern Gulf of Mexico. *In*: Proceedings of the 3rd International Coral Reef Symposium. Miami, FL: University of Miami, Volume 1. p. 31–37.

Hoegh-Guldberg O, Mumby PJ, Hooten AJ, Steneck RS, Greenfield P, Gomez E, Harvell CD, Sale PF, Edwards AJ, Caldeira K, et al. 2007. Coral reefs under rapid climate change and ocean acidification. Science. 318:1737–1742.

Hübscher C, Dullo C, Flögel S, Titschack J, Schönfeld J. 2010. Contourite drift evolution and related coral growth in the eastern Gulf of Mexico and its gateways. Int J Earth Sci. 99:191–206.

Hughes TP, Baird AH, Bellwood DR, Card M, Connolly SR, Folke C, Grosberg R, Hoegh-Guldberg O, Jackson JBC, Kleypas J, et al. 2003. Climate change, human impacts, and the resilience of coral reefs. Science. 301:929–933.

Ingersoll E. 1881. Proceedings of the United States National Museum. Washington, DC: Smithsonian Institution.

International Energy Agency. 2013. Climate change. Available from: http://www.iea.org/newsroomandevents/news/2013/march/name,36350,en.html via the Internet. Accessed 31 January, 2014.

Jaap WC. 1979. Observation on zooxanthellae expulsion at Middle Sambo Reef, Florida Keys. Bull Mar Sci. 29:414–422.

Jaap WC. 1985. An epidemic zooxanthellae expulsion during 1983 in the lower Florida Keys coral reefs: Hyperthermic etiology. *In*: Proceedings of the 5th International Coral Reef Congress. Tahiti, Volume 6. p. 143–148.

Jaap WC, Hallock P. 1990. Coral reefs. *In:* Myers RL, Ewel JJ, editors. Ecosystems of Florida. Orlando: University of Central Florida Press. p. 574–616.

Jaap WC, Lyons WG, Dustan P, Halas J. 1989. Stony coral (Scleractinia and Milleporina) community structure at Bird Key Reef, Ft. Jefferson National Monument, Dry Tortugas, Florida. Florida Marine Research Publication No. 46. St. Petersburg, FL: Florida Marine Research Institute.

Jaap WC, Sargent FJ. 1993. The status of the remnant population of *Acropora palmate* (Lamarck, 1816) at Dry Tortugas National Park, Florida, with a discussion of possible causes of changes since 1881. *In:* Proceedings of the Colloquium on Global Aspects of Coral Reefs: Health, hazards, and history. Miami: University of Miami. p. 101–105.

Jaap WC, Mallison D, Hine A, Muller P, Jarrett B, Wheaton J. 2000. Deep reef communities: Tortugas Banks and Pulley Ridge. *In*: Proceedings of the First International Symposium on Deep Sea Corals. Halifax, NS. Abstract. p. 12–13.

Jaap WC, Szmant A, Jaap K, Dupont J, Clarke R, Somerfield P, Ault J, Bohnsack JA, Kellison SG, Kellison GT, et al. 2008. A perspective on the biology of Florida Keys coral reefs. *In:* Riegl BM, Dodge RE, editors. Coral reefs of the USA. Berlin: Springer. p. 75–126.

Jaap WC, Watkins HE. 2013. Contributions of the Gulfstream Natural Gas pipeline to fish habitat in the eastern Gulf of Mexico. Technical Report to Gulfstream Natural Gas Co. LLC. St. Petersburg: Lithophyte Research.

Jarrett BD, Hine AC, Halley RB, Naar DF, Locker SD, Neumann AC, Twichell D, Hu C, Donahue BT, Jaap WC, et al. 2005. Strange bedfellows: A deep hermatypic coral reef superimposed on a drowned barrier island; Southern Pulley Ridge, SW Florida platform margin. Mar Geol. 214:295–307.

Joyce EA Jr, Williams J. 1969. Rationale and pertinent data. Memoirs of the Hourglass Cruises. Volume I, pt 1. St. Petersburg: FMRI.

Koenig CC, Coleman FC. 2011. Protection of grouper and red snapper spawning in shelf-edge marine reserves of the northeastern Gulf of Mexico: Demographics, movements, survival and spillover effects. MARFIN Project final report. Project Number: NA07NMF4330120.

Landsberg JH. 2002. The effects of harmful algal blooms on aquatic organisms. Rev Fish Sci. 10:113–390.

Lee TN, Williams RE, McGowan M, Szmant AF, Clarke ME. 1992. Influence of gyres and wind-driven circulation on transport of larvae and recruitment in the Florida Keys coral reefs. Cont Shelf Res. 12:97–1002.

Lee TN, Clarke ME, Williams E, Szmant AF, Berger T. 1994. Evolution of the Tortugas gyre and its influence on recruitment in the Florida Keys. Bull Mar Sci. 54:621–646.

Lee TN, Williams E, Johns E, Wilson D, Smith NP. 2002. Transport processes linking South Florida coastal ecosystems. *In:* Porter J, Porter K, editors. The Everglades, Florida Bay, and coral reefs of the Florida Keys: An ecosystem sourcebook. Boca Raton: CRC Press. p. 309–342.

Lirman D, Schopmeyer S, Manzello D, Grammer LJ, Precht WF, Muller-Karger F, Banks K, Bartels E, Bourque A, Byrne J, et al. 2011. Severe 2010 cold-water event caused unprecedented mortality to corals of the Florida Reef Tract and reversed previous survivorship patterns. PLoS ONE. 6(8): e23047. Available from: http://dx.doi.org/10.1371/journal.pone.0023047 via the Internet. Accessed 31 January, 2014.

Longley WH, Hildebrandt FF. 1941. Systematic catalogue of the fishes of Tortugas, Florida with observations on color, habitats, and local distribution. Pap Tortugas Lab. 34:1–331.

Ludwick JC, Walton WR. 1957. Shelf-edge, calcareous prominences in the northeastern Gulf of Mexico. Am Assoc Pet Geol Bull. 41:2054–2101.

Mallinson D, Hine AC, Hallock P, Locker S, Shinn E, Naar D, Donahue B, Weaver D. 2003. Development of small carbonate banks on the south Florida platform margin: Response to sea level and climate change. Mar Geol. 199:45–63.

Morris JA Jr, Whitfield PE. 2009. Biology, ecology, control and management of the invasive Indo-Pacific lionfish: An updated integrated assessment. NOAA Technical Memorandum NOS-NCCOS, 99. Beaufort, NC: NOAA, National Ocean Service Center for Coastal Fisheries and Habitat Research.

Muller-Karger FE. 2000. The spring 1998 northeastern Gulf of Mexico (NEGOM) cold water event: Remote sensing evidence for upwelling and for eastward advection of Mississippi water (or: How an errant Loop Current anticyclone took the NEGOM for a spin). Gulf Mex Sci. 18:55–67.

Mumby P, Steneck RS. 2008. Coral reef management and conservation in light of rapidly evolving ecological paradigms. Trends Ecol Evol. 10:555–563.

Mullins HT, Gardulski AF, Hinchey EJ, Hine AC. 1988a. The modern carbonate ramp of central west Florida. J Sed Petrol. 58:273–290.

Mullins HT, Gardulski AF, Hine AC, Melillo AJ, Wise SW Jr, Applegate J. 1988b. Three-dimensional sedimentary framework of the carbonate ramp slope of central west Florida: A sequential seismic stratigraphic perspective. Geol Soc Am Bull. 100:514–533.

National Marine Fisheries Service. 2005. 2005 Status of U.S. fisheries: Report to Congress. Available from: http://www.nmfs.noaa.gov/sfa/statusoffisheries/SOSmain.htm via the Internet. Accessed 31 January, 2014.

Norse EA, Brooke S, Cheung WWL, Clark MR, Ekeland I, Froese R, Gjerde KM, Haedrich RL, Heppell SS, Morato T, et al. 2012. Sustainability of deep-sea fisheries. Mar Policy. 36:307–320.

Pandolfi JM, Connolly SR, Marshall DJ, Cohen AL. 2011. Projecting coral reef futures under global warming and acidification. Science. 333:418–422.

Parker RO Jr, Ross SW. 1986. Observing reef fishes from submersibles off North Carolina. NE Gulf Sci. 8:31–49.

Paul JH, Hollander D, Coble P, Daly KL, Murasko S, English D, Basso J, Delaney J, McDaniel L, Kovach CW, et al. 2013. Toxicity and mutagenicity of Gulf of Mexico waters during and after the Deepwater Horizon oil spill. Environ Sci Technol. 47(17):9651–9659. Available from: http://pubs.acs.org/doi/abs/10.1021/es401761h via the Internet. Accessed 31 January, 2014.

Pinsky ML, Worm B, Fogarty MJ, Sarmiento JL, Levin SA. 2013. Marine taxa track local climate velocities. Science. 341(6151):1239–1242.

Porter JW. 1976. Autotrophy, heterotrophy, and resource portioning in Caribbean reef-building corals. Am Nat. 110:731–742.

Quattrini AM, Ross SW. 2006. Fishes associated with North Carolina shelf-edge hardbottoms and initial assessment of a proposed marine protected area. Bull Mar Sci. 79:137–163.

Quattrini AM, Etnoyer PJ, Doughty C, English L, Falco R, Remon N, Rittinghousec M, Cordesa E. 2014. A phylogenetic approach to octocoral community structure in the deep Gulf of Mexico. Deep Sea Res Pt II Top Stud Oceanogr. 99:92–102.

Rabalais NN, Díaz RJ, Levin LA, Turner RE, Gilbert D, Zhang J. 2010. Dynamics and distribution of natural and human-caused hypoxia. Biogeosciences. 7:585–619.

Reed JK, Weaver DC, Pomponi SA. 2006. Habitat and fauna of deep-water *Lophelia pertusa* coral reefs off the southeastern U.S.: Blake Plateau, Straits of Florida, and Gulf of Mexico. Bull Mar Sci. 78:343–375.

Reich CD, Poore RZ, Hickey TD. 2013. The role of Vermetid gastropods in the development of the Florida Middle Ground, northeast Gulf of Mexico. J Coast Res Spec Issue. 63:46–57.

Roberts CM. 2002. Deep impact: The rising toll of fishing in the deep sea. Trends Ecol Evol. 17:242–245.

Roberts HH, Rouse LJ Jr, Walker ND, Hudson H. 1982. Cold water stress in Florida Bay and Northern Bahamas: A product of winter frontal passages. J Sed Pet. 52:145–155.

Ross SW, Quattrini AM. 2007. The fish fauna associated with deep coral banks off the southeastern United States. Deep-Sea Res I Oceanogr Res Pap. 54:975–1007.

Sager WW, Schroeder WW, Laswell JS, Davis KS, Rezak R, Gittings SR. 1992. Mississippi-Alabama outer continental shelf topographic features formed during the late Pleistocene-Holocene transgression. Geo-Mar Lett. 12:41–48.

Sale PF. 2008. Management of coral reefs: Where we have gone wrong and what we can do about it. Mar Pollut Bull. 56:805–809.

Schomer SN, Drew RD. 1982. An ecological characterization of the lower Everglades, Florida Bay, and the Florida Keys. FWS/OBS-82/58.1. Washington, DC: U.S. Fish and Wildlife Service.

Schroeder WW. 2002. Observations of *Lophelia pertusa* and the surficial geology at a deepwater site in the northeastern Gulf of Mexico. Hydrobiologica. 471:29–33.

Schroeder WW, Brooke SD. 2011. Habitat-forming deepwater scleractinian corals in the Gulf of Mexico. *In:* Buster NA, Holmes CW, editors. Gulf of Mexico origin, waters, and biota: Volume 3, Geology. College Station, TX: Texas A&M University. p. 355–363.

Shinn EA, Jaap WC. 2005. Field guide to the major organisms and processes building reefs and islands of the Dry Tortugas: The Carnegie Dry Tortugas Laboratory Centennial Celebration (1905–2005). St. Petersburg, FL: U.S. Geological Survey and Florida Fish and Wildlife Conservation Commission.

Simon JL, Dauer DM. 1977. Reestablishment of a benthic community following natural defaunation. *In:* Coull BC, editor. Ecology of marine benthos. Columbia, SC: University of South Carolina Press. p. 139–158.

Smith GB. 1975. The 1971 red tide and its impact on certain communities in the mid-eastern Gulf of Mexico. Environ Lett. 9:141–152.

Smith GB. 1976. Ecology and distribution of eastern Gulf of Mexico reef fishes. Florida Marine Research Publication No. 19. St. Petersburg, FL: Florida Marine Research Institute.

Smith GB. 1979. Relationship of eastern Gulf of Mexico reef-fish communities to the species equilibrium theory of insular biogeography. J Biogeogr. 6:49–61.

Smith GB, Austin HM, Bertone SA, Hasting RW, Ogren LH. 1975. Fishes of the Florida Middle Ground with comments on ecological zoogeography. Florida Marine Research Publication No. 9. St. Petersburg, FL: FMR.

Snelgrove PVR. 1999. Getting to the bottom of marine biodiversity: Sedimentary habitats. BioScience. 49:129–138.

Squires DF. 1964. Fossil coral thickets in Wairapa, New Zealand. J Paleontol. 38:904–915.

Steidinger KA, Ingle RM. 1972. Observations on the 1971 summer red tide in Tampa Bay. Environ Lett. 3:271–278.

Stiles ML, Harrouild-Kolieb E, Faure P, Ylitalo-Ward H, Hirshfield MF. 2007. Deep sea trawl fisheries of the southeastern U.S. and Gulf of Mexico. Washington, DC: Oceana.

Sulak KJ, Brooks RA, Luke KE, Norem AD, Randall MT, Quaid AJ, Yeargin GE, Miller JM, Harden WM, Caruso JH, et al. 2007. Demersal fishes associated with *Lophelia pertusa* coral and hard-substrate biotopes on the continental slope, northern Gulf of Mexico. *In:* George RY, Cairns SD, editors. Conservation and adaptive management of seamount and deep-sea coral ecosystems. Miami, FL: University of Miami. p. 65–92.

Thompson MJ, Schmidt TW. 1977. Validation of the species/time random count technique sampling fish assemblages at Dry Tortugas. *In:* Proceedings of the Third International Coral Reef Symposium. Miami, FL: University of Miami, Volume 1. p. 283–288.

Tilmant JT. 1984. Reef fish. *In:* Jaap W, editor. The ecology of the south Florida coral reefs: A community profile. U.S. Fish and Wildlife Service FWS/OBS-82/08. Washington, DC: U.S. Fish and Wildlife Service. p. 52–63.

Tolimieri N. 1998. Contrasting effects of microhabitat use on large-scale abundance in two families of Caribbean reef fishes. Mar Ecol Prog Ser. 167:227–239.

Veron JEN. 2011. Ocean acidification and coral reefs: An emerging big picture. Diversity. 3:262–274.

Walsh JJ, Jolliff JK, Darrow BP, Lenes JM, Milroy SP, Remsen A, Dieterle DA, Carder KL, Chen FR, Vargo GA, et al. 2006. Red tides in the Gulf of Mexico: Where, when, and why? J Geophys Res. 111:C11003.

Weaver DC, Dennis GD, Sulak KJ. 2002. Northeastern Gulf of Mexico coastal and marine ecosystem program: Community structure and trophic ecology of demersal fishes on the Pinnacles reef tract. Final synthesis report. U.S. Department of the Interior, Geological Survey USGS BSR-2001-0008; Minerals Management Service, Gulf of Mexico OCS Region, OCS Study MMS 2002-034.

Wolanski E. 2001. Oceanographic processes of coral reefs: Physical and biological links in the Great Barrier Reef. Boca Raton: CRC Press.

Woodward-Clyde and Continental Shelf Associates, Inc. 1983. Southwest Florida Shelf ecosystem study: Year 1 Report. MMS Contract 14-12-0001-29142.

Zhang IK, Hollowed AB, Lee J-B, Kim D-H. 2011. An IFRAME approach for assessing impacts of climate change on fisheries. J Mar Sci. 68:1318–1328.

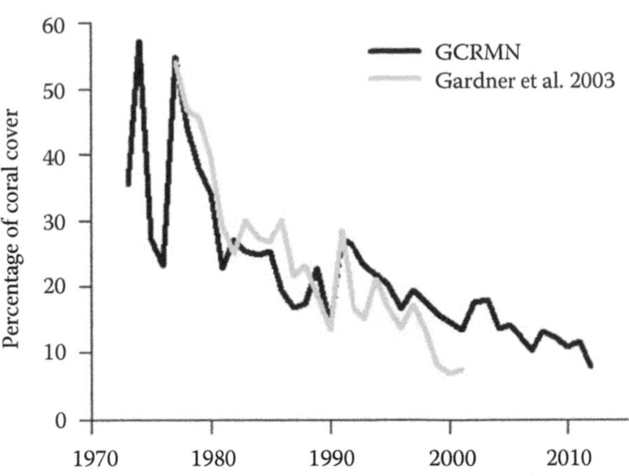

Figure 1.2 Decline in percentage of coral cover on Caribbean coral reefs from 1963 to the present based on data compiled by the IUCN (i.e., the International Union for the Conservation of Nature's Tropical Americas Coral Reef Resilience Workshop) with yearly averages weighted by the area surveyed per study and compared to Gardner et al. (2003) (yearly averages weighted by the inverse of a study's sample variance). GCRMN, Global Coral Reef Monitoring Network. Credit: © IUCN (International Union for the Conservation of Nature).

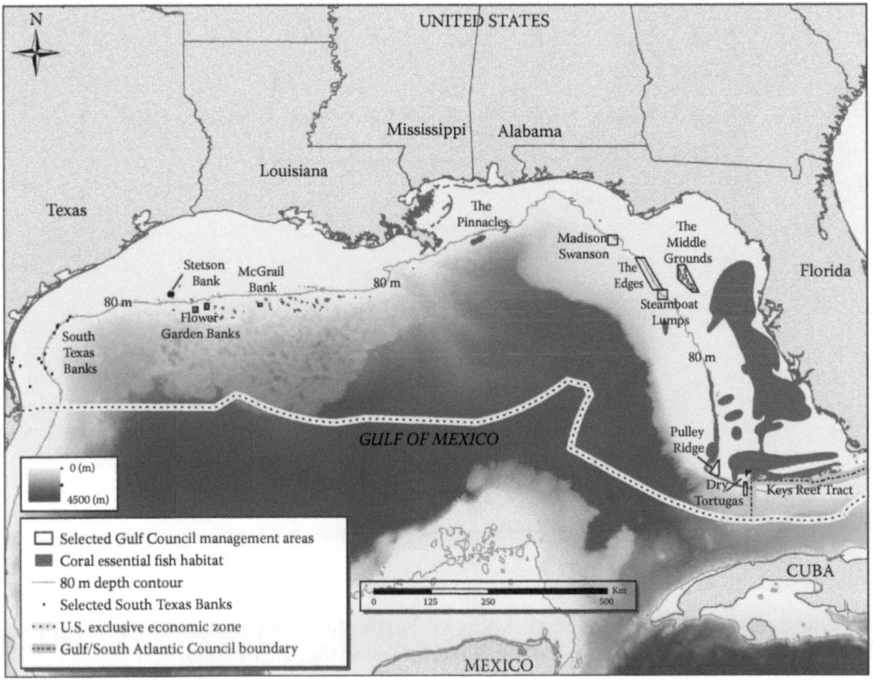

Figure 3.1 Map of the Gulf of Mexico exclusive economic zone (yellow dashed line) depicting Gulf Council–designated fishery management areas (black boxes) and other known areas of importance with shallow and mesophotic corals. The orange area is the 2005 coral essential fish habitat, the thin gray line is the 80 m depth contour, and the blue dashed line indicates the jurisdictional boundary between the Gulf Council and the South Atlantic Fishery Management Council.

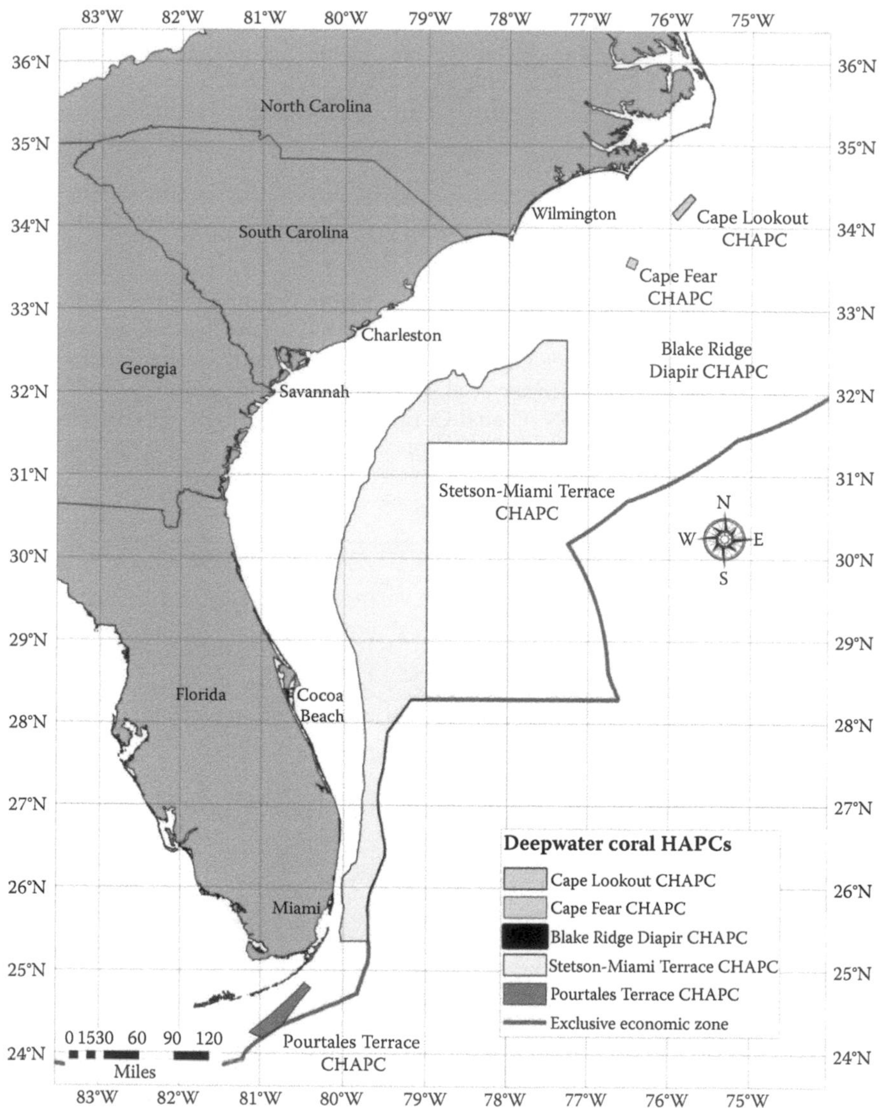

Figure 4.1 The deepwater Coral Habitat Areas of Particular Concern (CHAPCs) implemented by SAFMC through Comprehensive Ecosystem-Based Amendment 1 in 2009.

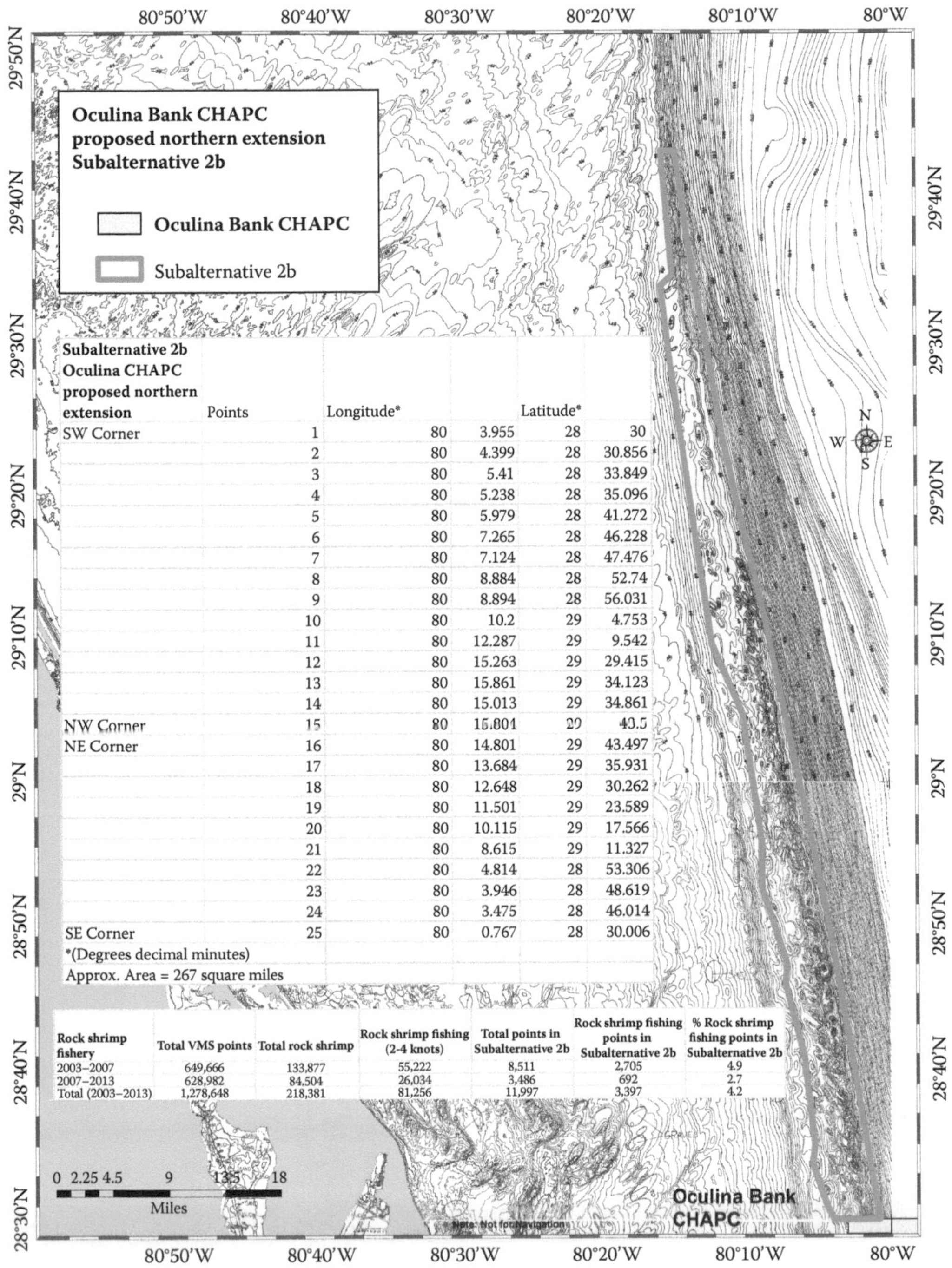

Subalternative 2b Oculina CHAPC proposed northern extension	Points	Longitude*		Latitude*	
SW Corner	1	80	3.955	28	30
	2	80	4.399	28	30.856
	3	80	5.41	28	33.849
	4	80	5.238	28	35.096
	5	80	5.979	28	41.272
	6	80	7.265	28	46.228
	7	80	7.124	28	47.476
	8	80	8.884	28	52.74
	9	80	8.894	28	56.031
	10	80	10.2	29	4.753
	11	80	12.287	29	9.542
	12	80	15.263	29	29.415
	13	80	15.861	29	34.123
	14	80	15.013	29	34.861
NW Corner	15	80	15.801	29	43.5
NE Corner	16	80	14.801	29	43.497
	17	80	13.684	29	35.931
	18	80	12.648	29	30.262
	19	80	11.501	29	23.589
	20	80	10.115	29	17.566
	21	80	8.615	29	11.327
	22	80	4.814	28	53.306
	23	80	3.946	28	48.619
	24	80	3.475	28	46.014
SE Corner	25	80	0.767	28	30.006
*(Degrees decimal minutes)					
Approx. Area = 267 square miles					

Rock shrimp fishery	Total VMS points	Total rock shrimp	Rock shrimp fishing (2-4 knots)	Total points in Subalternative 2b	Rock shrimp fishing points in Subalternative 2b	% Rock shrimp fishing points in Subalternative 2b
2003–2007	649,666	133,877	55,222	8,511	2,705	4.9
2007–2013	628,982	84,504	26,034	3,486	692	2.7
Total (2003–2013)	1,278,648	218,381	81,256	11,997	3,397	4.2

Figure 4.2 The SAFMC Preferred Alternative for a northern expansion of the Oculina Bank HAPC included in Coral Amendment 8 (under development). The chart incorporates VMS data from the rock shrimp fishery, 2003–2013.

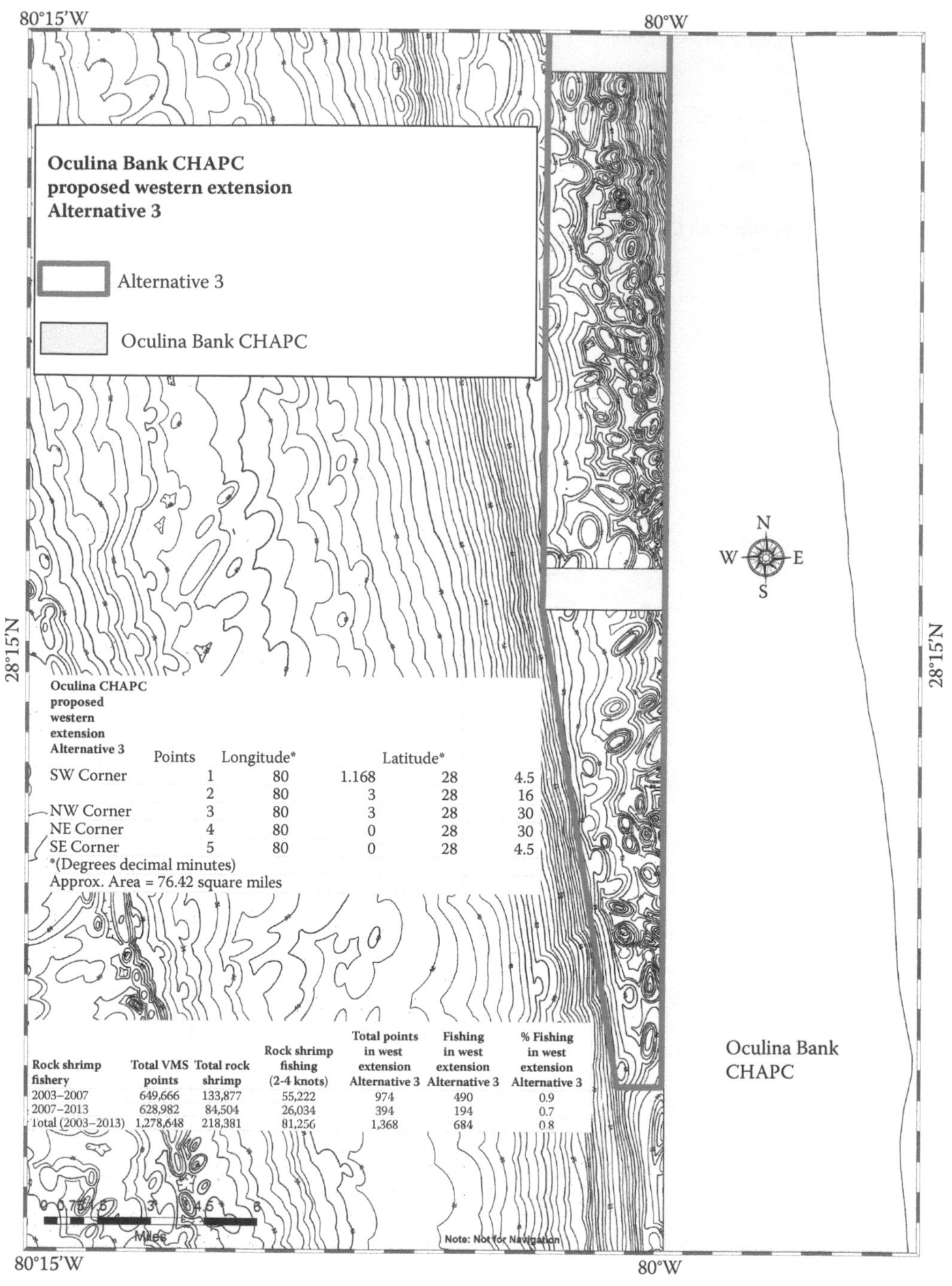

Oculina CHAPC proposed western extension Alternative 3

	Points	Longitude*		Latitude*	
SW Corner	1	80	1.168	28	4.5
	2	80	3	28	16
NW Corner	3	80	3	28	30
NE Corner	4	80	0	28	30
SE Corner	5	80	0	28	4.5

*(Degrees decimal minutes)
Approx. Area = 76.42 square miles

Rock shrimp fishery	Total VMS points	Total rock shrimp	Rock shrimp fishing (2-4 knots)	Total points in west extension Alternative 3	Fishing in west extension Alternative 3	% Fishing in west extension Alternative 3
2003–2007	649,666	133,877	55,222	974	490	0.9
2007–2013	628,982	84,504	26,034	394	194	0.7
Total (2003–2013)	1,278,648	218,381	81,256	1,368	684	0.8

Figure 4.3 The SAFMC Preferred Alternative for a western expansion of the Oculina Bank HAPC included in Coral Amendment 8 (under development). The chart incorporates VMS data from the rock shrimp fishery, 2003–2013.

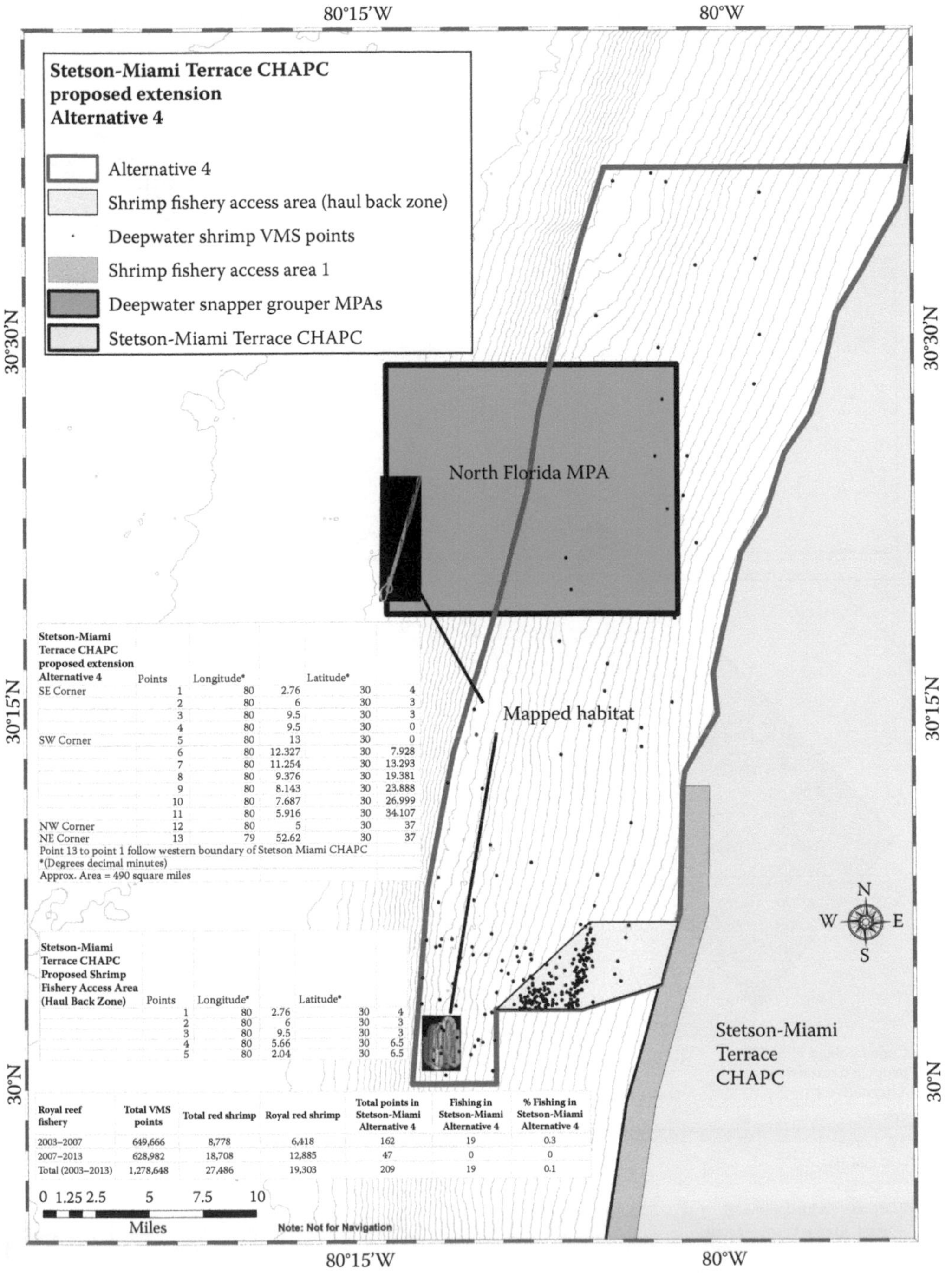

Stetson-Miami Terrace CHAPC proposed extension Alternative 4	Points	Longitude*		Latitude*	
SE Corner	1	80	2.76	30	4
	2	80	6	30	3
	3	80	9.5	30	3
	4	80	9.5	30	0
SW Corner	5	80	13	30	0
	6	80	12.327	30	7.928
	7	80	11.254	30	13.293
	8	80	9.376	30	19.381
	9	80	8.143	30	23.888
	10	80	7.687	30	26.999
	11	80	5.916	30	34.107
NW Corner	12	80	5	30	37
NE Corner	13	79	52.62	30	37

Point 13 to point 1 follow western boundary of Stetson Miami CHAPC
*(Degrees decimal minutes)
Approx. Area = 490 square miles

Stetson-Miami Terrace CHAPC Proposed Shrimp Fishery Access Area (Haul Back Zone)	Points	Longitude*		Latitude*	
	1	80	2.76	30	4
	2	80	6	30	3
	3	80	9.5	30	3
	4	80	5.66	30	6.5
	5	80	2.04	30	6.5

Royal reef fishery	Total VMS points	Total red shrimp	Royal red shrimp	Total points in Stetson-Miami Alternative 4	Fishing in Stetson-Miami Alternative 4	% Fishing in Stetson-Miami Alternative 4
2003–2007	649,666	8,778	6,418	162	19	0.3
2007–2013	628,982	18,708	12,885	47	0	0
Total (2003–2013)	1,278,648	27,486	19,303	209	19	0.1

Figure 4.4 The SAFMC Preferred Alternative for a western expansion of the Stetson-Miami Terrace CHAPC included in Coral Amendment 8 (under development). The chart incorporates VMS data from the rock shrimp fishery 2003–2013 indicating royal red fishery activity. MPA, marine protected area.

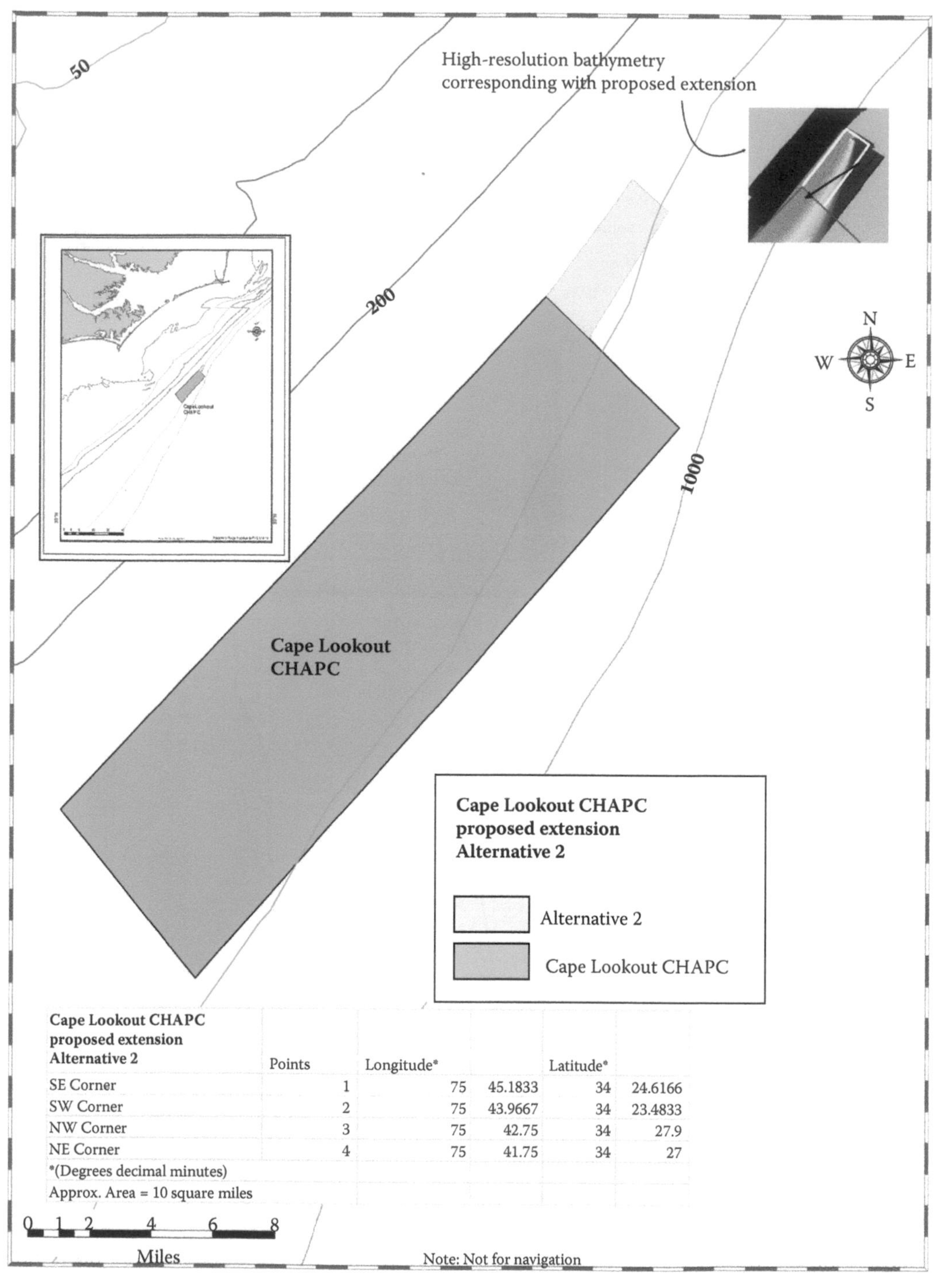

Figure 4.5 The SAFMC Preferred Alternative for a northern expansion of the Cape Lookout CHAPC included in Coral Amendment 8 (under development).

Figure 5.2 Representative habitats found on Pourtalès Terrace from ROV dive images. (a) Habitat: deepwater *L. pertusa* coral mound; live coral colony (~1 m diameter), black coral (*Leiopathes* sp.), and mora codling (*Laemonema melanurum*); CHAPC, Site 25, 509 m. (b) Habitat: flat rock pavement with low relief ledges; big roughy (*G. darwinii*), anemones (*Actinoscyphia* sp.), various gorgonians, hydroids, and crinoids; CHAPC, Mound #311 Wall, Site 23, 302 m. (c) Habitat: mound-slope, rock pavement; dense cover of Stylasteridae corals and numerous demosponge species; HAPC, Mound #311, Site 22, 233 m. (d) Habitat: flat pavement with sediment; blueline tilefish (*C. microps*), stylasterid corals; CHAPC, Alligator Bioherm #4, Site 18, 183 m. (e) Habitat: mound-wall, upper slope and edge of mound; dense cover of Stylasteridae coral; MPA, NW Mound, Site 14, 191 m. (f) Habitat: soft bottom; tripod fish (*B. grallator*); unprotected, Pourtalès Escarpment, Site 21, 836 m. (g) Multibeam sonar of CHAPC, Sinkhole site (Site 24) and *Lophelia* mound site (Site 25); dive tracks are shown by solid red lines. (h) Multibeam sonar of CHAPC, Alligator Bioherm #3 (Site 20).

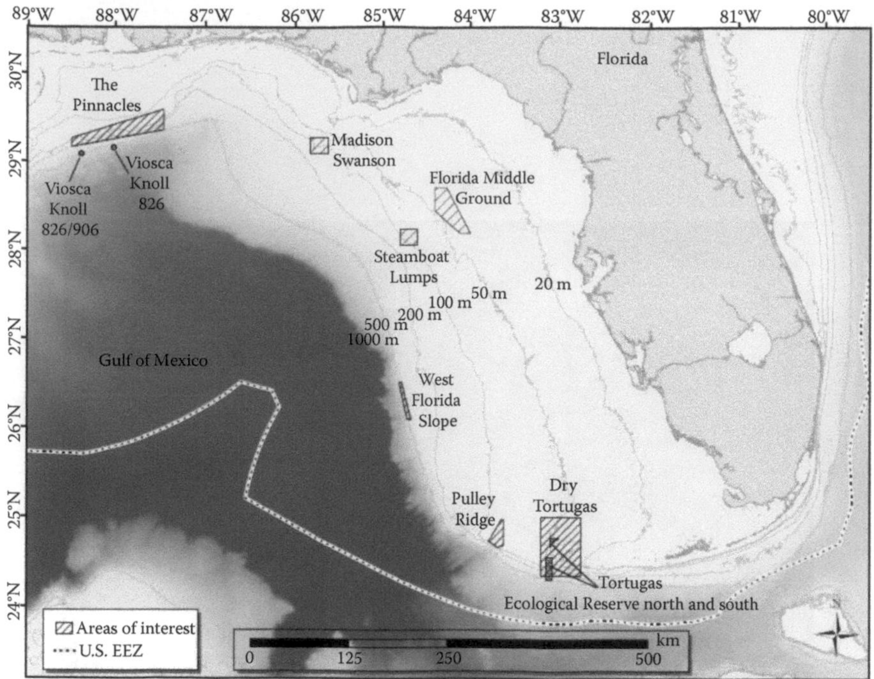

Figure 6.1 Eastern Gulf of Mexico areas of interest: the Dry Tortugas, Pulley Ridge, Florida Middle Grounds, Madison Swanson, Steamboat Lumps, the Pinnacles, and Viosca Knolls. EEZ, exclusive economic zone. (From Cartography by Mark Mueller, Gulf of Mexico Fishery Management Council.)

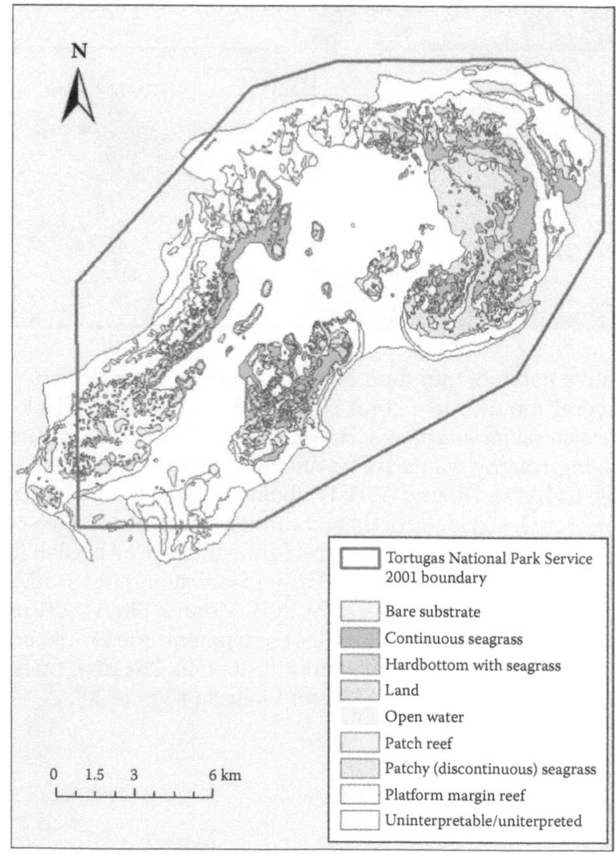

Figure 6.4 Dry Tortugas benthic habitat map. (From Florida Fish and Wildlife Conservation Commission cartography.)

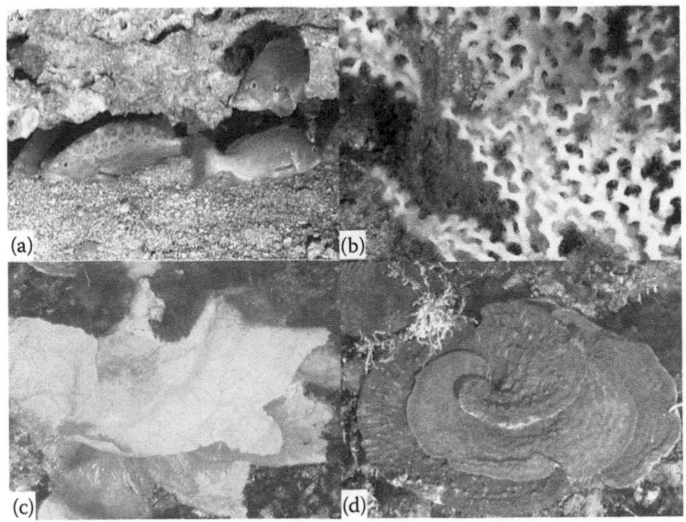

Figure 6.5 (a) Scamp grouper, *M. phenax*, under a ledge, north Pulley Ridge. (b) *O. tenella*, colonial azooxanthellate coral, north Pulley Ridge. (c) Green alga, *Anadyomene menziessi*, south Pulley Ridge. (d) Swirling plates of *L. cucullata*, zooxanthellate coral, south Pulley Ridge. (Photographs courtesy of Sustainable Seas Expedition, DeepWorker submersible.)

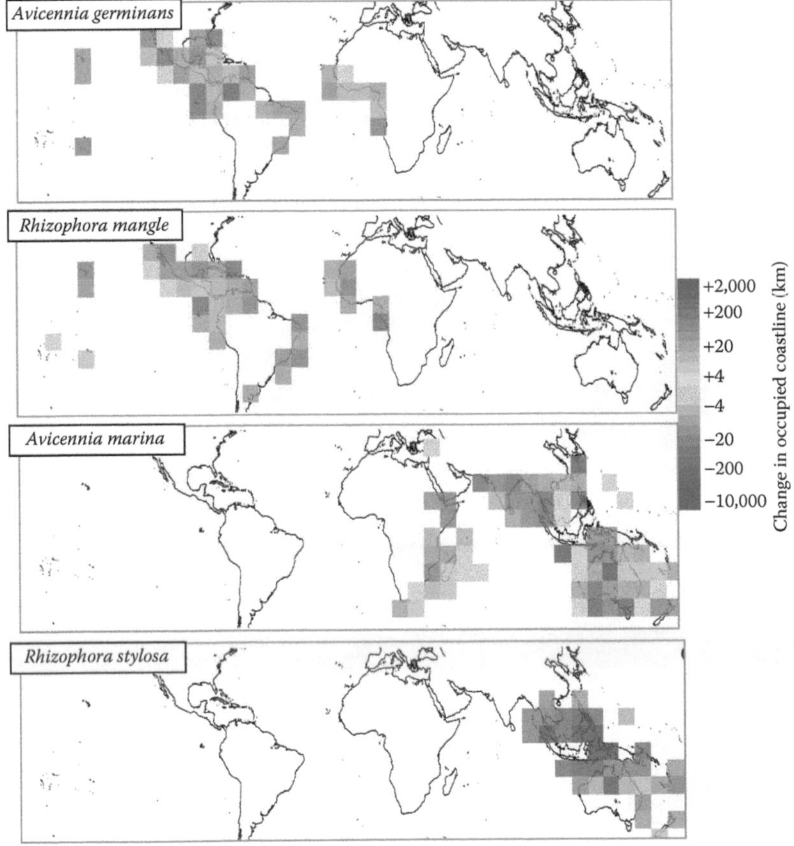

Figure 10.2 Change in predicted occupancy for four prominent mangrove species projected for 2080 relative to current occupancy. (From Record, S., Charney, N.D., Zakaria, R.M., Ellison, A.M., *Ecosphere*, 4, art34, 2013. With permission.)

00.5 1 2 3 4 5 6 7 8 9 10 Km

Figure 10.4 Mosaic of habitats for the area surrounding Big Pine Key, Florida Keys. Terrestrial and shoreline habitats are uplands (gray) and mangrove (brown). Aquatic habitats are aligned north to south between the islands, with continuous seagrass beds (dark green) and patchy seagrass beds (light green). Coral patch reefs and the continuous offshore reef tract are shown in red. The Overseas Highway is shown as a black line running east to west.

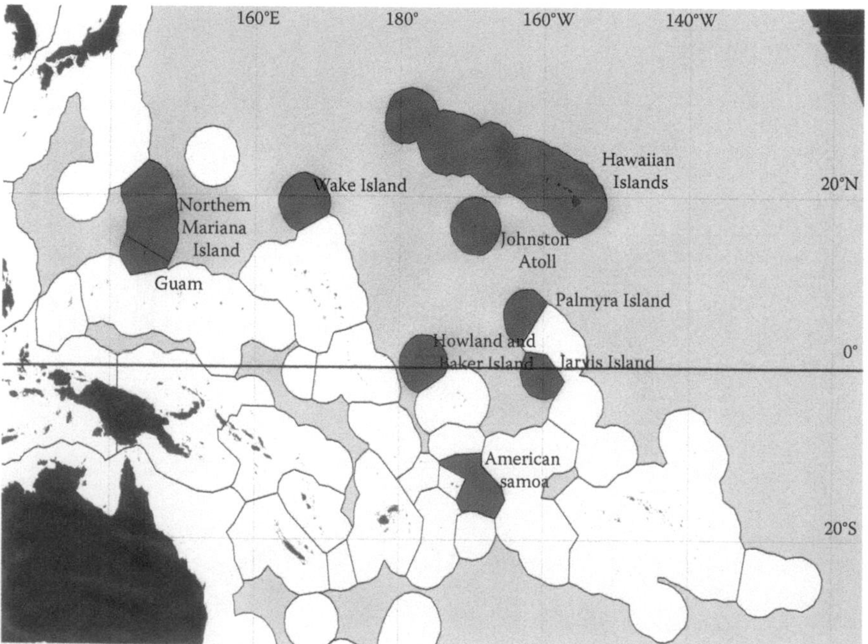

Figure 11.1 Map of the Western Pacific region showing the exclusive economic zones (EEZs). The red EEZs (3–200 nmi) are under the fishery management jurisdiction of the Western Pacific Regional Fishery Management Council.

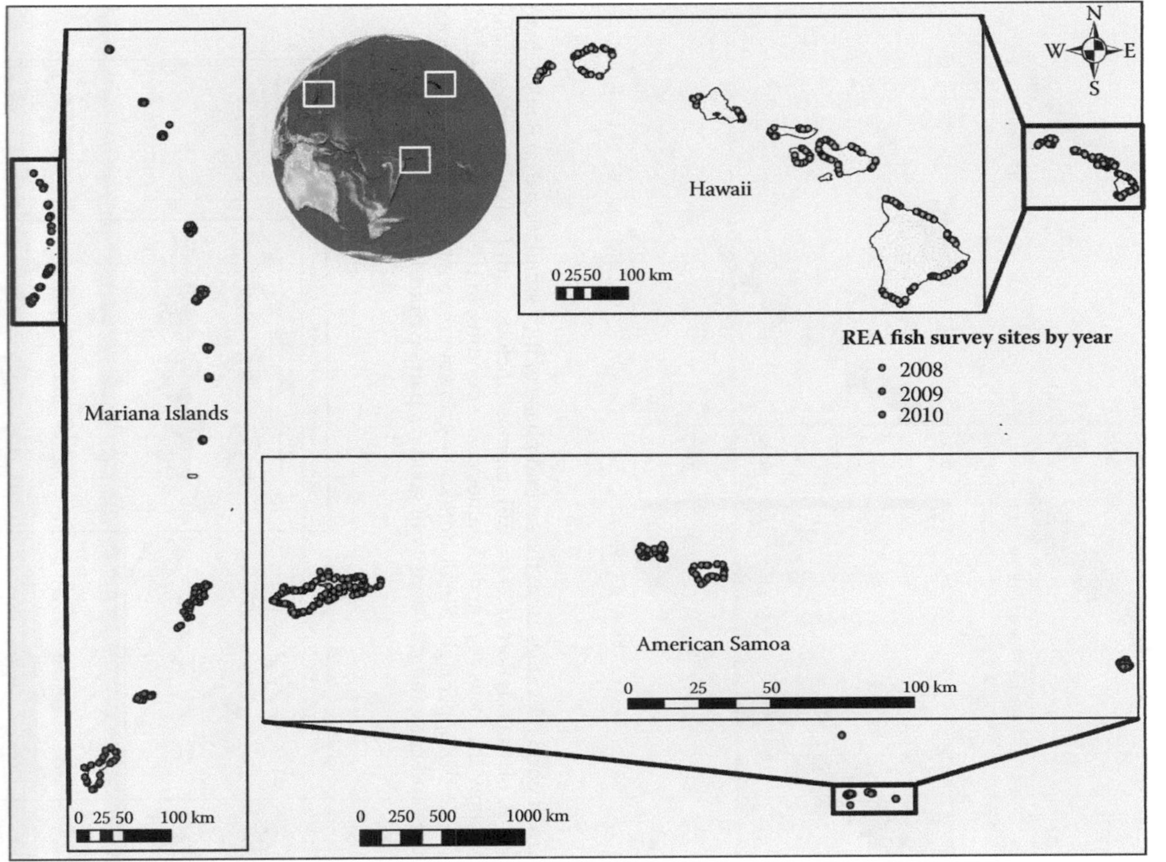

Figure 11.3 Map of the randomly selected sampling locations of the NOAA-CRED Rapid Assessment and Monitoring Program that generated the standing biomass estimates from 2008 to 2010. The sampling sites shown are only for biomass data used in the analysis. Biomass data exist for the Northwestern Hawaiian Islands and Pacific Remote Island Areas. (Map provided by Kaylyn McCoy of NOAA-CRED. With permission.)

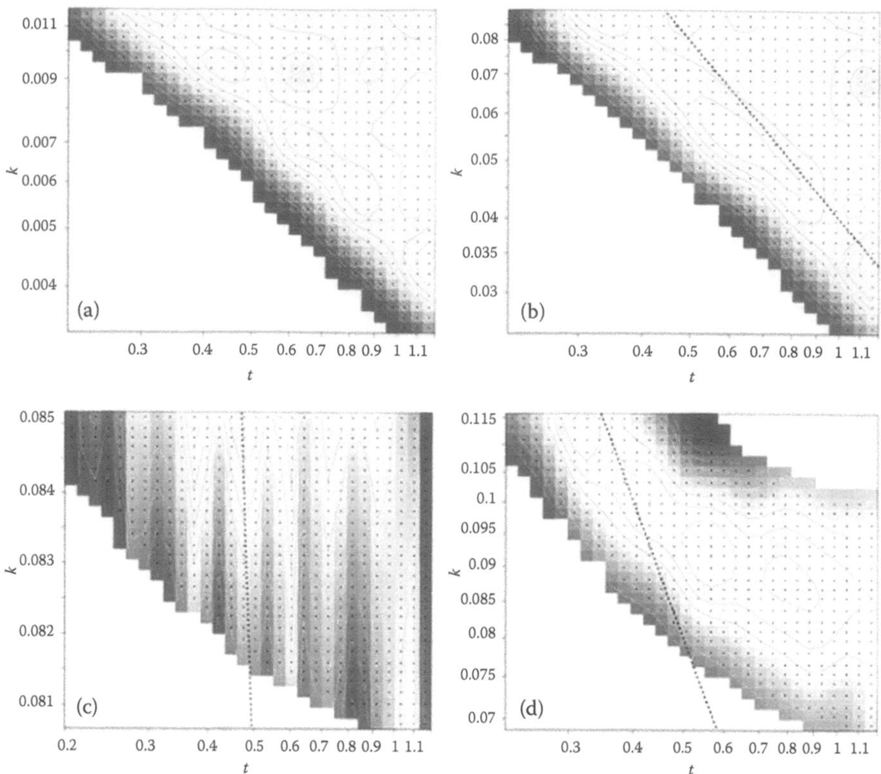

Figure 11.7 Pairs of good *r*, *k* values from the second phase of four simulations: (a) $F=0.01$, biomass ignored; (b) $F=0.10$, biomass ignored; (c) $F=0.01$, biomass included; (d) $F=0.10$, biomass included. The dotted lines show the locus of r^* and k^* values corresponding to the real MSY. In panel (a), the line is off the scale to the upper right. Single dots show grid squares containing at least one good *r*, *k* pair. Contour lines and color indicate density of good *r*, *k* pairs (white: high; red: low).

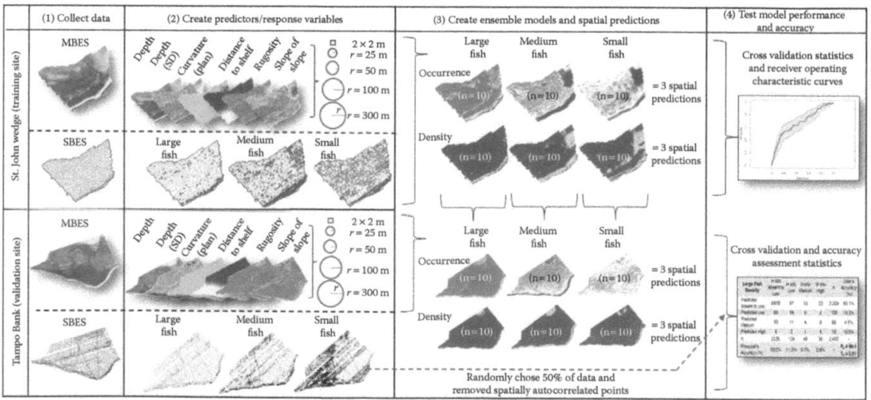

Figure 12.2 Example of analytical steps and data types in the multiscale seascape ecology approach to spatial predictive modeling. MBES, multibeam echosounder for mapping the seafloor; SBES, singlebeam echosounder for mapping location and body size of fish. (Adapted from Costa, B., Taylor, C.J., Kracker, L., Battista, T., and Pittman, S.J., *PLoS ONE*, 2013.)

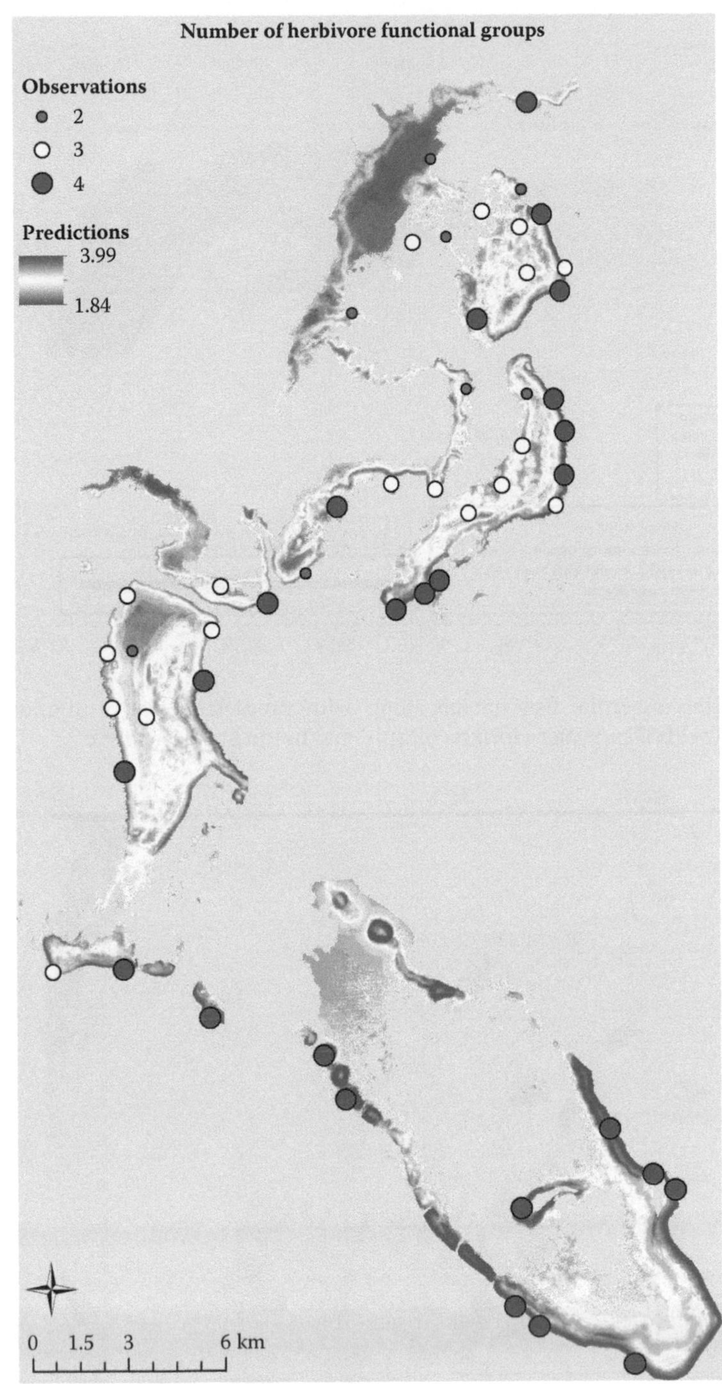

Figure 12.3 Predictive map of richness in herbivorous fish functional groups across coral reef ecosystems in Fiji as one indicator in a suite of resilience indicators. Predictive models were developed using random forest on the basis of high-resolution satellite data. (Adapted from Knudby, A., Jupiter, S., Roelfsema, C., Lyons, M., and Phinn, S., *Remote Sens*, 5, 1311–1334, 2013.)

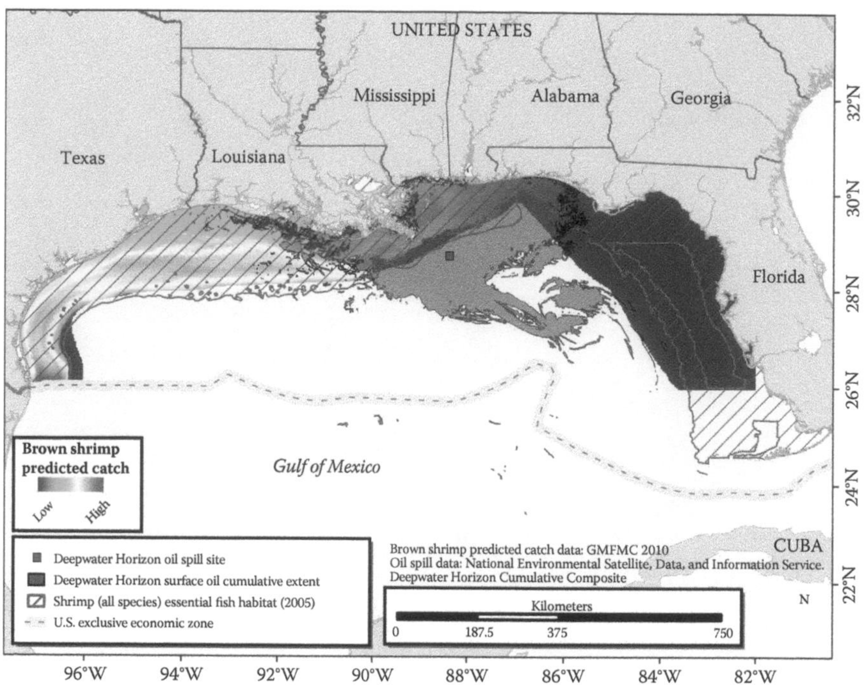

Figure 13.2 Shrimp essential fish habitat along with predicted relative abundance for brown shrimp, overlain with Deepwater Horizon oil spill maximum surface extent.

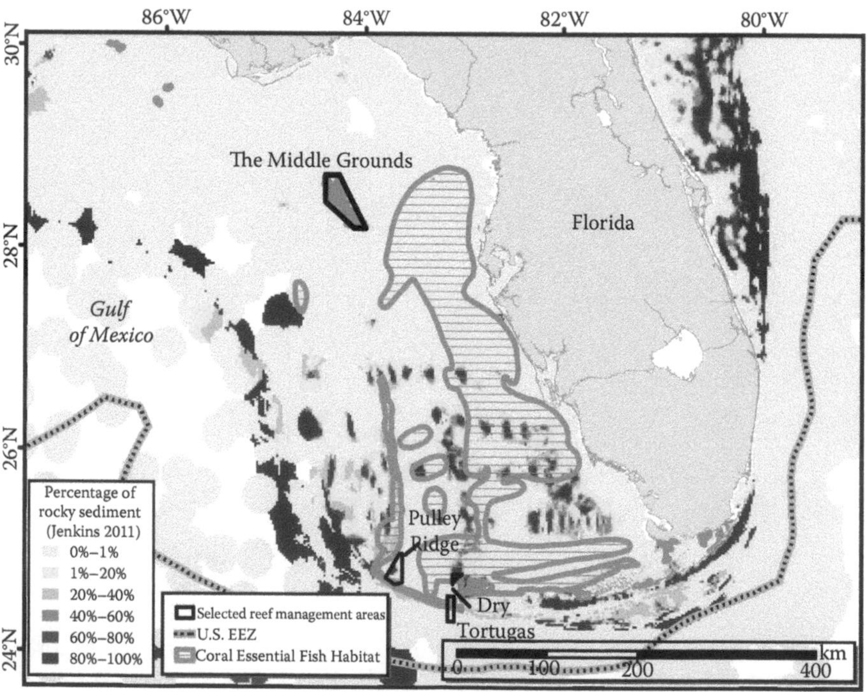

Figure 13.3 Coral essential fish habitat spatial representation (blue hatching) and percentage of rocky bottom sediments as described in Jenkins 2011. Note: white indicates no data.

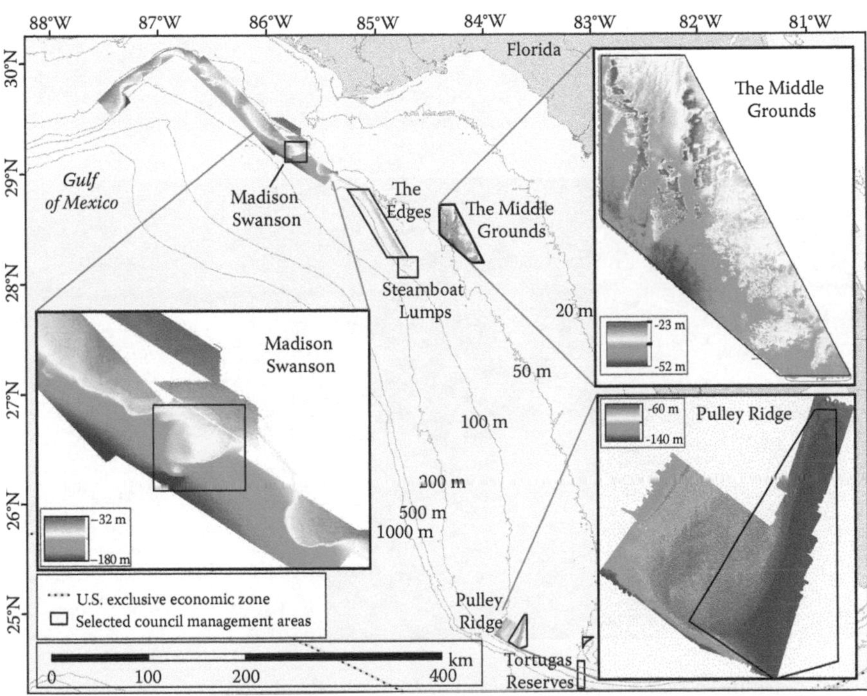

Figure 13.4 Overview map showing distribution of multibeam echo sounder imagery collected by multiple USF/USGS surveys, with selected Gulf Council fishery management areas in the eastern Gulf of Mexico. The Middle Grounds, Madison Swanson, and Pulley Ridge are expanded for detail. Depth range color ramps are specific to each inset area and cannot be directly compared.

Figure 14.2 A Caribbean patch reef by day (a) and by night (b) showing surrounding seagrasses, a grazed "halo," and movements of day-active fishes including parrotfishes and night-active grunts (Haemulidae) and *Diadema* sea urchins.

chapter seven

Coral community development on offshore platforms in the Gulf of Mexico

What we now know

Paul W. Sammarco

Contents

Introduction

It is well known that demersal and reef-associated fish require hard bottom as habitat for population establishment and growth—for fish to settle as larvae, grow into adulthood, and reproduce (Orth et al. 1984, Choat and Bellwood 1991). These life stages are further enhanced when the hard bottom has a complex three-dimensional structure, affording habitat for feeding, breeding, and refuge from predators and competitors (Doherty 1991, Syms and Jones 2000). Scleractinian corals on hardbottom habitat enhance three-dimensional structure (Nagelkerken 1977, Luckhurst and Luckhurst 1978, Thresher 1983, Bell and Galzin 1984, Roberts and Ormond 1987, Jones 1991), which is one of the reasons why coral reefs possess such high reef fish diversity and abundance.

Previously, little information was available on the sessile, epibenthic communities living on oil and gas platforms in the northern Gulf of Mexico (Gallaway and Lewbel 1981, Driessen 1989, Bright et al. 1991, Adams 1996, K. Deslarzes [Geo-Marine Inc.] pers comm). Moreover, the available data were limited in scope. Recently, additional information has become available, indicating that scleractinian corals have colonized these platforms in the first 10–30 m of water (Bright et al. 1991, K. Deslarzes pers. comm.).

Artificial reefs have long been known to have substantial fish populations associated with them (Grossman et al. 1997). In fact, there are now many artificial reef programs in the United States, where a variety of structures have been introduced into coastal waters to facilitate fish community development (Fikes 2013). However, there has been ongoing controversy as to whether these structures "attract" or "produce" fish (Wilson et al. 2001, Broughton 2012), that is, whether fish are attracted to them and are transient, or whether fish colonize these structures, remain there, become residents, grow, and reproduce there. It would appear that these structures both attract and produce. For example, some demersal fish such as red snapper (Lutjanidae) frequent the platforms but do not stay at any one platform in particular (Stanley and Wilson 1997, Franks 2000, Fabi et al. 2002, Gallaway et al. 2009). In contrast, other demersal and reef-associated fish, such as pomacentrids, balistids, acanthurids, and scarids, settle on the platforms as larvae, grow into juveniles, develop into adults, become residents there, and reproduce there (Love et al. 1994, Rooker et al. 1997). Clearly, these structures are producing fish.

Over the past 15–20 years, a large amount of quantitative data has been collected on the relationships between the scleractinian coral populations that now exist on oil and gas platforms in the northern Gulf of Mexico. Information has also been gathered on how these satellite coral populations relate to those on the NOAA (U.S. National Oceanographic and Atmospheric Agency) Flower Garden Banks National Marine Sanctuary. The purpose of this chapter is to provide an overview of current knowledge regarding coral populations on oil and gas platforms in this region and comment on their potential effects on fish biodiversity and population maintenance. In light of these effects, implications for platform decommissioning after these structures have completed their useful life as oil and gas production units will also be discussed.

Background

Natural coral reefs are not common in the northern Gulf of Mexico (Ludwick 1964, Parker et al. 1983, Rezak et al. 1985, Kennicutt et al. 1995, Dufrene 2005). There are perhaps three sets of shallow, biogenic, actively accreting reefs. The first set occurs in the western Gulf of Mexico off Tampico, Mexico—the Islas de Lobos (Jordan-Dahlgren and Rodriguez-Martinez 2003, Tunnell et al. 2007), the second set of reefs is in the east—the NOAA Florida Keys National Marine Sanctuary (Porter 2001), and the third set is southeast of Houston, Texas—the NOAA Flower Garden Banks National Marine Sanctuary (Rezak et al. 1985). There are many other banks on or just off the continental shelf of the northern Gulf of Mexico (Lugo-Fernandez et al. 2001, Schmahl et al. 2003, Schmahl and Hickerson 2006), some of which bear live scleractinian corals. These are deeper carbonaceous reefs, probably drowned during the rise in sea level approximately 14,000–20,000 years ago (Rezak et al. 1985, Montaggioni and Braithwaite 2009). Some of those in shallower water have become algal ridges or algal-sponge zones (e.g., Geyer Bank) (Rezak et al. 1985, Dennis and Bright 1988, Cancelmo 2008, pers. obs). Thus, for tens of thousands of years, hardbottom in shallow water has been scarce in this region as the bottom has been dominated primarily by soft sediment and, for the most part, still is.

The expansion of oil and gas exploration and production activities from land, offshore onto the continental shelf of the Gulf of Mexico during the 1940s (US Geological Survey 1995, Priest 2009) changed the distribution and abundance of shallow hardbottom habitats in the northern Gulf of Mexico. Production platforms were introduced into Gulf of Mexico waters and eventually spread across most of the continental shelf except for the waters around Florida (Sammarco et al. 2004). Since the 1940s, 40,000 oil/gas wells have been drilled on the continental shelf for resource extraction (Francois 1993). The number of platforms on the shelf has been decreasing and is currently estimated to be 3200 (Knott 1995, Dauterive 2000, J. Sinclair [U.S. Department of the Interior, Bureau of Safety and Environmental Enforcement], pers. comm.). Substantial hard substrate has been absent for thousands of years in this region but is now being provided by these platforms (Curray 1965a,b, Frost 1977, Blum et al. 1998, 2001, Schroeder et al. 2000), which resemble islands in that they rise from the bottom, enter shallow water (the euphotic zone), and break through the surface. These artificial substrates or habitats have become very productive and are now both producing demersal and reef-associated fish (Love et al. 1994, Rooker et al. 1997) and attracting pelagic fish (Stanley and Wilson 1997, Franks 2000, Fabi et al. 2002, Gallaway et al. 2009), and are therefore sought after by recreational fishers.

Coral populations on platforms in the northern Gulf of Mexico

Species richness and abundance

Numerous surveys have shown that offshore platforms in the northern Gulf of Mexico serve as substratum for the growth and development of hermatypic (reef-building) corals as well as other Caribbean and Gulf of Mexico reef organisms (Sammarco et al. 2004). These include some protected, threatened, or endangered species, such as sea turtles (Sonnier et al. 1976, Plotkin et al. 1993, Renaud and Carpenter 1994) and seafans (Gorgonacea; Gittings et al. 1992, Williamson et al. 2011), and scleractinian corals. The platforms also host ahermatypic (non-reef-building) corals (Scarborough and Kendall 1994, Gass and Roberts 2006), including the invasive Indo-Pacific species *Tubastraea coccinea* (Lesson 1829) (Fenner 2001, Ferreira 2003, Fenner and Banks 2004, Paula and Creed 2004, 2005, Sammarco et al. 2004, 2012a), which colonized the Caribbean ~65 years ago (Cairns 2000, Humann and DeLoach 2002) and is now common in the western Atlantic Ocean.

Patterns of coral species richness on oil and gas platforms around the East and West Flower Garden Banks have now been determined through scuba diver surveys down to 37 m depth and by ROV (remotely operated vehicle) surveys down to 138 m depth. One striking feature is a peak of species richness directly at the Flower Garden Banks, which then dissipates with radial distance (Sammarco et al. 2012a). This is the first indication that most of the corals settling and growing on the platforms are most likely derived from the Flower Garden Banks.

The density pattern of hermatypic corals (i.e., the number of corals occurring on platforms per unit area in the northern Gulf of Mexico) is very similar to the species richness pattern, with a peak in coral density at or near the Flower Garden Banks. Densities decrease radially as one moves away from these banks, suggesting that the Flower Garden Banks are the source of coral larvae on the platforms (Sammarco et al. 2012a). This pattern was observed for three species of coral: *Madracis decactis* (Lyman 1859), *Montastraea cavernosa* (Linnaeus), and *Oculina diffusa* (Lamarck 1816). This distribution of coral was skewed toward the east, downcurrent from the Flower Garden Banks. This may be caused by the westerly boundary current at the edge of the continental shelf drawing coral larvae from the Flower Garden Banks to the east, where they encounter these platforms and settle.

Studies have also confirmed that coral community development, in terms of both species richness and abundance, increases with the age of the platform (Sammarco et al. 2004). Thus, platforms that have been deployed for a longer time appear to be more environmentally valuable with respect to coral community development than those that have been deployed for a shorter time. The break point in age for coral community development appears to be about 15 years, after which coral community development accelerates. The lack of coral community development prior to this time could be due to the presence of antifouling and anticorrosion coatings on the platforms, which degrade with time, or the fact that other sessile epibenthic organisms need to settle on the platform first.

Genetic affinities

The most reliable way to determine whether the corals found on the platforms in the northern Gulf of Mexico are actually derived from the Flower Garden Banks is through analysis of colonies using molecular genetics (Mettler et al. 1988, Futuyma 1998). A variety of molecular (amplified fragment length polymorphisms AFLPs) (Vos et al. 1995, Mueller and Wolfenbarger 1999, Bernatchez and Duchesne 2000, Sunnucks 2000) and statistical analytical techniques, including AMOVA (Excoffier et al. 1992), AFLPop (v. 1.0) (Duchesne and Bernatchez 2002), and Structure (Pritchard et al. 2000), have shown that the genetic affinity or genetic relatedness of corals such as *M. decactis* on the platforms is highest on those platforms around the Flower Garden Banks and declines with distance from the banks, particularly to the east as far as Terrebonne Bay, Louisiana (Atchison 2005, Atchison et al. 2008, Sammarco et al. 2012b). This observation of declining genetic affinity to the east is important because the prevailing circum-Gulf boundary current is westerly in this region (Sturges and Blaha 1976, Sturges 1993) and is most likely carrying coral larvae to the east (Sturges and Lugo-Fernandez 2005, Sammarco et al. 2012a,b; also see Lugo-Fernandez 1998, Lugo-Fernandez et al. 2001), strongly indicating that the Flower Garden Banks are the source of the coral larvae for *M. decactis* and other similar hermatypic corals (Figure 7.1). Examination of the genetic affinity between populations of corals on each of the platforms to each other, as well as to the Flower Garden Banks, indicates that those populations closest to each other are not significantly different from each other (Atchison 2005, Atchison et al. 2008). Coral populations on platforms further from one another are significantly different from each other. Genetic distance increases with spatial distance between the platforms, reaching a maximum at ~90 km.

In contrast, the genetic affinities of populations of *T. coccinea* on the platforms were relatively constant across the continental shelf regardless of distance from the Flower Garden Banks (Sammarco et al. 2012b), demonstrating that there was no trend from east to west or *vice versa*. This suggested that these banks were not the larval source for the *T. coccinea* populations on the platforms. More likely, populations from other parts of the Gulf of Mexico, probably to the west and south, were seeding these areas via the western boundary current (Figure 7.2).

Recruitment

Coral recruitment on the platforms in the northern Gulf of Mexico, even around the Flower Garden Banks, is rare (Baggett and Bright 1985, Sammarco 2013, Sammarco unpublished data) and levels of larval settlement are very low, as shown when recruitment levels on the platforms are compared with those on the banks (Brazeau et al. 2005, 2008, 2011, Sammarco unpublished data). Comparison of the results of coral larval settlement experiments involving the Flower Garden Banks and findings from the Helix Experiment (an earlier meso-scale coral settlement experiment performed on the Great Barrier Reef in Australia) (Sammarco and Andrews 1988, 1989, Andrews et al. 1989, Gay and Andrews 1994) indicates that the low levels of recruitment on the platforms may be the result of a distance effect. That is, once the coral larvae are released from the source reef (i.e., the Flower Garden Banks), they may be retained around the reef by meso-scale eddies in the lee of the reef extending hundreds of meters to kilometers out from the reef. Once the larvae are advected beyond this radius and outside the influence of these eddies, they may be dispersed largely through diffusion (Okubo 1994). Very few coral larvae may actually interact with the platforms and settle (Sammarco 2013), particularly as the platforms are not solid in structure (like a reef or an island) and their pilings are reticulate and most

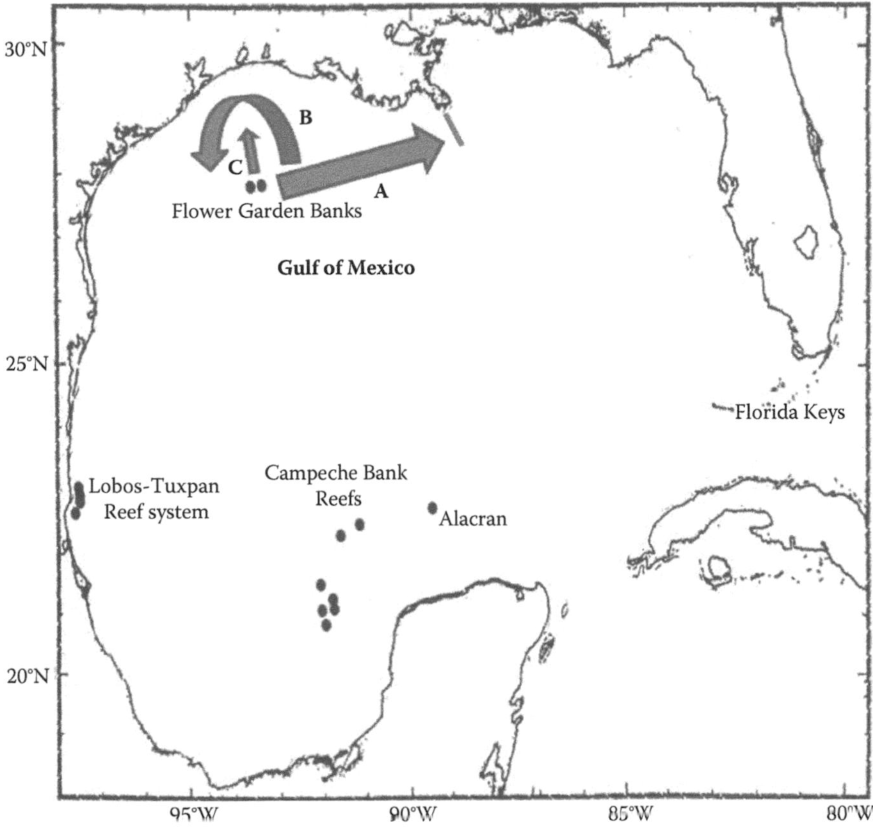

Figure 7.1 Inferred larval dispersal patterns for *M. decactis* and *M. cavernosa* based on the distribution and abundance patterns of both species and genetic affinities of populations on offshore oil and gas platforms and the Flower Garden Banks. (A) Larval dispersal by the western boundary current of the northern Gulf of Mexico. (B) Larval dispersal by large-scale eddies and coastal transport described by Lugo-Fernandez et al. (From Lugo-Fernandez, A., Deslarzes, K.J.P., Price, J.M., Boland, G.S., and Morin, M.V., *Cont Shelf Res*, 21, 47–67, 2001.) (C) Extension of coral populations to the north from these banks, most likely driven by B.

likely only generate small eddies in the wake of currents (Sammarco et al. in press). This rarity of settlement on the platforms makes the coral populations there relatively fragile, because, with these very low levels of recruitment, they would require a long time to regenerate after mass mortality as recolonization would happen at a very slow pace (Slobodkin 1961, MacArthur and Wilson 1967).

The Mississippi River as a geographic barrier

Most corals are stenohaline and exhibit limited tolerance to variations around normal salinity (35‰) (Levinton 1982). Near its most central northern point, the Gulf of Mexico receives an extraordinarily high discharge of freshwater from the Mississippi River (Arnold et al. 2000), one of the largest rivers in the world, which discharges an average of 555 km^3 of water per year (Gross 1977). The coral populations on the platforms either side of the Mississippi River mouth appear visually identical with a very simple coral

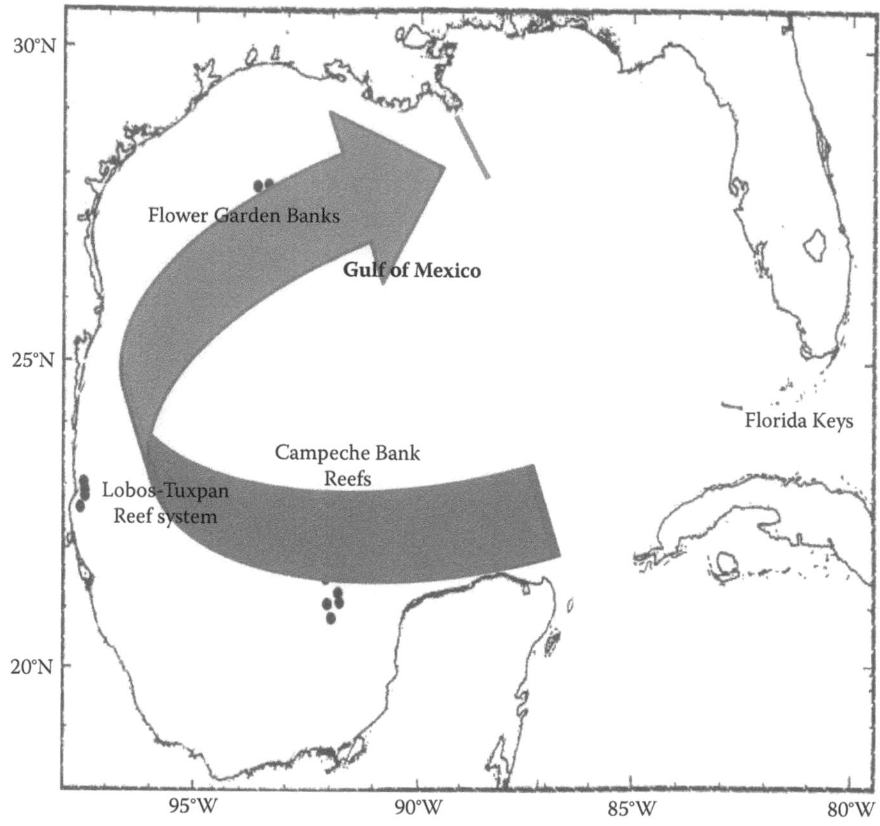

Figure 7.2 Inferred larval dispersal pattern for *T. coccinea*, based on the distribution and abundance patterns and genetic affinities of populations on offshore oil and gas platforms and the Flower Garden Banks. In this case, the western boundary current is presumed to be the primary mechanism.

community structure and similar benthic communities. However, molecular genetics research has confirmed that there is no genetic affinity between the coral populations of the dominant species of *M. decactis* and *T. coccinea* on the two side of the river mouth (Sammarco et al. 2012b). This strongly indicates that the Mississippi River is a formidable barrier to larval dispersal in this region and appears to separate the eastern from the western coral populations (Wooten et al. 1988, Near et al. 2001, Soltis et al. 2006, Wiley and Chapman 2010).

This genetic finding also implies that the coral populations in the east are not derived from the Flower Garden Banks or from other western populations, but rather are probably receiving larvae from the Caribbean and the Loop Currents flowing up from the south (Figure 7.3) (Sturges and Lugo-Fernandez 2005). It is also possible that storm-driven currents may be advecting coral larvae from the southern Gulf of Mexico during severe storms, north-northeast to the eastern side of the Mississippi River (Lugo-Fernandez and Gravois 2010). Although there is little genetic affinity between *M. decactis* populations on different platforms on the same side of the river mouth (Figures 7.1 through 7.3; Sammarco et al. 2012b), genetic affinity is much higher between *T. coccinea* populations on either side of the river mouth, implying that larval dispersal and resultant recruitment is more

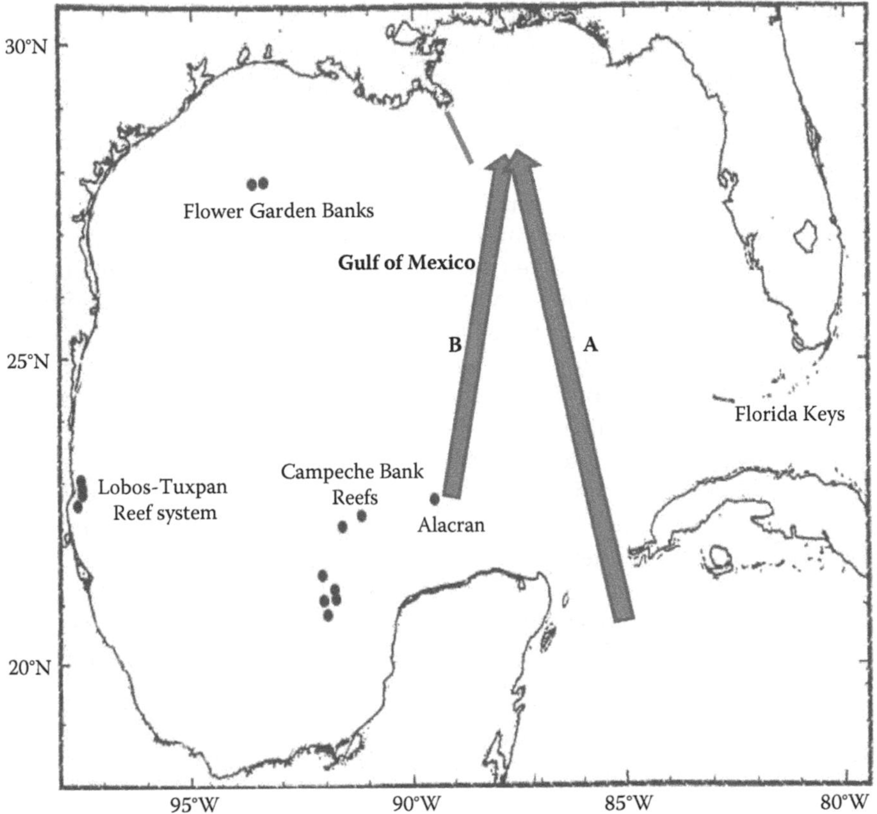

Figure 7.3 Inferred larval dispersal patterns for *M. decactis* and *T. coccinea* on the east side of the Mississippi River mouth, based on the distribution and abundance patterns of both species and genetic affinities of populations on offshore oil and gas platforms and the Flower Garden Banks. (A) Larval dispersal by the Caribbean Current and the Loop Current, characteristic of the Gulf of Mexico. (B) Larval dispersal possibly by severe storm–associated currents, driving larvae from the southern Gulf of Mexico to the NE region. (Based on the findings of Lugo-Fernandez, A. and Gravois, M., *Cont Shelf Res*, 30, 1226–1240, 2010.)

effective in this species than in *M. decactis*. This may also be the reason why *T. coccinea* has been such a successful invasive species in the western Atlantic, as its dispersal and recruitment abilities may surpass those of native species.

Closing comments: A new view of decommissioning and the "rigs-to-reefs" program

The information collected on coral populations that have successfully colonized oil and gas platforms in the northern Gulf of Mexico has changed our view of these structures and may also influence their decommissioning. At present, platforms must be removed from the sea within 12 months of the termination of a lease within a lease block, a regulation issued at a time when little was known about these structures and their function as artificial reefs in the Gulf of Mexico. In fact, with 3200 structures deployed at present, they represent the largest artificial reef system in the world. Our knowledge of the development of marine benthic, demersal, and pelagic communities living on and associated with these artificial reefs has now greatly expanded. On older platforms, particularly structures in place for over 15 years, these communities resemble those on developed Gulf of Mexico and Caribbean reefs. In addition, the platforms may help to stabilize the Flower Garden Banks by providing refuge for small populations of corals that could help regenerate populations on these banks in the event of mass mortality (Sammarco et al. 2004, 2012a). It is now evident that, from an environmental perspective, many platforms would be more valuable in than out of the water, toppled in place as permitted by the "rigs-to-reefs" program or towed to a designated location and toppled there (Dauterive 2000). Active rigs-to-reefs programs are sponsored by the federal government (http://www.bsee.gov/Exploration-and-Production/Decomissioning/FAQ/) jointly with the states of Louisiana (http://www.wlf.louisiana.gov/fishing/artificial-reef-program), Texas (http://www.tpwd.state.tx.us/landwater/water/habitats/artificial_reef/rigs-to-reefs.phtml), Mississippi (http://www.dmr.state.ms.us/joomla16/index.php/marine-fisheries/artificial-reef/75-rigs-to-reef), and Alabama (http://www.outdooralabama.com/fishing/saltwater/fisheries/artificial-reefs/unocal.cfm). There is growing awareness of these issues and active debate regarding the fate of platforms and their potential roles as longer-term artificial reefs, as exemplified by the "Save The Blue" program based in Houston, Texas (http://savetheblue.silverleafcms.com/resources/).

It would most likely benefit the Gulf of Mexico if these platforms were surveyed for their benthic populations, particularly corals and other protected taxa, as well as for demersal, reef-associated, and pelagic fish communities prior to decommissioning and removal. Such surveys could be carried out on an individual basis before decommissioning decisions were taken. In this way, an environmental value could be assigned to the platform based on factors such as location, proximity to safety fairways, hypoxic zones, and so on.

Not all platforms have well-developed reef communities (Sammarco et al. 2004, 2012a). Some platforms in place for under 15 years will most likely be less developed and thus less environmentally valuable than others. Others may be sited too far north on the shelf of the Gulf of Mexico, on the inside of the continental shelf, and exposed to particularly high sediment and nutrient loads (the hypoxic zone) derived from the Mississippi River, or winter temperatures of less than 18°C (Roberts et al. 2008), below which coral growth and survival is minimal (Levinton 1982). Sediment is also known to deter coral planulae from settling successfully (Sammarco 1980, Hodgson 1990, Hunte and Wittenberg 1992). Based on the distribution and abundance of corals on the platforms in the northern Gulf of Mexico, I estimate that 10%–20% of the platforms on the continental shelf in this region may qualify for preservation for environmental purposes. A protocol should be developed by which platforms would be assessed for their reef communities before decisions were taken on whether or not they should be removed or left in place after oil and gas extraction has ceased.

In the joint federal and state rigs-to-reefs programs, platforms can be either toppled in place or towed to a designated rigs-to-reefs site on the shelf and toppled there. This program was designed to facilitate the development of fisheries, but there is currently no evidence to indicate that these rigs-to-reefs structures promote the development of increased coral or benthic communities, which would in turn promote fish population growth (Sammarco et al. in press). The program is probably working, but not to its maximum efficiency. Therefore, the form of the toppled structures and how effective they are in attracting and/or producing fish at depth should be reviewed. Supplying additional smaller-scale habitat in the toppled platforms may promote demersal, reef-associated, and possibly fish population development.

Acknowledgments

Many thanks to the numerous people, who are far too many to individually name here, who helped me with these studies. Thanks to Mr. Mark Mueller and the Gulf of Mexico Fishery Management Council for making it possible for me to attend their meeting on the importance of coral reefs and artificial reefs as fish habitat. Thanks also to Steve Bortone for encouraging me to write down my thoughts on the relationship between oil and gas platforms and coral populations. My research on this subject was funded by the U.S. Department of the Interior—Minerals Management Service/Bureau of Ocean Energy Management. Support for attendance at this conference was provided by the Gulf of Mexico Fisheries Management Council. Thanks to all.

References

Adams CL. 1996. Species composition, abundance and depth zonation of sponges (phylum Porifera) on an outer continental shelf gas production platform, northwestern Gulf of Mexico. Final report and MSc thesis, Center for Coastal Studies, Texas A&M University, Corpus Christi, TX.

Andrews JC, Gay S, Sammarco PW. 1989. Models of dispersal and recruitment of coral larvae around an isolated reef (Helix Reef) in the central Great Barrier Reef. *In:* Proceedings of the 6th International Coral Reef Congress, James Cook University of North Queensland, 1988, Volume 2. Townsville, Australia: James Cook University of North Queensland Press. p. 469–474.

Arnold JG, Muttiah RS, Srinivasan R, Allen PM. 2000. Regional estimation of base flow and groundwater recharge in the Upper Mississippi River Basin. J Hydrol. 227:21–40.

Atchison AD. 2005. Offshore oil and gas platforms as stepping stones for expansion of coral communities: A molecular genetics analysis. MSc Thesis, Department of Oceanography and Coastal Sciences, Louisiana State University, Baton Rouge, LA.

Atchison AD, Sammarco PW, Brazeau DA. 2008. Genetic connectivity in corals on the Flower Garden Banks and surrounding oil/gas platforms, Gulf of Mexico. J Exp Mar Biol Ecol. 365:1–12.

Baggett LS, Bright TJ. 1985. Coral recruitment at the East Flower Garden Bank (northwestern Gulf of Mexico). *In*: Proceedings of the Fifth International Coral Reef Congress, Tahiti, French Polynesia, Volume 4. Moorea, French Polynesia: Antenne Museum – EPHE. p. 379–384.

Bell JD, Galzin R. 1984. Influence of live coral cover on coral-reef fish communities. Mar Ecol Prog Ser. 15:265–274.

Bernatchez L, Duchesne P. 2000. Individual-based genotype analysis in studies of parentage and population assignment: How many loci, how many alleles? Can J Fish Aquat Sci. 57:1–12.

Blum MD, Misner TJ, Collins ES, Scott DB. 1998. Rapid sea level rise and highstand (+2 m) during the middle Holocene, central Texas coast. Abstr Geol Soc Am. 30:7.

Blum MD, Misner TJ, Collins ES, Scott DB, Morton RA, Aslan A. 2001. Middle Holocene sea-level rise and highstand at 12 m, central Texas coast. J Sed Res. 71:4.

Brazeau DA, Sammarco PW, Atchison AD. 2008. Genetic structure in coral recruitment: Evidence of extreme patchiness in settlement. *In*: Proceedings of the 11th International Coral Reef Symposium. Fort Lauderdale, FL, July 2008. Abstract.

Brazeau DA, Sammarco PW, Atchison AD. 2011. Genetic structure in coral recruitment: Evidence of extreme patchiness in settlement. Aquat Biol. 12:55–67.

Brazeau DA, Sammarco PW, Gleason DF. 2005. A multi-locus genetic assignment technique to assess local recruitment of *Agaricia agaricites* on coral reefs. Mar Biol. 147:1141–1148.

Bright TJ, Gittings SR, Zingula R. 1991. Occurrence of Atlantic reef corals on offshore platforms in the northwestern Gulf of Mexico. Northeast Gulf Sci. 12:55–60.

Broughton C. 2012. Office of National Marine Sanctuaries science review of artificial reefs. US Department of Commerce, NOAA Office of National Marine Sanctuaries, Conservation Series, ONMS-12-05.

Cairns SD. 2000. Revision of the shallow-water azooxanthellate Scleractinia of the western Atlantic. Stud Nat Hist Caribb Reg. 75:1–240.

Cancelmo J. 2008. Texas coral reefs. Corpus Christi, TX: Texas A&M University. p. 87–95.
Choat JH, Bellwood DR. 1991. Reef fishes: Their history and evolution. *In:* Sale PF, editor. The ecology of fishes on coral reefs. San Diego, CA: Academic Press. p. 39–68.
Curray JR. 1965a. Late quaternary history, continental shelves of the United States. *In:* Wright HE Jr, Frey DG, editors. The quaternary of the United States. Princeton, NJ: Princeton University Press. p. 723–735.
Curray JR. 1965b. Sediments and history of Holocene transgression, continental shelf, northwest Gulf of Mexico: API Project 51. *In*: Proceedings of American Association of Petroleum Geologists. Tulsa, OK: American Association of Petroleum Geologists. p. 221–266.
Dauterive L. 2000. Rigs to reefs policy, progress, and perspective. OCS Study MMS 2000-073. New Orleans, LA: US Department of the Interior, Minerals Management Service, Gulf of Mexico OCS Region.
Dennis GD, Bright TJ. 1988. Reef fish assemblages on hard banks in the northwestern Gulf of Mexico. Bull Mar Sci. 43:280–307.
Doherty PJ. 1991. Spatial and temporal patterns in recruitment. *In:* Sale PF, editor. The ecology of fishes on coral reefs. San Diego, CA: Academic Press. p. 261–293.
Driessen PK. 1989. Offshore oil platforms: Mini-ecosystems. *In:* Reggio VC Jr, editor. Petroleum structures as artificial reefs: A compendium. Fourth International Conference on Artificial Habitats for Fisheries, rigs-to-reefs special session, November 1987, Miami, FL. OCS Study/ MMS 89-0021. New Orleans, LA: US Department of the Interior, Minerals Management Service. p. 3–6.
Duchesne P, Bernatchez L. 2002. AFLPOP: A computer program for simulated and real population allocation using AFLP markers. Mol Ecol Notes. 2:380–383.
Dufrene TA. 2005. Geologic variability and Holocene sedimentary record on the northern Gulf of Mexico inner to mid-continental shelf. MSc Thesis, Department of Oceanography and Coastal Sciences, Louisiana State University, Baton Rouge, LA.
Excoffier L, Smouse PE, Quattro JM. 1992. Analysis of molecular variation inferred from metric distances among DNA haplotypes: Application to human mitochondrial DNA restriction data. Genetics. 131:479–491.
Fabi G, Grati F, Luccetti A, Trovarelli L. 2002. Evolution of a fish assemblage around a gas platform in the northern Adriatic sea. ICES J Mar Sci. 59:S309–S315.
Fenner D. 2001. Biogeography of three Caribbean corals (Scleractinia) and the invasion of *Tubastraea coccinea* into the Gulf of Mexico. Bull Mar Sci. 69:1175–1189.
Fenner D, Banks K. 2004. Orange cup coral *Tubastraea coccinea* invades Florida and the Flower Garden Banks, northwestern Gulf of Mexico. Coral Reefs. 23:505–507.
Ferreira CEL. 2003. Non-indigenous corals at marginal sites. Coral Reefs. 22:498.
Fikes R. 2013. Artificial reefs of the Gulf of Mexico: A review of Gulf State programs & key considerations. National Wildlife Federation. Available from: http://www.nwf.org/News-and-Magazines/Media-Center/Reports/Archive/2013/11-08-13-Review-of-Gulf-of-Mexico-Artificial-Reefs.aspx via the Internet. Accessed 29 January, 2013.
Francois DK. 1993. Federal offshore statistics: 1992. Leasing, exploration, production, and revenues as of December 31, 1992. MMS 93-0066. Herndon, VA: US Department of the Interior, Minerals Management Service.
Franks J. 2000. A review: Pelagic fishes at petroleum platforms in the northern Gulf of Mexico; diversity, interrelationships, and perspective. Pêche thonière et dispositifs de concentration de poissons, Caribbean-Martinique, 15–19 October, 1999. Available from: http://archimer.ifremer.fr/doc/00042/15301/ via the Internet. Accessed 21 August, 2013.
Frost SH. 1977. Oligocene reef coral biogeography Caribbean and western Tethys. *In:* Werger MJA, editor. Proceedings of the 2nd International Symposium on Corals and Fossil Coral Reefs. Paris: Editions du B.R.G.M. p. 342–352.
Futuyma DJ. 1998. Evolutionary biology. 3rd ed. Sunderland, MA: Sinauer Associates.
Gallaway BJ, Lewbel GS. 1981. The ecology of petroleum platforms in the northwestern Gulf of Mexico: A community profile. FWS 10BS-82/87, Open File Report 82-03. Washington, DC: USFWS Office of Biology Services.

Gallaway BJ, Szedlmeyer ST, Gazey WJ. 2009. A life history review for red snapper in the Gulf of Mexico with an evaluation of the importance of offshore petroleum platforms and other artificial reefs. Rev Fish Sci. 17:48–67. Available from: http://dx.doi.org/10.1080/10641260802160717 via the Internet. Accessed 21 August, 2013.

Gass SE, Roberts JM. 2006. The occurrence of the cold-water coral *Lophelia pertusa* (Scleractinia) on oil and gas platforms in the North Sea: Colony growth, recruitment and environmental controls on distribution. Mar Pollut Bull. 52:549–559. Available from: http://dx.doi.org/10.1016/j.marpolbul.2005.10.002 via the Internet. Accessed 23 August, 2013.

Gay SL, Andrews JC. 1994. The effects of recruitment strategies on coral larvae settlement distributions at Helix Reef. *In:* Sammarco PW, Heron ML, editors. The biophysics of marine larval dispersal. Washington, DC: American Geophysical Union. p. 73–88.

Gittings SR, Boland GS, Deslarzes DJP, Combs CL, Holland BS, Bright TJ. 1992. Mass spawning and reproductive viability of reef corals at the East Flower Garden Bank, northwest Gulf of Mexico. Bull Mar Sci. 51:420–428.

Gross MG. 1977. Oceanography: A view of the earth. 2nd ed. Englewood Cliffs, NJ: Prentice-Hall.

Grossman GD, Jones GP, Seaman WJ Jr. 1997. Do artificial reefs increase regional fish production? A review of existing data. Fisheries. 22:17–23.

Hodgson G. 1990. Sediment and the settlement of larvae of the reef coral *Pocillopora damicornis*. Coral Reefs. 9:41–43.

Humann P, DeLoach N. 2002. Reef coral identification: Florida, Caribbean, Bahamas. Jacksonville, FL: New World Publisher.

Hunte W, Wittenberg M. 1992. Effects of eutrophication and sedimentation on juvenile corals. Mar Biol. 114:625–631.

Jones GP. 1991. Post-recruitment processes in the ecology of coral reef fish populations: A multifactorial perspective. *In:* Sale PF, editor. The ecology of fishes on coral reefs. San Diego, CA: Academic Press. p. 294–330.

Jordan-Dahlgren E, Rodriguez-Martinez RE. 2003. The Atlantic coral reefs of Mexico. *In:* Cortes J, editor. Latin American coral reefs. Amsterdam: Elsevier. p. 131–158.

Kennicutt MC II, Schroeder WW, Brooks JM. 1995. Temporal and spatial variations in sediment characteristics on the Mississippi-Alabama continental shelf. Cont Shelf Res. 15:1–18.

Knott D. 1995. US lessons for UK platform disposal. Oil Gas J. 93:17–18.

Levinton JS. 1982. Marine ecology. Englewood Cliffs, NJ: Prentice-Hall.

Love M, Hyland J, Ebeling A, Herrlinger T, Brooks A, Imamura E. 1994. A pilot study of the distribution and abundances of rockfishes in relation to natural environmental factors and an offshore oil and gas production platform off the coast of Southern California. Bull Mar Sci. 55:1062–1085.

Luckhurst BE, Luckhurst K. 1978. Analysis of the influence of the substrate variables on coral reef fish communities. Mar Biol. 49:317–323.

Ludwick JC. 1964. Sediments in the northeastern Gulf of Mexico. *In:* Miller RL, editor. Papers in marine geology, Shephard commemorative volume. New York: Macmillan.

Lugo-Fernandez A. 1998. Ecological implications of hydrography and circulation to the Flower Garden Banks, northwest Gulf of Mexico. Gulf Mex Sci. 16:144–160.

Lugo-Fernandez A, Deslarzes KJP, Price JM, Boland GS, Morin MV. 2001. Inferring probable dispersal of Flower Garden Banks coral larvae (Gulf of Mexico) using observed and simulated drifter trajectories. Cont Shelf Res. 21:47–67.

Lugo-Fernandez A, Gravois M. 2010. Understanding impacts of tropical storms and hurricanes on submerged bank reefs and coral communities in the northwestern Gulf of Mexico. Cont Shelf Res. 30:1226–1240.

MacArthur RH, Wilson EO. 1967. The theory of island biogeography. Princeton, NJ: Princeton University Press.

Mettler LI, Gregg TG, Schaffer HE. 1988. Population genetics and evolution. 2nd ed. Englewood Cliffs, NJ: Prentice-Hall.

Montaggioni LF, Braithwaite CJR. 2009. Quaternary coral reef systems: History, development processes and controlling factors. Developments in marine geology. Volume 5. Amsterdam: Elsevier.

Mueller UG, Wolfenbarger LL. 1999. AFLP genotyping and fingerprinting. Trends Ecol Evol. 14:389–394.

Nagelkerken W. 1977. The distribution of the graysby *Petrometopon cruentatum* (Lacepede) on the coral reef at the southwest coast of Curacao. *In*: Proceedings of the Third International Coral Reef Symposium, Miami, Florida, USA, 1976. Volume 1. Miami, FL: Rosenstiel School of Atmospheric Science, University of Miami. p. 311–315.

Near TJ, Page LM, Mayden RL. 2001. Intraspecific phylogeography of *Percina evides* (Percidae: Etheostomatinae): An additional test of the central highlands pre-Pleistocene vicariance hypothesis. Mol Ecol. 10:2235–2240.

Okubo A. 1994. The role of diffusion and related physical processes in dispersal and recruitment of marine populations. *In:* Sammarco PW, Heron ML, editors. The biophysics of marine larval dispersal. Washington, DC: American Geophysical Union. p. 5–34.

Orth RJ, Heck KL, van Montfrans J. 1984. Faunal communities in seagrass beds: A review of the influence of plant structure and prey characteristics on predator-prey relationships. Estuaries. 7:339–350.

Parker RO, Colby DR, Willis TD. 1983. Estimated amount of reef habitat on a portion of the US South Atlantic and Gulf of Mexico continental shelf. Bull Mar Sci. 33:935–940.

Paula AF, Creed JC. 2004. Two species of the coral *Tubastraea* (Cnidaria, Scleractinia) in Brazil: A case of accidental introductions. Bull Mar Sci. 74:175–183.

Paula AF, Creed JC. 2005. Spatial distribution and abundance of non-indigenous coral genus *Tubastraea* (Cnidaria, Scleractinia) around Ilha Grande, Brazil. Braz J Biol. 65:661–673. Available from: http://dx.doi.org/10.1590/S1519-69842005000400014 via the Internet. Accessed 31 August, 2013.

Plotkin PT, Wicksten MK, Amos AF. 1993. Feeding ecology of the loggerhead sea turtle *Caretta caretta* in the northwestern Gulf of Mexico. Mar Biol. 115:1–5.

Porter J, editor. 2001. The Everglades, Florida Bay, and coral reefs of the Florida Keys: An ecosystem sourcebook. Boca Raton, FL: CRC Press.

Priest T. 2009. The offshore imperative: Shell Oil's search for petroleum in postwar America. College Station, TX: Texas A&M University.

Pritchard JK, Stephens M, Donnelly P. 2000. Inference of population structure using multi-locus genotype data. Genetics. 155:945–959.

Renaud ML, Carpenter JA. 1994. Movements and submergence patterns of loggerhead turtles (*Caretta caretta*) in the Gulf of Mexico determined through satellite telemetry. Bull Mar Sci. 55:1–15.

Rezak R, Bright TJ, McGrail DW. 1985. Reefs and banks of the northwestern Gulf of Mexico. New York: Wiley.

Roberts CM, Ormond RFG. 1987. Habitat complexity and coral reef fish diversity and abundance on Red Sea fringing reefs. Mar Ecol Prog Ser. 41:1–8.

Roberts HH, Rouse LJ, Walker ND, Hudson JH. 2008. Cold-water stress in Florida Bay and Northern Bahamas: A product of winter cold-air outbreaks. J Sed Res. 52:145–155. Available from: http://dx.doi.org/10.1306/212F7EFA-2B24-11D7-8648000102C1865D via the Internet. Accessed 29 August, 2013.

Rooker JR, Dokken QR, Pattengill CV, Holt GJ. 1997. Fish assemblages on artificial and natural reefs in the Flower Garden Banks National Marine Sanctuary, USA. Coral Reefs. 16:83–92.

Sammarco PW. 1980. *Diadema* and its relationship to coral spat mortality: Grazing, competition, and biological disturbance. J Exp Mar Biol Ecol. 45:245–272.

Sammarco PW. 2013. Corals on oil and gas platforms near the Flower Garden Banks: Population characteristics, recruitment, and genetic affinity. OCS Study BOEM 2013. New Orleans, LA: US Department of the Interior, Bureau of Ocean Energy Management, Gulf of Mexico OCS Region.

Sammarco PW, Andrews JC. 1988. Localized dispersal and recruitment in Great Barrier Reef corals: The Helix experiment. Science. 239:1422–1424.

Sammarco PW, Andrews JC. 1989. The Helix experiment: Differential localized dispersal and recruitment patterns in Great Barrier Reef corals. Limnol Oceanogr. 34:896–912.

Sammarco PW, Atchison AD, Boland GS. 2004. Expansion of coral communities within the northern Gulf of Mexico via offshore oil and gas platforms. Mar Ecol Prog Ser. 280:129–143.

Sammarco PW, Atchison AD, Boland GS, Sinclair J, Lirette A. 2012a. Geographic expansion of hermatypic and ahermatypic corals in the Gulf of Mexico, and implications for dispersal and recruitment. J Exp Mar Biol Ecol. 436–437:36–49. Available from: http://dx.doi.org/10.1016/j.jembe.2012.08.009 via the Internet. Accessed 20 November, 2013.

Sammarco PW, Brazeau DA, Sinclair J. 2012b. Genetic connectivity in scleractinian corals across the northern Gulf of Mexico: Oil/gas platforms, and relationship to the Flower Garden Banks. PLoS One. 7(4):e30144. Available from: http://dx.doi.org/10.1371/journal.pone.0030144 via the Internet. Accessed 20 November, 2013.

Sammarco PW, Lirette A, Tung YF, Genazzio M, Sinclair J. In press. Coral community development on "rigs-to-reefs" vs. standing oil/gas platforms: Artificial reefs in the Gulf of Mexico. ICES J Mar Sci.

Scarborough A, Kendall JJ Jr. 1994. An indication of the process: Offshore platforms as artificial reefs in the Gulf of Mexico. Bull Mar Sci. 55:1086–1098.

Schmahl GP, Hickerson EL. 2006. Ecosystem approaches to the identification and characterization of a network of reefs and banks in the Northwestern Gulf of Mexico. Eos Trans Am Geophys Union. 87(36):Suppl.

Schmahl GP, Hickerson EL, Weaver DC. 2003. Biodiversity associated with topographic features in the northwestern Gulf of Mexico. *In*: Proceedings of the 22nd Annual Gulf of Mexico Information Transfer Meeting, Gulf of Mexico OCS Region, Kenner, LA. New Orleans, LA: US Department of the Interior, Minerals Management Service. p. 84–88.

Schroeder DM, Ammann AJ, Harding JA, MacDonald LA, Golden WT. 2000. Relative habitat value of oil and gas production platforms and natural reefs to shallow water fish assemblages in the Santa Maria Basin and Santa Barbara Channel, California. *In:* Browne DR, Mitchell KL, Chaney HW, editors. Proceedings of the 5th California Islands Symposium. Camarillo, CA: US Department of the Interior, Minerals Management Service, Pacific OCS Region. p. 493–498.

Slobodkin LB. 1961. Growth and regulation of animal populations. New York: Holt, Rinehart, and Winston.

Soltis DE, Morris AB, MacLachlan JS, Manos PS, Soltis PS. 2006. Comparative phylogeography of unglaciated eastern North America. Mol Ecol. 15:1261–1293. Available from: http://dx.doi.org/10.1111/j.1365-294X.2006.03061.x via the Internet. Accessed 29 August, 2013.

Sonnier F, Teerling J, Hoese HD. 1976. Observations on the offshore reef and platform fish fauna of Louisiana. Copeia. 1976(1):105–111.

Stanley DR, Wilson CA. 1997. Seasonal and spatial variation in the abundance and size distribution of fishes associated with a petroleum platform in the Northern Gulf of Mexico. Can J Fish Aquat Sci. 54:1166–1176.

Sturges W. 1993. The annual cycle of the western boundary current in the Gulf of Mexico. J Geophys Res. 98:18053–18068.

Sturges W, Blaha JP. 1976. A western boundary current in the Gulf of Mexico. Science. 92:367–369.

Sturges W, Lugo-Fernandez A, editors. 2005. Circulation in the Gulf of Mexico: Observations and models. Geophysical monograph series. Volume 161. Washington, DC: American Geophysical Union.

Sunnucks P. 2000. Efficient genetic markers for population biology. Trends Ecol Evol. 15:199–203.

Syms C, Jones GP. 2000. Disturbance, habitat structure, and the dynamics of a coral-reef fish community. Ecology. 81:2714–2729. Available from: http://dx.doi.org/10.1890/0012-9658(2000)081[2714:DHSATD]2.0.CO;2 via the Internet. Accessed 21 August, 2013.

Thresher RE. 1983. Environmental correlates of the distribution of planktivorous fishes in the One Tree Reef lagoon. Mar Ecol Prog Ser. 10:137–145.

Tunnell JW, Chavez EA, Withers K, editors. 2007. Coral reefs of the southern Gulf of Mexico. Corpus Christi, TX: Texas A&M University Press.

US Geological Survey. 1995. Areas of historical oil and gas exploration and production in the conterminous United States. SuDoc I 19.76:95–75. Washington, DC: US Geological Survey.

Vos P, Hogers R, Bleeker M, Reijans M, van de Lee T, Hornes M, Frijters A, Pot J, Peleman J, Kuiper M, et al. 1995. AFLP: A new technique for DNA fingerprinting. Nucleic Acids Res. 23:4407–4414.

Wiley BA, Chapman RW. 2010. Population structure of spotted seatrout, *Cynoscion nebulosus*, along the Atlantic coast of the US. *In:* Bortone SA, editor. Biology of the spotted seatrout. Boca Raton, FL: CRC Press. p. 31–40.

Williamson EA, Strychar KB, Withers K, Sterba-Boatwright B. 2011. Effects of salinity and sedimentation on the gorgonian coral, *Leptogorgia virgulata* (Lamarck 1815). J Exp Mar Biol Ecol. 409:331–338. Available from: http://dx.doi.org/10.1016/j.jembe.2011.09.014 via the Internet. Accessed 21 August, 2013.

Wilson J, Osenberg CW, St. Mary CM, Watson CA, Lindberg WJ. 2001. Artificial reefs, the attraction-production issue, and density dependence in marine ornamental fishes. Aquar Sci Conserv. 3:95–105.

Wooten MC, Scribner KT, Smith MH. 1988. Genetic variability and systematics of *Gambusia* in the southeastern United States. Copeia. 1988(2):283–289.

chapter eight

A strategic approach to address fisheries impacts on deep-sea coral ecosystems

Thomas F. Hourigan

Contents

Introduction

New research over the last two decades has revealed complex habitats formed by corals and sponges in deeper waters around the world. While less well known than shallow-water tropical coral reefs, the three-dimensional features formed by many deep-sea corals and sponges provide habitat for numerous species and represent an important component of deep-sea ecosystems (Roberts et al. 2009, Hogg et al. 2010). The speed with which these new discoveries have been translated into conservation action, both in U.S. waters and internationally, has been unprecedented. The primary target of these conservation efforts has been the impacts of commercial fisheries.

Though far less diverse than fish communities of shallow-water tropical coral reefs (Roberts et al. 2009), fishes, including a number of commercially and recreationally harvested species, are commonly found in deep-sea coral and sponge habitats (e.g., Reed 2002, Costello et al. 2005, Stone 2006, Ross and Quattrini 2007). While certain fish species are reliably associated with coral habitats (Ross and Quattrini 2007), many other fish–coral associations appear to be facultative (Auster 2005). Certain deepwater fish and coral species may favor the same rocky habitats (Tissot et al. 2006). Seamounts are prime locations for many deep-sea corals, as well as features over which orange roughy (*Hoplostethus atlanticus*), oreos (family Oreosomatidae), and other fishes aggregate in large numbers and are targeted by trawl fisheries (Clark and Koslow 2007). The contribution of intact coral or sponge habitats to fishery production is generally unknown. In the United States, this has complicated identification of these habitats as essential fish habitat. Foley et al. (2010) recently modeled the effects of deep-sea coral habitat decline on commercial fish stocks of redfish (*Sebastes* spp.) off Norway. They concluded that both the carrying capacity and growth rate of redfish were functions of *Lophelia* reef habitat.

Shallow-water tropical coral reefs are subject to numerous anthropogenic threats, particularly overfishing and destructive fishing, coastal development and land-based sources of pollution, and coral bleaching associated with climate change (Burke et al. 2011). In contrast, deep-sea corals do not contain symbiotic zooxanthellae and therefore cannot bleach; their habitats are predominantly further offshore and minimally impacted by land-based activities. Both shallow-water and deep-sea coral reefs are vulnerable to ocean acidification, which is predicted to reduce coral's ability to produce calcium carbonate skeletons (Guinotte et al. 2006), but the options for managing these impacts are limited.

Managers do have the ability to address one major threat to deep-sea coral ecosystems—fishing. Because deep-sea corals are generally long-lived, slow growing, and fragile, they and their associated communities are vulnerable to impacts from activities that physically disturb the corals. The most widespread of these activities is fishing using bottom-tending fishing gears, that is, gears that contact the bottom (Freiwald et al. 2004, Roberts et al. 2009). Many commercial fisheries exploit species whose distribution extends to 400–500 m depths. Significant expansion of bottom fisheries below 500 m (occasionally extending as deep as 2000 m) has mostly occurred in the last four decades, and has a record of unsustainable harvests (Sissenwine and Mace 2007, Pitcher et al. 2010). Studies from around the world have reported severe damage to deep-sea coral communities from bottom trawling (e.g., Koslow et al. 2001, Krieger 2001, Fosså et al. 2002, Reed 2002, Grehan et al. 2005, Koenig et al. 2005, Stone 2006, Clark and Koslow 2007, Waller et al. 2007, Althaus et al. 2009, Clark and Rowden 2009, Du Preeze and Tunnicliffe 2011). Benn et al. (2010) estimated that the extent of bottom trawling on the deep seafloor in the northeast Atlantic was an order of magnitude greater than the total extent of all the other bottom-contact activities. In U.S. waters, bottom-trawl fisheries were identified as the most serious threat to deep-sea coral habitats in Alaska, the U.S. west coast, and northeast and southeast regions (Hourigan et al. 2007). Other bottom-tending gear, including traps, bottom-set longlines, and gillnets, can also damage deep-sea corals (Krieger 2001, Mortensen et al. 2005, Baer et al. 2010, Sampaio et al. 2012). These gears may be used preferentially in steep and rocky habitats (i.e., areas of high rugosity) that are inaccessible for trawling, thereby representing the primary fishing gear damaging corals in such areas. In certain cases, longline gear may sweep a large area of the bottom during retrieval, damaging benthic organisms (Welsford and Kilpatrick 2008). Compared to bottom-trawl gear, however, these fixed fishing gears weigh less (i.e., exert less force on the substrate), contact a smaller physical area of the seafloor, and in many regions are used for fishing over a smaller geographic area, so their overall contribution to coral injury is likely to be less (Stone and Shotwell 2007). Heifetz et al. (2009) surveyed deep-sea corals in the central Aleutian Islands and found that areas fished with bottom trawls contained a significantly higher proportion of damaged corals and sponges than did areas that were only fished with fixed gear (pots and longlines).

In this chapter, I provide a brief review of U.S. and international efforts to address fisheries impacts on deep-sea coral and sponge habitats with a focus on developing best practices. Also identified are areas where additional progress appears needed. In U.S. waters, these habitats occur predominantly in the exclusive economic zone (EEZ), where the U.S. National Oceanic and Atmospheric Administration (NOAA), in partnership with the regional Fishery Management Councils, is responsible for managing fisheries. I show how NOAA has incorporated best practices to address the impacts of fisheries in its 2010 Strategic Plan for Deep-Sea Coral and Sponge Ecosystems (NOAA 2010a). Finally, I review regional research priorities that have been identified to support implementation of deep-sea coral and sponge conservation under the strategic plan and how NOAA's Deep Sea Coral Research and Technology Program is providing the science needed to address these priorities.

Addressing fishing impacts on deep-sea coral habitats: The national and international context

A new age of exploration catalyzes international conservation efforts

In 1984, the Oculina Bank Habitat Area of Particular Concern off Florida became the world's first area established specifically to protect deepwater azooxanthellate coral habitat (Reed 2002). Its early protection was probably due, in part, to its occurrence in relatively shallow water (70–100 m) and to its use by large spawning aggregations of commercially valuable gag (*Mycteroperca microlepis*) and scamp (*M. phenax*) serranid groupers. New exploration and discoveries in the 1990s revealed the extent and biological richness of other deepwater coral habitats (e.g., *Lophelia pertusa* reefs in the northeast Atlantic and corals on seamounts in the Pacific), as well as evidence of damage from fishing gear. This resulted in protection from trawling in selected areas within the EEZs of numerous countries (reviewed by Davies et al. 2007, Roberts et al. 2009). The majority of these areas protected relatively discrete individual features.

An ecosystem approach

In contrast to these relatively small, targeted closures, in 1986, NOAA's Fisheries Service, on the recommendation of the Western Pacific Fishery Management Council, prohibited the use of bottom trawls and bottom-set nets in the EEZ of the U.S. Pacific Islands (51 FR 27413) (WestPac 1986). The avoidance of habitat degradation was one of the reasons given for this prohibition. The Western Pacific Fishery Management Council's Precious Coral Fishery Management Plan (WestPac 1979) previously identified bottom trawling by foreign fishing fleets as having a major potential adverse impact on precious coral beds.

Further deep-sea benthic conservation on this scale was not seen again until 2001, when New Zealand closed 19 seamounts to all bottom trawling and dredging (Brodie and Clark 2004), and in 2007 complemented these protections with the designation of 17 large Benthic Protected Areas (Fisheries [Benthic Protection Areas] Regulations 2007, Spear and Cannon 2012, Rieser et al. 2013). Bottom trawling was banned in the Benthic Protected Areas, and other commercial fishing operations were monitored by independent observers. Together these measures protect 32% of the New Zealand EEZ from bottom trawling, including 52% of large seamounts in New Zealand waters.

In a similar manner, Australia expanded the 1999 Tasmanian Seamounts Marine Reserve, established to a large extent in response to documented impacts on coral habitat by the orange roughy trawl fishery (Koslow et al. 2001), to create a regionwide South-east Commonwealth Marine Reserve Network in 2007 (Director of National Parks 2013). This is now part of an Australian Commonwealth marine reserve system, which, as of 2012, included a total area of 3.1 million km^2 managed primarily for biodiversity conservation, including extensive deepwater habitats (Department of the Environment 2013). These measures sought to protect benthic habitats from fishing impacts on a scale commensurate with an ecosystem approach to management.

In the United States, essential fish habitat provisions of the Magnuson-Stevens Fisheries Conservation and Management Act (MSA) underlie the Pacific and North Pacific Fishery Management Councils decisions to develop regionwide approaches to protect benthic biogenic habitats, including deep-sea corals, from the adverse impacts of fishing gears. Extensive deepwater areas were afforded protections in 2007 (reviewed by Lumsden et al. 2007, Hourigan 2009), including known coral habitats (e.g., Aleutian "coral gardens" composed of dense assemblages of stylasterid corals, octocorals, and sponges). In 2010, NOAA and the South Atlantic Fishery Management Council established five deepwater Coral Habitat Areas of Particular Concern, including the largest marine protected area along the U.S. east coast, protecting the most extensive and best-developed *Lophelia* reef habitats in U.S. waters (SAFMC 2010).

Central to these conservation measures were the following approaches (Hourigan 2009):

- Area-based management measures addressed multiple deepwater biogenic habitats (e.g., corals, sponges, and bryozoans) over large geographic regions.
- Particularly vulnerable areas, especially seamounts and major identified deep-sea coral and sponge habitats, were protected from impacts by all bottom-tending gear.
- Where survey data were lacking, geomorphological "proxies" (e.g., canyons, banks, or mounds) that might be expected to be associated with vulnerable biogenic habitats were identified and protected from bottom trawling.
- The current "footprint" of bottom-trawl and dredge fisheries (i.e., the geographic area where most of the fishing using these mobile bottom-tending gears occurred) was defined in partnership with the fishing community, and expansion of these fisheries into deeper waters was prevented until they can be surveyed to identify potentially vulnerable habitats.
- Fisheries observers in Alaska and the west coast, and vessel monitoring systems in southeastern United States provided key information to inform adaptive management and enforcement.

Protecting vulnerable marine ecosystems on the high seas

Multilateral efforts to address the impacts of deep-sea fisheries in areas beyond national jurisdictions (i.e., the high seas) began in 2004 and led to a 2006 United Nations General Assembly (UNGA) Resolution (61/105) calling upon states and regional fishery management organizations and arrangements to protect vulnerable marine ecosystems from significant adverse impacts from deep-sea bottom fishing. UNGA Resolution 61/105 specifically identified "cold-water corals" as an example of a vulnerable marine ecosystem. Subsequent reviews by the UN Secretary General in 2009 and 2011 identified implementation progress, and additional resolutions were adopted in 2009 and 2011 (UNGA Resolutions 64/72 and 66/68). The United Nations Food and Agriculture Organization developed "International Guidelines for the Management of Deep-Sea Fisheries in the High Seas" (FAO 2009) to help guide implementation of UNGA Resolution 61/105.

Recent reviews (e.g., Rogers and Gianni 2010, Weaver et al. 2011) of progress in implementing UN General Assembly Resolutions 61/105 and 64/72, including efforts to prevent significant adverse impacts on vulnerable marine ecosystems, have highlighted both progress and continuing challenges. The Commission for the Conservation of Antarctic Marine Living Resources and most regional fishery management organizations have adopted area-based closures to certain bottom-tending fishing gears to protect vulnerable marine ecosystems, although the extent and type of closures varied. Several regional fishery management organizations have attempted to define the historical area where bottom-fishing has been conducted and "freeze the footprint" of such fishing as an interim measure until management measures are developed. Often, however, efforts to establish the historical fishing footprint have suffered from a lack of geographic specificity on where fishing occurred and lack of information on fishing effort (Penney 2010, Weaver et al. 2011).

In most cases, deep-sea corals and sponges were the primary benthic taxa used as indicators of vulnerable marine ecosystems, but the thresholds (e.g., amount of coral bycatch) that precipitated management action were often different among regional fishery management organizations. Auster et al. (2010) and others have identified concerns with relying on "move-on rules" to respond to evidence of unforeseen encounters with vulnerable marine ecosystems while fishing, and recommended stronger *a priori* impact assessments, along with closed areas, gear restrictions, and fishing effort controls to prevent significant adverse impacts on such ecosystems. A number of workshops (e.g., FAO 2011, Weaver et al. 2011) and authors (Davies and Guinotte 2011, Ross and Howell 2013, Vierod et al. 2014) have highlighted the value of predictive models that can help identify potential vulnerable marine ecosystems. Ardron et al. (in press) have synthesized many of these recommendations and best practices into a step-by-step approach for identifying vulnerable marine ecosystems for protection.

NOAA's strategic approach

In 2010, NOAA released the "Strategic plan for deep-sea coral and sponge ecosystems: research, management, and international cooperation" (NOAA 2010a). The strategic plan identifies goals, objectives, and approaches for NOAA to improve the understanding, conservation, and management of deep-sea coral and sponge ecosystems. Managing fishing threats to these ecosystems is a major focus.

For the purposes of the strategic plan, NOAA defined the term "deep-sea corals" as azooxanthellate corals (i.e., corals that do not depend upon symbiotic algae and light for energy) generally occurring at depths below 50 m. Of particular ecological importance and conservation concern are "structure-forming deep-sea corals," those colonial deep-sea coral species that provide vertical structure above the seafloor that can be utilized by other species and are most likely to be damaged by interactions with fishing gear (Lumsden et al. 2007). Structure-forming deep-sea corals include both branching stony corals that form a structural framework (e.g., *L. pertusa*) as well as individual colonies of corals, such as gorgonians and other octocorals, black corals, gold corals, and lace corals. The NOAA strategic plan also recognized that other biogenic habitats, particularly deep-sea sponge habitats, can play similar ecological roles and face similar threats as deep-sea coral habitats (Hogg et al. 2010).

Three of the NOAA strategic plan's six objectives under the conservation and management goal address the adverse impacts of fishing gear on deep-sea coral and sponge habitats (Table 8.1). The objectives were developed with public participation and in consultation with the Fishery Management Councils, and build directly on the experience of the North Pacific, Pacific, and South Atlantic Fishery Management Councils and evolving international best practices. Conservation relies on long-term, year-round closures to damaging fishing gears as a principal management measure.

While recent U.S. efforts to protect benthic habitats—including deep-sea corals—from fishing impacts have generally focused on requirements to protect essential fish habitat (MSA Sec. 305(b)), the strategic plan integrates all applicable legal authorities provided to NOAA through the Magnuson-Stevens Act and the National Marine Sanctuaries Act. In addition to essential fish habitat authorities, National Standard 9 (MSA Sec. 301(a)(9)) requires that conservation and management measures minimize bycatch to the extent practicable. Deep-sea corals and sponges that are harvested in a fishery but not sold or kept for personal use are considered bycatch. To date, only the North Pacific Fishery Management Council's 2007 habitat closures explicitly cited this authority.

The 2007 reauthorization of the Magnuson-Stevens Act recognized the importance of deep-sea coral habitats, authorizing NOAA to establish a Deep Sea Coral Research and Technology Program (MSA Sec. 408) and providing new discretionary authority to protect deep-sea coral habitats in their own right (MSA Sec. 303(b)(2)). NOAA has encouraged use of this discretionary authority to advance the agency's and Fishery Management Councils' conservation objectives. The New England and Mid-Atlantic Fishery Management Councils are currently exploring its use to protect deep-sea coral areas in U.S. northeastern waters.

The overall decision process for managing bottom-tending gear impacts is summarized in Figure 8.1 and described below, along with selected issues that have been raised in implementing the fishing objectives.

Table 8.1 Summary of Conservation and Management Objectives from NOAA's Strategic Plan for Deep-Sea Coral and Sponge Ecosystems

Goal	Objectives
Conservation and management	1. Protect areas containing known deep-sea coral or sponge communities from impacts of bottom-tending fishing gear. 2. Protect areas that may support deep-sea coral and sponge communities where mobile bottom-tending fishing gear has not been used recently, as a precautionary measure. 3. Develop regional approaches to further reduce interactions between fishing gear and deep-sea corals and sponges. 4. Enhance conservation of deep-sea coral and sponge ecosystems in National Marine Sanctuaries and Marine National Monuments. 5. Assess and encourage avoidance or mitigation of adverse impacts of nonfishing activities on deep-sea coral and sponge ecosystems. 6. Provide outreach and coordinated communications to enhance public understanding of these ecosystems.

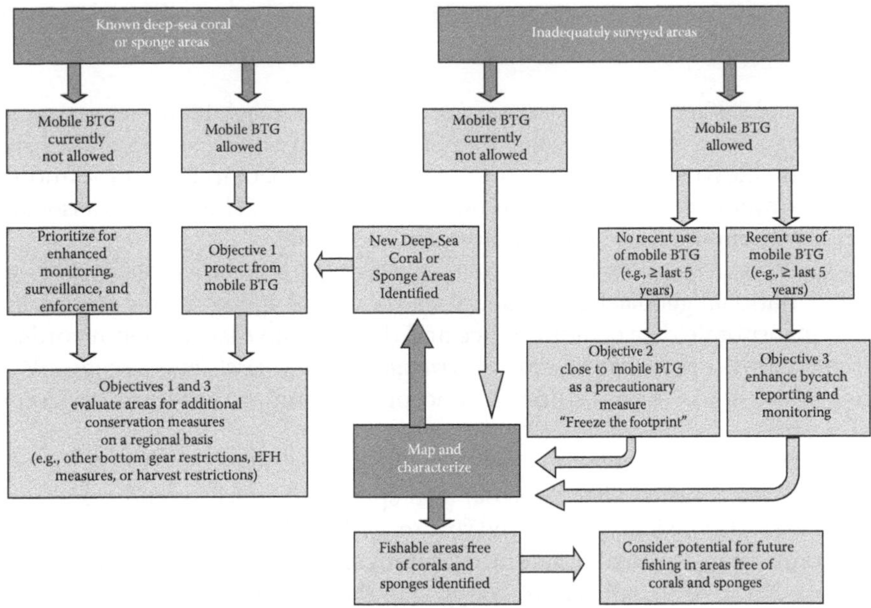

Figure 8.1 NOAA's precautionary approach to manage bottom-tending gear, especially mobile bottom-tending gear and other adverse impacts of fishing on deep-sea coral and sponge ecosystems, as described in NOAA's Strategic Plan for Deep-Sea Coral and Sponge Ecosystems. BTG, bottom-tending gear; EFH, essential fish habitat.

Objective 1. Protect areas containing known deep-sea coral or sponge communities from impacts by bottom-tending gear

This objective recognizes the intrinsic value of biogenic habitats and promotes their protection from gear impacts, particularly through spatial closures to bottom-tending gear. Where deep-sea coral and sponge habitats are identified, NOAA will work with the Fishery Management Councils to monitor fisheries and ensure that these habitats are adequately protected from impacts by fishing gear.

Key implementation issues

While there is growing consensus that important areas of deep-sea coral habitat should be protected (e.g., Freiwald et al. 2004, Roberts et al. 2009), there is less consensus on when the coral fauna in a particular area should be considered a conservation priority. Nearly everyone would recognize the need to protect an extensive, deepwater coral reef, such as *L. pertusa* reefs (Norway, Gulf of Mexico) that represent thousands of years of coral growth, or the Aleutian Island "coral gardens" with their dense coral and sponge aggregations teeming with fish (Stone 2006, Stone and Shotwell 2007). However, corals are a diverse and widespread group of organisms, varying in size, density, and habitat. While most species are probably vulnerable to physical disturbance, there is no intention that every area with deep-sea corals should be closed to bottom-tending, or even bottom-trawl, gear.

NOAA's strategic plan does not identify national "rules" for when an area should be considered a priority for protective measures. Areas should be identified on a case-by-case basis, recognizing that different regions have different characteristic fauna, fisheries, and needs. For deep-sea corals, factors that should be considered in these decisions include the following:

- *Characteristics of the deep-sea coral species*: Species that reach larger sizes and are more robust may contribute to greater habitat complexity. For example, Du Preeze and Tunnicliffe (2011) surveyed sites off northern British Columbia and found that over half of primnoid corals over 30 cm tall had rockfishes (*Sebastes* spp.) associated with them, while small corals had no associated rockfishes.
- *The size of the reef or coral aggregation, or density of structure-forming deep-sea corals*: NOAA's national geodatabase of deep-sea corals and sponges uses pioneering methods to incorporate data on abundance and density into distribution records.
- *The co-occurrence of multiple species of structure-forming corals (and sponges)*: While biological diversity is often a difficult concept to operationalize, this may represent a useful proxy.
- *The occurrence of rare species or species with limited distributions*: NOAA has identified the ivory tree coral (*Oculina varicosa*) as a species of concern, and other vulnerable species may also be identified in the future. In 2012, NOAA received a petition to list 43 Alaskan deep-sea corals under the Endangered Species Act (Center for Biological Diversity 2012). Although NOAA found that the petition did not present substantial scientific or commercial information indicating that a listing was warranted (Federal Register 2013: 78 FR 10601), threats to these and other species will continue to be reviewed.
- *The importance of the ecological function provided by the deep-sea corals as habitat*: For example, framework-forming colonial corals, such as *L. pertusa*, are likely to provide habitat for other species. Moreover, these corals are more vulnerable to physical damage from fishing gear, compared to noncolonial stony coral species. In certain cases, the functional role of certain taxa may be poorly understood. Pennatulacean octocorals (sea pens and their relatives) provide a particularly challenging case. Unlike most other corals, which are predominantly associated with hard substrata, sea pens generally occur in soft sediments, where they can form dense aggregations. Such habitats are also well suited for bottom trawling, and so interactions with fisheries are common (e.g., NMFS 2013). The potential role of sea pen aggregations as habitat for other species has also been equivocal. Brodeur (2001) observed schools of Pacific Ocean perch rockfish (*Sebastes alutus*) associated with the dense aggregations of a pennatulacean (*Halipteris willemoesi*) in the Bering Sea, but few other associations with managed species have been documented. Baillon et al. (2012) reported that several species of sea pens appeared to provide shelter for redfish larvae (*Sebastes* spp.). NOAA is beginning to evaluate the potential contribution of sea pens as habitat for other species.
- *The extent to which the area is sensitive to human-induced environmental degradation*: The degree of threat from fishing impacts, either alone or in combination with other human activities, is an important consideration in deciding on management measures and their potential conservation effectiveness.
- *The "intactness" of a particular habitat*: The more intact or pristine the habitat, the more likely it will retain its full suite of natural ecological functions. Such habitats may therefore be a higher priority for protection.
- *The likelihood of occurrence of deep-sea corals in unsurveyed areas based on the results of coral habitat suitability models or similar methods*: Since few deep-sea areas have been adequately surveyed, models or other proxies may be needed for conservation decisions.

These criteria, while very general, have significant overlap with criteria developed internationally to identify "[e]cologically and biologically significant areas" in the deep sea (CBD 2008) and vulnerable marine ecosystems (FAO 2009). Both these international criteria identify "cold-water corals" as examples of important, vulnerable taxa. Many of the steps recommended for the identification of vulnerable marine ecosystems by Ardron et al. (in press) are also applicable to identifying and prioritizing areas for protection from fishing impacts within the EEZ.

Objective 2. As a precautionary measure, protect areas that may support deep-sea coral and sponge communities where mobile bottom-tending fishing gear has not been used recently

The second objective aims to limit fishing efforts that employ the most damaging fishing gears to only those areas where they have traditionally been used, to protect unfished areas likely to support deep-sea coral or sponge habitats. This approach is also known as "freezing the footprint" of the fishery. It recognizes that protecting known deep-sea coral and sponge habitats is not enough, since most of the seafloor within fishable depths has not been surveyed for vulnerable biogenic habitats. Untrawled or lightly trawled areas are more likely to contain undamaged biogenic habitats (Clark and Rowden 2009, Heifetz et al. 2009). This precautionary approach reverses the burden of proof required to allow fishing in these areas, limiting expansion of bottom-trawl or dredge fishing into new areas until research surveys identify habitats suitable for such gear.

The precautionary approach to "freeze the footprint" of bottom trawling to protect deep-sea corals and sponges was first advanced by nongovernmental organizations in the context of international and domestic conservation (Oceana 2004, Roberts and Hirshfield 2004, Shester and Ayers 2005). As noted above, it formed a central approach in interim measures applied by regional fishery management organizations and the Commission for the Conservation of Antarctic Marine Living Resources to protect vulnerable marine ecosystems. It was also a key component of habitat protection measures implemented by the North Pacific, Pacific, and South Atlantic Fishery Management Councils. Both the New England and Mid-Atlantic Fishery Management Councils are currently considering variations on a "freeze the footprint" of bottom trawling as one component of protections for deep-sea coral habitats.

Key implementation issues

Accurate, fine-scale information on fishing effort is critical in implementing a successful "freeze the footprint" approach (Weaver et al. 2011, Ardron et al. in press). Fishing information aggregated in large geographic blocks may not adequately reflect historic fishing effort, both overestimating the "footprint" and precluding determination of lightly fished areas that may still contain intact deep-sea coral or sponge communities (Clark and Rowden 2009, Rogers and Gianni 2010). In the Aleutian Islands, access to industry-provided tow-by-tow information allowed significant refinement of the existing trawl footprint. In contrast, on the U.S. west coast such fine-scale information was not available, and the precautionary closure westward of the 700 fathom isobath (1280 m) was significantly deeper than most of the historic trawl footprint (NMFS 2013). This highlights the importance of involving participants from the fishing community early in the planning efforts (Spear and Cannon 2012).

In the case of the South Atlantic Fishery Management Council, identification of the "footprint" of the deepwater shrimp fishery was facilitated by a fine-scale analysis of vessel monitoring system data (SAFMC 2010). Information on permitted fishing vessel location, bearing, speed, and distance from shore, especially if linked to observer data, can provide an independent indication of the fishing footprint, especially when analyzed in association with bathymetry and other environmental information.

Rieser et al. (2013) reviewed examples of benthic protection zones from catch-share fisheries in Alaska, New Zealand, and the southern Indian Ocean. They concluded that many areas identified by the fishing industry in a "freeze the footprint" approach were areas where the fishing industry was not interested in fishing. They recommend greater application of analytical methods based on fisheries-independent data and modeling to help determine the boundaries of bottom-trawl closures rather than relying solely on recommendations from the fishing industry.

Objective 3. Develop regional approaches to further reduce interactions between fishing gear and deep-sea corals and sponges

Objectives 1 and 2 identified NOAA's national approach to protect areas from impacts of fishing gear where there were known deep-sea coral or sponge habitats, or as a precautionary measure, where mobile bottom-tending gear had not yet been widely used. However, in many cases the most important management opportunities may lie in areas within the current fishing footprint, but where there is limited information on the deep-sea coral or sponge communities that occur there. NOAA's strategic plan encouraged each region to evaluate additional management options, ranging from additional area-based measures (e.g., closures or gear restrictions) to nonspatial measures (e.g., fishing effort reductions and modification of gear design or gear type). Objective 3 places particular emphasis on improving information on bycatch of corals and sponges and applying this information to management. Bycatch information gathered by independent fishery observers, where available, is likely to be most useful. Fishermen may be reluctant to self-report coral bycatch if they believe the information could be used to support potential closures to their fishing grounds.

Building on the comprehensive, essential fish habitat measures implemented through the North Pacific and Pacific Fishery Management Councils, as well as the South Atlantic Fishery Management Council's Comprehensive Ecosystem-based Amendment and deepwater Coral Habitat Areas of Particular Concern (SAFMC 2010), NOAA seeks to place deep-sea coral and sponge conservation in a regional ecosystem context. Regionally tailored management approaches also allow for regular review of protection measures and adaptive management. NOAA has encouraged regional Fishery Management Councils to incorporate new information on deep-sea coral and sponge habitats in required 5-year Essential Fish Habitat Reviews and to consider whether additional conservation and management measures are needed for their protection.

Key implementation issues

For certain fisheries, bycatch analysis may be of particular value in both identifying potential areas with vulnerable corals or sponges and understanding the impact of fisheries on these resources within the current fishing footprint (Shester and Ayers 2005, Kenchington et al. 2009). Bottom trawls or bottom-set longlines and gillnets often become entangled with corals and sponges (Stone and Shotwell 2007) during fishing in habitats where these taxa occur. These gears are poor sampling tools for most sessile benthic organisms, and the weight, volume, or abundance of bycatch of corals or sponges landed on deck is likely a very imperfect indication of the benthic community (Heifetz et al. 2009, Auster et al. 2010).

However, fishing vessels reach many more areas than can be sampled by fishery-independent means, and bycatch may provide the only information available for these areas. In addition, the Magnuson-Stevens Act requires that bycatch of corals or sponges be minimized to the extent practicable—providing a direct link between observed bycatch and required management action.

Practical use of bycatch information for deep-sea coral and sponge management in the United States has been hampered by the lack of independent observers on many fishing vessels, and when there are observers, lack of training in how to distinguish taxa. Experience in New Zealand (e.g., Tracey et al. 2011) indicates that proper training of fishery observers can result in accurate identification of many deep-sea coral taxa, and improved information on the distribution and abundance of these taxa, and help quantify the interactions of commercial fisheries with these taxa. NOAA's Deep Sea Coral Research and Technology Program is currently supporting the development of coral and sponge guides and training to enhance the capacity of fisheries observers to identify, quantify, and report deep-sea coral and sponge bycatch.

Science to support management: NOAA's Deep Sea Coral Research and Technology Program

NOAA's Deep Sea Coral Research and Technology Program was established by the 2007 Magnuson-Stevens Act reauthorization (Sec. 408) to identify and map the locations of deep-sea corals, monitor activity in locations where deep-sea corals are known or likely to occur, and submit information to the councils. The program integrates expertise and resources available across NOAA to provide scientific information needed to conserve and manage deep sea coral ecosystems.

The Deep Sea Coral Research and Technology Program is distinguished by its focus on developing spatially explicit information directly relevant to management, with a primary focus on information needed to inform fisheries management. The program conducts 3-year regional mapping and research initiatives, developed in consultation with the regional Fishery Management Councils. Appropriations for the program began in 2009, and research initiatives have been conducted in the southeastern United States (South Atlantic Fishery Management Council Region, 2009–2011) and the west coast (Pacific Fishery Management Council Region, 2010–2012), and initiated in Alaska (North Pacific Fishery Management Council Region, 2012–2014) and northeastern United States (New England and Mid-Atlantic Fishery Management Council Regions, 2013–2015). Research in the U.S. Pacific Islands is planned for 2015–2017.

These 3-year field efforts are complemented by smaller-scale projects in most U.S. regions to analyze existing information that can inform management. Projects to date have included analyzing and mapping the distribution and intensity of bottom-fishing activities, "mining" older research for the locations of deep-sea corals and sponges, and developing genetic techniques to improve identification of coral and sponge bycatch in fisheries, among others (NOAA 2012). These large and small efforts are brought together by a national capability to manage and disseminate data and information, including the first national geodatabase of deep-sea coral and sponge locations. The overall vision is for an integrated program where analysis of existing information helps target future field research, effective research and mapping protocols developed in one region are applied in subsequent regions, and the combined results are used by managers to advance deep-sea conservation.

Regional research priorities

Before beginning field research in a region, the program has held workshops to identify information needs and management-targeted research priorities. The workshops bring together academic scientists, managers, and other federal and state partners. The top priorities identified in each of these regions are shown in Table 8.2. While each workshop included rich discussion of local needs and approaches, the overall priorities to inform management are remarkably consistent across different regions.

1. *Map and characterize habitats*: Mapping the location of deep-sea coral habitats is invariably a top priority. In every region, past surveys have explored only a tiny portion of the potentially suitable habitat for deep-sea corals. Multibeam sonar mapping is key for nearly all work. While bathymetry and backscatter measured using multibeam sonar may reveal mounds associated with major deep-sea stony coral reefs (e.g., *Lophelia* bioherms in the southeastern United States), locating most other coral habitats and all further characterization require additional photo or video surveys.
2. *Habitat associations*: In most regions, identifying and understanding the association of other species, particularly of managed fisheries species that can inform essential fish habitat determinations, with coral and sponge habitats was the second highest research priority.
3. *Habitat suitability modeling*: Developing and validating habitat suitability models for coral and sponge distributions was a priority, both to identify conservation targets and to better focus future research surveys. The program has supported the development of regionwide models in each region, as well as finer-scale models at subregional scales. Benthic surveys are being conducted using towed cameras and remotely operated vehicles to validate and further refine these models.
4. *Biology of biogenic habitat species*: Understanding the life history and ecology of deep-sea corals and sponges was also a recurrent theme. Key aspects that can inform the potential for the recovery of damaged habitats include: age and growth rates, reproductive modes, trophic dynamics, environmental tolerances, and population connectivity.
5. *Anthropogenic impacts*: Finally, in each region, documenting the geographic extent of bottom-contact fishing effort, the extent and severity of damage by different fishing gears, and measures of recovery, if any, were identified as a priority.

Translating science to management

In its first 4 years, NOAA's Deep Sea Coral Research and Technology Program has contributed to significant management efforts in both the Atlantic and Pacific (NOAA 2012):

- Field research in the southeast (2009–2011) helped the South Atlantic Fishery Management Council delineate allowable fishing zones within the new deepwater Coral Habitat Areas of Particular Concern. It also discovered previously unknown deepwater coral habitats formed by *O. varicosa* and *L. pertusa* outside currently protected areas (NOAA 2012). The council has proposed expanding the boundaries of the Oculina Bank Habitat Area of Particular Concern and two of the deepwater Coral Habitat Areas of Particular Concern (SAFMC 2013).

Table 8.2 Research Needs To Inform Management of Deep-Sea Coral and Sponge Ecosystems

Rank	Southeast United States 2009	U.S. west coast 2010	Alaska 2010	Northeast United States[a] 2011
1	*Map*: Complete multibeam sonar surveys for proposed deepwater coral MPAs	*Map and characterize*: Identify species distribution, abundance, densities, and diversity throughout the California Current large marine ecosystem	*Map and characterize*: Mine existing information to expand understanding of deep-sea coral and sponge distribution; conduct new mapping	*Map, model and characterize*: Evaluate existing information; develop predictive models; conduct new mapping
2	*Habitat associations*: Describe types, strengths, and patterns of habitat associations of other species with deep-sea corals and sponges	*Habitat associations*: Determine the ecological roles of deep-sea corals and sponges, that is, the nature of associations between deep-sea corals and sponges and other species (invertebrates and fishes)	*Habitat associations*: Determine the functions corals and sponges provide for managed species at different life stages	*Biology*: Determine biology, biodiversity, and ecology of deep-sea corals and sponges
3	*Map and characterize*: Explore and characterize sites of interest identified within the coral HAPCs after completing multibeam sonar surveys	*Biology*: Understand the basic biology of deep-sea corals, including taxonomy, age structure, growth, reproduction, population connectivity, and life histories	*Habitat suitability modeling*: Construct predictive models, test and refine	*Anthropogenic impacts*: Determine natural and human impacts on deep-sea coral and sponge ecosystems; understanding disturbance effects and recovery; economic and social values associated with deep-sea corals
4	*Anthropogenic impacts*: Identify and assess impact on areas from fishing and nonfishing activities	*Habitat suitability modeling*: Understand abiotic and biotic requirements and suitability; geologic context and drivers of deep-sea coral habitat; variables to be used as habitat proxies; and water column physical, chemical, and biological parameters of deep-sea coral habitats	*Biology*: Determine the population characteristics (demographics) of deep-sea corals and sponges, including growth rates, life history and reproductive traits, trophic dynamics, environmental tolerances, and connectivity	

(*continued*)

Table 8.2 (Continued) Research Needs To Inform Management of Deep-Sea Coral and Sponge Ecosystems

Rank	Southeast United States 2009	U.S. west coast 2010	Alaska 2010	Northeast United States[a] 2011
5	*Habitat suitability modeling:* Develop modeling tools to predict where corals are likely to occur	*Anthropogenic impacts:* Understand anthropogenic and natural impacts on deep-sea coral ecosystems; develop baseline conditions (indices of health and condition at both species and community level)	*Anthropogenic impacts:* Understand fishing impacts, bycatch, and closed areas	

Notes: Priorities were developed and ranked by participants at regional workshops organized by NOAA's Deep Sea Coral Research and Technology Program. (From NOAA deep-sea coral and sponge ecosystems exploration and research priorities workshop for the U.S. South Atlantic, Wilmington, NC, NOAA, p. 16, 2010b; NOAA deep-sea coral and sponge ecosystems explora-tion and research priorities workshop for the U.S. West Coast, Portland, OR, NOAA, p. 18, 2010c; NOAA deep-sea coral and sponge ecosystems exploration and research priorities workshop for Alaska Anchorage, AK, NOAA, p. 23, 2011; NOAA deep-sea coral and sponge ecosystems exploration and research priorities workshop for the Northeast U.S. Highlands, NJ, NOAA, p. 16, 2012a.).

HAPC, Habitat Area of Particular Concern; MPA, marine protected area.

[a] Priorities for the northeast region were not ranked.

- Extensive new information and research on coral and sponge habitats, associated fishes, and bycatch off the U.S. west coast are being used in the Pacific Fishery Management Council's 5-Year Groundfish Essential Fish Habitat Review (GEFHRC 2012, NMFS 2013) and management proposals.
- Program-funded surveys are being used to evaluate potential expansion of National Marine Sanctuary boundaries off the west coast and in the Gulf of Mexico.
- Information from the national geodatabase and predictive deep-sea coral modeling are being used by the New England and Mid-Atlantic Fishery Management Councils in essential fish habitat and deep-sea coral management proposals, and by the Mid-Atlantic Regional Council on the Ocean as input for regional marine spatial planning.

Conclusions

Both deep-sea corals and their tropical shallow-water counterparts form biogenic habitats that are hotspots of biological diversity in the sea and whose future is integrally linked to fisheries. Shallow-water coral reefs support subsistence, commercial, and recreational fisheries that provide food for millions of people. As described by other authors in this volume, overfishing has been a major driver of tropical coral reef decline, interacting synergistically with local nonfishing impacts and bleaching associated with global climate change. With some exceptions, deep-sea coral habitats have not been directly targeted for their fish resources but have nevertheless suffered unintended, but extensive, collateral damage from commercial fisheries using bottom trawls and other bottom-contact fishing gear. It is likely that many deep-sea coral habitats on the continental shelves have been irreparably damaged where bottom-trawl fisheries are active. Intact deeper habitats provide scope for collaborative efforts that will enhance the sustainability of bottom fisheries and the conservation of deep-sea coral ecosystems.

In the United States, area-based conservation approaches established in NOAA's Strategic Plan for Deep-Sea Coral and Sponge Ecosystems are being initiated by regional Fishery Management Councils to protect deep-sea coral and sponge habitats from fishing impacts. Stakeholders from both the commercial fisheries and conservation communities have been actively involved in developing management alternatives. These efforts are supported by new information and products from mapping, surveys, predictive modeling, and research by NOAA's Deep Sea Coral Research and Technology Program. Required 5-year reviews of essential fish habitat and designations of Habitat Areas of Particular Concern offer an opportunity for NOAA and regional Fishery Management Councils to apply this new information in an adaptive management framework.

Implementation is at an early stage, but the number of fisheries impacted by new management measures is expected to be small. Without knowing the extent to which harvested fish stocks benefit from intact deep-sea coral habitats (Auster 2005), it is difficult to predict whether these conservation efforts will result in net benefits or costs to the affected fisheries. At a minimum, the unmanaged expansion of destructive fishing techniques into the deep sea observed elsewhere around the world (Pitcher et al. 2010) and associated damage to deep-sea coral ecosystems will be limited in U.S. waters.

References

Althaus F, Williams A, Schlacher TA, Kloser RJ, Green MA, Barker BA, Bax NJ, Brodie P, Schlacher-Hoenlinger MA. 2009. Impacts of bottom trawling on deep-coral ecosystems of seamounts are long-lasting. Mar Ecol Prog Ser. 397:279–294.

Ardron JA, Clark MR, Penney AJ, Hourigan TF, Rowden AA, Dunstan PK, Watling L, Shank TM, Tracey DM, Dunn MR, et al. (in press). A systematic approach towards the identification and protection of vulnerable marine ecosystems. Mar Policy.

Auster PJ. 2005. Are deep-water corals important habitats for fishes? *In*: Freiwald A, Roberts JM, editors. Cold-water corals and ecosystems. Berlin: Springer-Verlag. p. 747–760.

Auster PJ, Gjerde K, Heupel E, Watling L, Grehan A, Rogers AD. 2010. Definition and detection of vulnerable marine ecosystems on the high seas: Problems with the "move-on" rule. ICES J Mar Sci. 68:254–264.

Baer A, Donaldson A, Carolsfeld J. 2010. Impacts of longline and gillnet fisheries on aquatic biodiversity and vulnerable marine ecosystems. DFO Can Sci Advis Sec Res Doc. 2010/012 vii + 78.

Baillon S, Hamel J-F, Wareham VE, Mercier A. 2012. Deep cold-water corals as nurseries for fish larvae. Front Ecol Environ. 10(7):351–356.

Benn AR, Weaver PPE, Billett DSM, van den Hove S, Murdock AP, Doneghan GB, Le Bas T. 2010. Human activities on the deep seafloor in the north east Atlantic: An assessment of spatial extent. PLoS ONE 5(9):e12730.

Brodeur RD. 2001. Habitat-specific distribution of Pacific ocean perch (*Sebastes alutus*) in Pribilof Canyon, Bering Sea. Cont Shelf Res. 21:207–224.

Brodie S, Clark MR. 2004. The New Zealand seamount management strategy: Steps towards conserving offshore marine habitat. *In*: Beumer JP, Grant A, Smith DC, editors. Aquatic protected areas: What works best and how do we know? Proceedings of the World Congress on Aquatic Protected Areas, Cairns, 2002. Australia: Australian Society for Fish Biology. p. 664–673.

Burke L, Reytar K, Spaulding M, Perry A. 2011. Reefs at risk revisited. Washington, DC: World Resources Institute.

Center for Biological Diversity. 2012. Petition to list 43 coral species under the Endangered Species Act [Internet]. Available from: http://www.biologicaldiversity.org/campaigns/alaska_corals/pdfs/43AlaskaCoralsESAPetition-08-20-2012.pdf. Accessed September, 2013.

Clark MR, Koslow JA. 2007. Impacts of fishing on seamounts. *In*: Pitcher TJ, Hart PJB, Morato T, Santos R, Clark M, editors. Seamounts: Ecology, fisheries and conservation. Blackwell Fisheries and Aquatic Resources Series 12. Oxford: Blackwell Scientific. p. 413–441.

Clark MR, Rowden AA. 2009. Effect of deepwater trawling on the macro-invertebrate assemblages of seamounts on the Chatham Rise, New Zealand. Deep Sea Res I. 56:1540–1544.

CBD (Convention on Biological Diversity). 2008. Decision IX/20 annex 2 [Internet]. Available from: http://www.cbd.int/decision/cop/default.shtml?id=11663. Accessed July, 2013.

Costello MJ, McCrea M, Freiwald A, Lundälv T, Jonsson L, Bett BJ, van Weering TCE, de Haas H, Roberts JM, Allen D, et al. 2005. Role of cold-water *Lophelia pertusa* coral reefs as fish habitat in the NE Atlantic. *In*: Friewald A, Roberts JM, editors. Cold-water corals and ecosystems. Berlin: Springer-Verlag. p. 771–805.

Davies AJ, Guinotte JM. 2011. Global habitat suitability for framework-forming cold-water corals. PLoS ONE 6(4):e18483.

Davies AJ, Roberts MJ, Hall-Spencer J. 2007. Preserving deep-sea natural heritage: Emerging issues in offshore conservation and management. Biol Conserv. 138:299–312.

Department of the Environment. 2013. Australian Government, Commonwealth marine reserves [Internet]. Available from: http://www.environment.gov.au/marinereserves/. Accessed September, 2013.

Director of National Parks. 2013. South-east Commonwealth Marine Reserves Network management plan 2013–2023. Canberra: Director of National Parks.

Du Preeze C, Tunnicliffe V. 2011. Shortspine thornyhead and rockfish (Scorpaenidae) distribution in response to substratum, biogenic structures and trawling. Mar Ecol Prog Ser. 425:217–231.

FAO (Food and Agricultural Organization). 2009. International guidelines for the management of deep-sea fisheries in the high seas. §42 [Internet]. Available from: http://www.fao.org/docrep/011/i0816t/i0816t00.htm. Accessed June, 2013.

FAO (Food and Agricultural Organization). 2011. Report of the FAO workshop on the implementation of the international guidelines for the management of deep-sea fisheries in the high seas: Challenges and ways forward, Busan, Republic of Korea, 10–12 May 2010. FAO Fisheries and Aquaculture Report. No. 948. Rome: FAO.

Fisheries (Benthic Protection Areas) Regulations [Internet]. 2007. New Zealand Legislation. Available from: http://www.legislation.govt.nz/regulation/public/2007/0308/latest/whole.html#DLM973968. Accessed September, 2013.

Foley N, Kahui V, Armstrong CW, van Rensburg TM. 2010. Estimating linkages between redfish and cold water coral on the Norwegian coast. Mar Res Econ. 25:105–120.

Fosså JH, Mortensen PB, Furevik DM. 2002. The deep-water coral *Lophelia pertusa* in Norwegian waters: Distribution and fishery impacts. Hydrobiologia. 471:1–12.

Freiwald A, Fosså JH, Grehan A, Koslow T, Roberts JM. 2004. Cold-water coral reefs. Cambridge, UK: UNEP-WCMC.

GEFHRC (Groundfish Essential Fish Habitat Review Committee). 2012. Pacific Coast groundfish 5-year review of essential fish habitat report to the Pacific Fishery Management Council. Phase 1: New information. September 2012.

Grehan AJ, Unnithan V, Olu-Le Roy K, Opderbecke J. 2005. Fishing impacts on Irish deep-water coral reefs: Making a case for coral conservation. *In*: Thomas J, Barnes P, editors. Benthic habitats and the effects of fishing. Bethesda, MD: American Fisheries Society. p. 819–832.

Guinotte JM, Cairns S, Freiwald A, Morgan L, George R. 2006. Will human-induced changes in seawater chemistry alter the distribution of deep-sea scleractinian corals? Front Ecol Environ. 4:141–146.

Heifetz J, Stone RP, Shotwell SK. 2009. Damage and disturbance to coral and sponge habitat of the Aleutian Archipelago. Mar Ecol Prog Ser. 397:295–303.

Hogg MM, Tendal OS, Conway KW, Pomponi SA, Van Soest RWM, Gutt J, Krautter M, Roberts JM. 2010. Deep-sea sponge grounds: Reservoirs of biodiversity. Biodiversity Series No. 32. Cambridge, UK: UNEP-WCMC.

Hourigan TF. 2009. Managing fishery impacts on deep-water coral ecosystems of the United States: Emerging best practices. Mar Ecol Prog Ser. 397:333–340.

Hourigan TF, Lumsden SE, Dorr G, Bruckner AW, Brooke S, Stone RP. 2007. Deep coral ecosystems of the United States: Introduction and national overview. *In*: Lumsden SE, Hourigan TF, Bruckner AW, Dorr G, editors. The state of deep coral ecosystems of the United States. NOAA Technical Memorandum CRCP-3. Silver Spring, MD: NOAA. p. 1–64.

Kenchington E, Cogswell A, Lirette C, Murillo-Pérez FJ. 2009. The use of density analyses to delineate sponge grounds and other benthic VMEs from trawl survey data. Serial No. N5626. NAFO SCR Doc. 09/6. Dartmouth, Nova Scotia: Northwest Atlantic Fisheries Organization.

Koenig CC, Shephard AN, Reed JK, Coleman FC, Brooke SD, Brusher J, Scanlon K. 2005. Habitat and fish populations in the deep-sea *Oculina* coral ecosystem of the western Atlantic. *In*: Thomas J, Barnes P, editors. Benthic habitats and the effects of fishing. Bethesda, MD: American Fisheries Society. p. 795–805.

Koslow JA, Gowlett-Holmes K, Lowry JK, O'Hara T, Poore GCB, Williams A. 2001. Seamount benthic macrofauna off southern Tasmania: Community structure and impacts of trawling. Mar Ecol Prog Ser. 213:111–125.

Krieger KJ. 2001. Coral (*Primnoa*) impacted by fishing gear in the Gulf of Alaska. *In*: Willison JHM, Hall J, Gass SE, Kenchington ELR, Butler M, Doherty P, editors. Proceedings of the First International Symposium on Deep-Sea Corals. Halifax, NS: Ecology Action Centre, Dalhousie University and N.S. Museum. p. 106–116.

Lumsden SE, Hourigan TF, Bruckner AW, Dorr G, editors. 2007. The state of deep coral ecosystems of the United States. NOAA Technical Memorandum CRCP-3. Silver Spring, MD: NOAA.

Mortensen PB, Buhl-Mortensen L, Gordon DC Jr., Fader GBJ, McKeown DL, Fenton DG. 2005. Effects of fisheries on deep-water gorgonian corals in the Northeast Channel, Nova Scotia (Canada). Am Fish Soc Symp. 41:369–382.

NMFS (National Marine Fisheries Service). 2013. Groundfish essential fish habitat synthesis report. Seattle: Northwest Fisheries Science Center. April 2013.

NOAA (National Oceanic and Atmospheric Administration). 2010a. NOAA strategic plan for deep-sea coral and sponge ecosystems: Research, management, and international cooperation. NOAA Technical Memorandum CRCP-11. Silver Spring, MD: NOAA.

NOAA (National Oceanic and Atmospheric Administration). 2010b. NOAA deep-sea coral and sponge ecosystems exploration and research priorities workshop for the U.S. South Atlantic. Wilmington, NC: NOAA, p. 16; Available from: http://data.nodc.noaa.gov/coris/library/NOAA/CRCP/other/other_crcp_publications/south_atlantic_dsc_research_priority_wrkshp_wilmington.pdf. Accessed February 2014.

NOAA (National Oceanic and Atmospheric Administration). 2010c. NOAA deep-sea coral and sponge ecosystems exploration and research priorities workshop for the U.S. West Coast. Portland, OR: NOAA, p. 18; Available from: http://data.nodc.noaa.gov/coris/library/NOAA/CRCP/other/other_crcp_publications/west_coast_deep_sea_coral_workshop_rpt.pdf. Accessed February 2014.

NOAA (National Oceanic and Atmospheric Administration). 2011. NOAA deep-sea coral and sponge ecosystems exploration and research priorities workshop for Alaska Anchorage, AK: NOAA, p. 23; Available from: http://docs.lib.noaa.gov/noaa_documents/CoRIS/Alaska_Deep-Sea_Coral_Workshop_Report_Final.pdf. Accessed February 2014.

NOAA (National Oceanic and Atmospheric Administration). 2012a. NOAA deep-sea coral and sponge ecosystems exploration and research priorities workshop for the Northeast U.S. Highlands, NJ: NOAA, p. 16; Available from: http://docs.lib.noaa.gov/noaa_documents/CoRIS/Deep-Sea_Coral_Workshop_Aug-2011-final.pdf. Accessed February 2014.

NOAA (National Oceanic and Atmospheric Administration). 2012b. Deep sea coral research and technology program: 2012 Report to Congress. Silver Spring, MD: NOAA.

Oceana. 2004. Rulemaking petition to the U.S. secretary of commerce to protect deep-sea coral and sponge habitat. Washington, DC: Oceana.

Penney AJ. 2010. Mapping of high seas bottom fishing effort data: Purposes, problems and proposals. SPRFMO Scientific Working Group. SWG-10-DW-02 Rev1. Wellington, New Zealand: South Pacific Regional Fisheries Management Organisation.

Pitcher TJ, Clark MR, Morato T, Watson R. 2010. Seamount fisheries: Do they have a future? Oceanography. 23(1):134–144.

Reed JK. 2002. Deep-water Oculina coral reefs of Florida: Biology, impacts, and management. Hydrobiologia. 471:43–55.

Rieser A, Watling L, Guinotte JM. 2013. Trawl fisheries, catch shares and the protection of benthic marine ecosystems: Has ownership generated incentives for seafloor stewardship? Mar Policy. 40:75–83.

Roberts S, Hirshfield M. 2004. Deep-sea corals: Out of sight, but no longer out of mind. Front Ecol Environ. 2(3):123–130.

Roberts JM, Wheeler A, Freiwald A, Cairns S. 2009. Cold-water corals: The biology and geology of deep-sea coral habitats. Cambridge, UK: Cambridge University Press.

Rogers AD, Gianni M. 2010. The implementation of UNGA resolutions 61/105 and 64/72 in the management of deep-sea fisheries on the high seas. Report prepared for the Deep-Sea Conservation Coalition. London, UK: International Programme on the State of the Ocean.

Ross RE, Howell KL. 2013. Use of predictive habitat modelling to assess the distribution and extent of the current protection of "listed" deep-sea habitats. Divers Distrib. 19(4):433–445.

Ross SW, Quattrini AM. 2007. The fish fauna associated with deep coral banks off the southeastern United States. Deep-Sea Res I. 54:975–1007.

SAFMC (South Atlantic Fishery Management Council). 2010. Comprehensive Ecosystem-based Amendment 1. June 2010. North Charleston, SC: SAFMC.

SAFMC (South Atlantic Fishery Management Council). 2013. Coral Amendment 8 to the Coral, Coral Reef, and Live/Hardbottom Fishery Management Plan of the South Atlantic: Modifications to habitat areas of particular concern and transit provisions. August 2013 (approved by Council September 2013). North Charleston, SC: SAFMC.

Sampaio Í, Braga-Henriques A, Pham C, Ocaña O, de Matos V, Morato T, Porteiro F. 2012. Cold-water corals landed by bottom longline fisheries in the Azores (north-eastern Atlantic). J Mar Biol Assoc UK. 92:1547–1555. Available from: http://dx.doi.org/10.1017/S0025315412000045 via the Internet. Accessed 6 January, 2014.

Shester G, Ayers J. 2005. A cost effective approach to protecting deep-sea coral and sponge ecosystems with an application to Alaska's Aleutian Islands region. *In*: Freiwald A, Roberts JM, editors. Cold-water corals and ecosystems. Berlin: Springer-Verlag. p. 1151–1169.

Sissenwine MP, Mace PM. 2007. Can deep water fisheries be managed sustainably? Report and documentation of the expert consultation on deep-sea fisheries in the high seas. FAO Fish Rep. 838. Rome: FAO. p. 61–111.

Spear B, Cannon J. 2012. Benthic protection areas: Best practices and recommendations. Sustainable Fisheries Partnership Report. Honolulu, HI: Sustainable Fisheries Partnership.

Stone RP. 2006. Coral habitat in the Aleutian Islands of Alaska: Depth distribution, fine-scale species associations, and fisheries interactions. Coral Reefs. 25:229–238.

Stone RP, Shotwell SK. 2007. State of deep coral ecosystems in the Alaska region: Gulf of Alaska, Bering Sea and the Aleutian Islands. *In*: Lumsden SE, Hourigan TF, Bruckner AW, Dorr G, editors. The state of deep coral ecosystems of the United States. NOAA Technical Memorandum CRCP-3. Silver Spring, MD: NOAA. p. 65–108.

Tissot BN, Yoklavich MM, Love MS, York K, Amend M. 2006. Benthic invertebrates that form habitat on deep banks off southern California, with special reference to deep sea coral. Fish Bull. 104:167–181.

Tracey D, Baird SJ, Sanders BM, Smith MH. 2011. Distribution of protected corals in relation to fishing effort and assessment of accuracy of observer identification. NIWA Client Report no. WLG2011-33 prepared for the Department of Conservation, Wellington. Wellington, New Zealand: National Institute of Water & Atmospheric Research.

Vierod ADT, Guinotte JM, Davies AJ. 2014. Predicting the distribution of vulnerable marine ecosystems in the deep sea using presence-background models. Deep Sea Res. II 99:6–18.

Waller R, Watling L, Auster P, Shank T. 2007. Anthropogenic impacts on the Corner Rise seamounts, north-west Atlantic Ocean. J Mar Biol Assoc UK. 87:1075–1076.

Weaver PPE, Benn A, Arana PM, Ardron JA, Bailey DM, Baker K, Billett DSM, Clark MR, Davies AJ, Durán Muñoz P, et al. 2011. The impact of deep-sea fisheries and implementation of the UNGA Resolutions 61/105 and 64/72. Report of an international scientific workshop. Southampton: National Oceanography Centre.

Welsford D, Kilpatrick R. 2008. Estimating the swept area of demersal longlines based on in-situ video footage. Document WG-FSA-08/58. Hobart, Australia: CCAMLR.

WestPac (Western Pacific Fishery Management Council). 1979. Fishery management plan for the precious coral fisheries (and associated non-precious corals) of the western Pacific Region. Honolulu, HI: WestPac.

WestPac (Western Pacific Fishery Management Council). 1986. Combined fishery management plan, environmental assessment and regulatory impact review for the bottomfish and seamount groundfish fisheries of the western Pacific Region. Honolulu, HI: WestPac.

chapter nine

Restoring deepwater coral ecosystems and fisheries after the Deepwater Horizon oil spill

Andrew N. Shepard

Contents

Introduction

The Deepwater Horizon oil spill (DWHOS) in 2010 was the largest marine oil spill in U.S. history, unprecedented in its extent and complexity of pollutants and potential impacts. The spill immediately affected deep (>50 m depth) coral ecosystems and, 3 years after the spill, is still threatening both shallow and deep coral communities. Monetary penalties for ecosystem restoration via settlements and litigation may exceed $20 billion and provide unprecedented regional opportunities to support ecosystem-based management (EBM) of Gulf of Mexico ecosystems, including coral ecosystems and associated fisheries. In contrast to ecosystem restoration of coastal habitats and resources, restoration of offshore habitats, where there is often little or no baseline information, is more complex and operationally challenging. This chapter describes the DWHOS restoration science programs and recommends actions they may support to develop EBM for restoring and sustaining deep (>50 m depth) coral ecosystems and fisheries. Recommended restoration actions include support for Gulf of Mexico integrated ecosystem assessment (IEA) to identify coral ecosystem restoration goals, metrics (indicators), and gaps in knowledge; improvement of fisheries-independent survey tools and recreational landings data to inform reef fish stock assessments; support for coral habitat characterization and mapping; support for long-term observing and monitoring; and actions to ensure ecosystem restoration projects and programs engage stakeholders (scientists, decision makers, and citizens) and result in useful outputs.

Oil impacts on coral ecosystems and fisheries

Deep (>50 m depth) coral ecosystems in the Gulf of Mexico consist of hard bottom with beds of soft and hard coral colonies, and bioherms formed by stony corals (Figure 9.1) (Schroeder et al. 2005, Brooke and Schroeder 2007, CSAI 2007). Regional Fishery Management Councils around the United States are imposing conservation measures intended to protect deep coral ecosystems from a variety of stressors including fisheries-related perturbations, climate change, and energy exploration and development activities. In 2010, for example, the South Atlantic Fishery Management Council established five deepwater coral Habitat Areas of Particular Concern (HAPCs) totaling over 627,000 ha (NOAA 2012). Federal regulations in the Gulf of Mexico require oil and gas development activities to avoid potentially biologically sensitive areas, including coral ecosystems (MMS 2009). The Gulf of Mexico Regional Fishery Management Council has designated several shelf-edge banks that support productive coral communities as marine managed areas (Coleman et al. 2004).

The DWHOS at the Macondo well site (Figure 9.1) began on 20 April 2010 after the rig exploded, lasted 87 days, and vented an estimated 4.9 million barrels (±10%) of oil into the deep gulf (FOSC 2011). In addition, the largest amount of dispersant ever used in U.S. waters was applied and included almost 5.3 million liters of Corexit 9527 by air and vessels to the sea surface slick, and over 2.9 million liters of Corexit 9500A at the well head to break up and disperse the oil (FOSC 2011, Kujawinski et al. 2011). During sea surface oil spills, highly water-soluble hydrocarbon components are lost to the atmosphere within hours to days; however, during the DWHOS "...gas and oil experienced a significant residence time in the water column with no opportunity for the release of volatile species to the atmosphere. Hence, water-soluble petroleum compounds dissolved into the water column to a much greater extent than is typically observed for surface spills" (Reddy et al. 2012:20233).

Large parts of the Gulf of Mexico, including both state and federal waters, were closed to fishing during the DWHOS in May–October 2010 (Figure 9.1) (NMFS 2012a). Impacts of

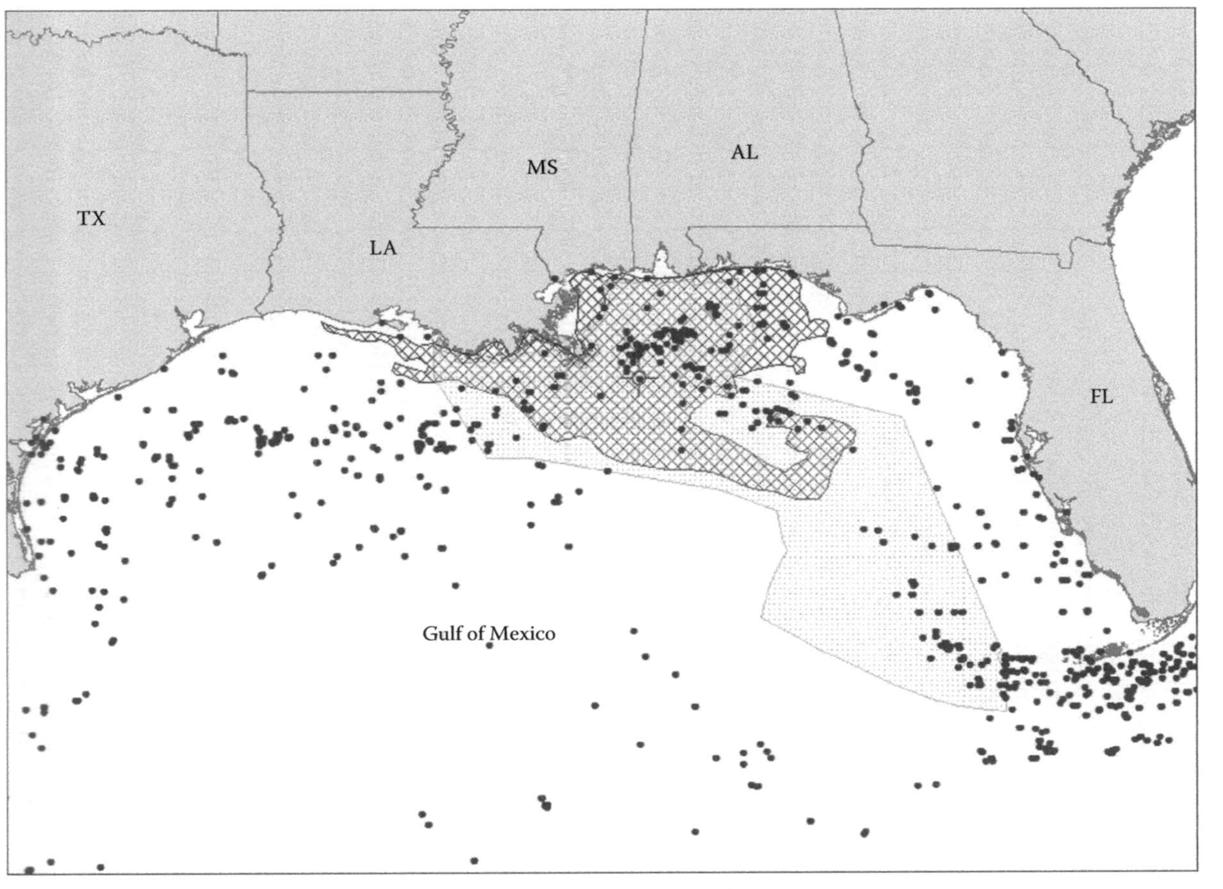

Figure 9.1 DWHOS spill extent (cross-hatched polygon; from NOAA Subsurface Monitoring Unit. 2013. BP Deepwater Horizon Oil Spill Cumulative NESDIS SAR Composite. Layer created by NESDIS Satellite Analysis Branch, 9/28/11 using ESRI ArcGIS 9.3. Available from: http://gomex.erma.noaa.gov/.), fishery closure (stippled polygon; from NMFS (National Marine Fisheries Service). 2011. ERMA Deepwater Gulf Response: Cumulative Emergency Fishery Closure. Layer created by SERO GIS Coordinator, April 2011, using ESRI ArcGIS 9.3. Available from: http://gomex.erma.noaa.gov/.), and known locations of habitat-forming soft and hard coral ecosystems at >50 m depth. (From Etnoyer, PJ. 2009. NOAA Gulf of Mexico Digital Atlas: Gulf of Mexico Deep Sea Coral Database. Layer created by P. Etnoyer, August 2009, using ESRI ArcGIS 9.3. Available from: http://gulfatlas.noaa.gov.)

the DWHOS on the recreational and commercial industries of Louisiana alone may have cost $285–$428 million in lost revenue by 2013, resulting in the loss of 2700–4000 jobs, and lost employee earnings of $68–$103 million (IEM 2010). While the Gulf Coast Claims Facility has paid out over $700 million to the gulfwide fishing industry, the long-term consequences of the oil spill on the fishing industry have yet to be assessed (NMFS 2012a). Notwithstanding the direct damages and impacts of the spill, gulf fisheries, wildlife, and associated habitats have been under stress for decades from environmental degradation (EPA 2011), overfishing and gear impacts (NMFS 2012b), invasive species (Schofield 2010), and climate change (Osgood 2008).

Oil spills have lethal and sublethal impacts on coral ecosystems. Many factors complicate assessment of DWHOS damage to Gulf of Mexico coral ecosystems, for example, the nature of the spill (e.g., size, location, composition, and timing), differences in resilience (ecosystem, species, organismal), acute versus chronic impacts, and synergistic disturbances (NOAA 2001, Oil Spill Commission 2011). In areas with natural hydrocarbon seepage like the northern gulf shelf and slope (MacDonald et al. 1995), corals may also adapt to background levels of oil and gas exposure (Al-Dahash and Mahmoud 2012). *Lophelia pertusa* bioherms and communities co-occur with cold seep communities in the northern gulf, although they settle and grow best in areas of inactive seepage on authigenic carbonate substrate, fed by pelagic versus seep-generated production (Becker et al. 2009). Oil pollution, including chemically enhanced fractions (e.g., with dispersants), may inhibit coral growth, reduce productivity, cause tissue loss, increase coral larvae mortality, reduce larval fitness, and impede larval settlement in preferred habitats (Loya and Rinkevich 1980, NOAA 2001, Yender et al. 2010, Goodbody-Gringley et al. 2013). Few studies, however, have considered lethal and sublethal effects on deep coral ecosystems below 50 m depth. Gass and Roberts (2006) reported that deepwater oil and gas activities may cause *L. pertusa* mortality and highlighted the lack of basic biological information for this species and associated communities.

Although post-DWHOS surveys did not detect DWHOS pollutants on shallow hermatypic coral reefs off Texas or Florida (Oil Spill Commission 2011, Goodbody-Gringley 2013), the areal extent of the spill covered shelf-edge and slope coral ecosystems from Florida to Louisiana (Figure 9.1) (Paul et al. 2013). White et al. (2012) provided evidence of impacts on a deep sea coral ecosystem 11 km from the DWHOS well site, including dead and dying corals and endemic invertebrates (brittle stars). Goodbody-Gringley et al. (2013) showed increased mortality and reduced settlement success of scleractinian coral larvae in the presence of specific components of the DWHOS spill (Louisiana sweet crude and dispersant).

In addition to coral protected areas in the Florida Keys, shallow (<50 m depth) and deeper mesophotic (50–150 m) coral reefs and beds across the Gulf of Mexico are designated as essential fish habitat for managed reef fisheries (GMFMC 2010) and HAPCs with constraints on fishing activities (NOAA 2013). Gulf regional fishery management plans cover 142 coral taxa and 31 reef-associated fish species including many demersal species of grouper, snapper, and shrimp, and pelagic species such as amberjack and tuna (GMFMC 2012). Loss of coral habitat (live cover and rugosity) in the Caribbean region is linked to declines in reef fisheries (Bellwood et al. 2004, Hoegh-Guldberg et al. 2007). Deep coral ecosystems in the northern gulf do not yet experience intense fishing pressure or related protections, although some endemic species such as red and golden crabs, royal red shrimp, and barrelfish have fisheries in other regions (Brooke and Schroeder 2007).

Oil pollutants may also have direct impacts on fish species associated with coral ecosystems. Peterson et al. (2003) noted that long-term exposure of fish embryos to weathered oil (three- to five-ringed polycyclic aromatic hydrocarbons [PAHs]) at ppb concentrations has population consequences through indirect effects on growth, deformities, and behavior, with long-term consequences on mortality and reproduction. Murawski et al. (in press) correlated proximity to the DWHOS well site with increased PAHs in livers and lesions on various species of reef fish.

Ecosystem restoration after Deepwater Horizon oil spill (DWHOS)

The purpose of the Oil Pollution Act of 1990 enacted after the *Exxon Valdez* oil spill is "…to make the environment and public whole for injury to or loss of natural resources and services as a result of a discharge or substantial threat of a discharge of oil (referred to as an 'incident')" (NOAA 1996:1). This objective is achieved through restoration, rehabilitation, replacement, or acquisition of equivalent natural resources or services that were injured or lost as a result of the spill. The size and complexity of the DWHOS coupled with the lack of baseline data on marine resources in the Gulf of Mexico is complicating efforts to assess injuries and seek compensation from responsible parties. This complexity is also reflected in the variety of claims, settlements, and litigation that may result in penalties potentially distributed to hundreds of programs and thousands of projects supporting gulf ecosystem and economic recovery. Penalties for both civil and criminal charges have been levied against a variety of responsible parties to assist this recovery, with much more pending future litigations or settlements (Figure 9.2). The funded programs are all tasked with supporting ecosystem restoration to some degree (Table 9.1).

Still pending as of September 2013, Clean Water Act (CWA) civil penalties may be the largest source of funding for ecosystem restoration. The act makes it unlawful to discharge oil into navigable waters or along shorelines. Fines range from $1100 to $4300 per barrel discharged

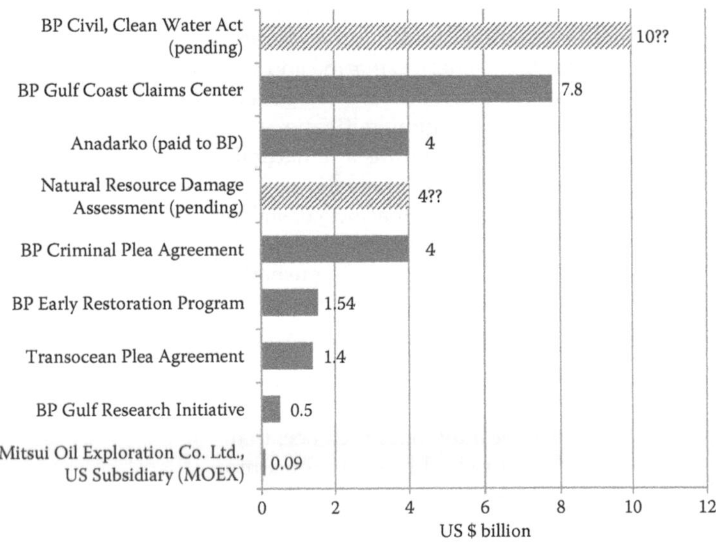

Figure 9.2 Breakdown of existing and pending ecosystem restoration funding from DWHOS civil and criminal penalties. Cross-hatched penalties are pending trials and amounts are unknown, conservative estimates as of March 2014.

Table 9.1 Ecosystem Restoration Programs Funded by DWHOS Penalties as of March 2014

Program	Ecosystem objectives	Funding (9/1/13)	Source
Gulf of Mexico Research Initiative (GOMRI)	Oil spill impacts research may support restoration applications	$500 million	BP
National Academy of Science (NAS)	Strategies and technologies for monitoring and protecting human and environmental health (gulf or other areas); oil/gas exploration safety	$500 million	BP and Transocean criminal settlements
North American Wetlands Conservation Fund (NAWCF)	Wetlands restoration and conservation projects located in gulf states or otherwise designed to benefit migratory bird species and other wildlife and habitat affected by oil spill	$100 million	BP settlement
National Fish and Wildlife Foundation (NFWF)	Natural resource restoration in AL, FL, MS, and TX; resilient coastal ecosystems and barrier island restoration/creation in LA, according to the LA Coastal Master Plan	$2.6 billion	BP and Transocean criminal settlements
Natural Resources Damage Assessment (NRDA), Early Restoration	Restoration activities aimed at returning natural resources or services to prespill baseline condition and compensating for interim losses, according to NRDA findings	$1.54 billion	BP
RESTORE Act, Section 1603, direct component	35% of $800 million from Transocean settlement; primarily for economic recovery; objectives include restoration and protection of the natural resources, ecosystems, fisheries, marine and wildlife habitats, beaches, and coastal wetlands of the Gulf Coast region; mitigation of damage to fish, wildlife, and natural resources; and implementation of a federally approved marine, coastal, or comprehensive conservation management plan, including fisheries monitoring (Department of the Treasury 2013)	$280 million	Transocean criminal settlement
RESTORE Act. Section 1603, comprehensive plan component	30% of $800 million from Transocean settlement; for projects and programs, using the best available science, that would restore and protect the natural resources, ecosystems, fisheries, marine and wildlife habitats, beaches, coastal wetlands, and economy of the Gulf Coast region (Department of the Treasury 2013)	$240 million	Transocean criminal settlement

Table 9.1 (Continued) Ecosystem Restoration Programs Funded by DWHOS Penalties as of March 2014

Program	Ecosystem objectives	Funding (9/1/13)	Source
RESTORE Act, Section 1603, spill impact component	30% of $800 million from Transocean settlement; projects, programs, and activities included in state expenditure plans must contribute to the overall economic and ecological recovery of the Gulf Coast, address the objectives in the Comprehensive Plan, and be preapproved by council (Department of the Treasury 2013)	$240 million	Transocean criminal settlement
RESTORE Act, Section 1604, NOAA ecosystem restoration science program	2.5% of $800 million from Transocean settlement for: marine and estuarine research, monitoring and ocean observing; data collection and stock assessments; pilot programs for fishery independent data and reduction of exploitation of spawning aggregations (Department of the Treasury 2013)	$20 million	Transocean criminal settlement
RESTORE Act, Section 1605, centers of excellence	2.5% of $800 million from Transocean settlement; for science and technology development in support of ecosystem restoration	$20 million	Transocean criminal settlement

(CWA 2002). The civil case against BP and other responsible parties began in February 2013 and may not end until 2015. Total penalties may reach $20 billion if the responsible parties are deemed criminally negligent and the amount of oil spilled during the 87-day event is estimated to be 4.9 million barrels (FOSC 2011). CWA penalties for oil discharge are normally deposited into the Oil Spill Liability Trust Fund and used to pay for oil pollution removal costs and natural resource damage assessments, among other activities. In the case of the DWHOS, however, the Resources and Ecosystems Sustainability, Tourist Opportunities, and Revived Economies of the Gulf Coast States Act (RESTORE Act) directs that 80% of CWA penalties are put into a Gulf Coast Restoration Trust Fund (Trust Fund) (Public Law 112-141 2012) and are divided among five "components" (Table 9.1) (Department of the Treasury 2013).

The Gulf Coast Ecosystem Restoration Council, established in Section 1603 of the RESTORE Act, released its initial comprehensive restoration plan in 2013 (GCERC 2013a) including Appendix A, a draft list of the types of projects that may be supported by the council, as obtained from council members (GCERC 2013b); <2% (i.e., eight) of 585 projects in Appendix A relate to restoring and sustaining offshore ocean habitats and resources. A single project addresses shallow coral reefs off Florida. When combined with early restoration program projects approved as of May 2013 by the NRDA Trustees (Deepwater Horizon Natural Resource Trustees 2012a,b, 2013), the types of projects proposed to date disproportionately deal with coastal restoration or lost human use and not the restoration of ocean ecosystems, despite the fact that the blowout occurred in 1200 m of water, 68 km from land, and the majority of unrecovered oil ended up in shelf and slope waters and sediments (FOSC 2011, Reddy et al. 2012).

Restoration options for shallow coral reefs resemble those for other coastal ecosystems in that direct interventions (e.g., transplants, removal of derelict fishing gear, debris cleanup) can be employed. For impacted corals in 1000–1500 m of water southwest of the DWHOS well site (White et al. 2012), direct interventions may not be practical. For deep coral ecosystems, where conventional restoration options falter, restoration may entail measures to monitor long-term recovery, protect and sustain ecosystem goods and services, and rebuild ecosystem resilience.

The natural resources damage assessment process primarily bases its restoration targets on equivalency analysis wherein the damaged habitat or resource is made whole by returning it to baseline conditions or by protecting an equivalent amount of like habitat or resource in another location (NOAA 1996). The lack of data and information from remote offshore habitats and resources may result in both inadequate assessment of damages and a lack of baseline information required for traditional restoration approaches. As indicated by the National Research Council (NRC 2012), these approaches severely limit restoration options and do "not make the public whole," as measured by the value of all the ecosystem services that flow from the affected areas. The NRC (2013) recommends an ecosystem services approach to assessment and restoration that aims to restore all the services provided by the impacted habitats and resources. For deep coral ecosystems, this vital information on their ecologic and economic importance is unknown and must be assessed both for restoration and the sustainability of related ecosystem goods and services.

Recommended deep coral ecosystem restoration actions

Recommended actions are organized into three groups of objectives intended to promote EBM for deep (>50 m depth) coral ecosystems and associated living resources, as well as other gulf ocean ecosystem goods and services (Table 9.2, Figure 9.3). Objective

Table 9.2 Recommended Objectives and Tasks to Support EBM Funded by Gulf Restoration Programs

Objectives	Actions
A: Identify priority needs and gaps in knowledge for comanagement of gulf deep coral ecosystems	1. Review EBM successes and lessons learned from other ecosystems.
	2. Engage stakeholders in developing EBM operational frameworks for coral ecosystem goods and services that identify science goals, gaps in science information, management priorities, and desired outcomes.
B: Conduct research, monitoring, and technology developments required to fill gaps	1. Expand surveys and research needed, as determined by gap analyses, to support management of activities that affect coral ecosystems.
	2. Develop and fund an expanded (applications, time and space scales) gulfwide observing and monitoring program.
	3. Develop and apply ecosystem models to support decision makers.
	4. Support and incentivize new technologies to improve fisheries efficiency and reduce the negative impacts of fishing activities on coral ecosystems.
	5. Determine socioeconomic value of coral ecosystem services.
C: Support information management and synthesis	1. Integrate data, derivative products, and research results to support EBM for coral ecosystems.
	2. Support synthesis of useful products for managers and communication to all stakeholders.

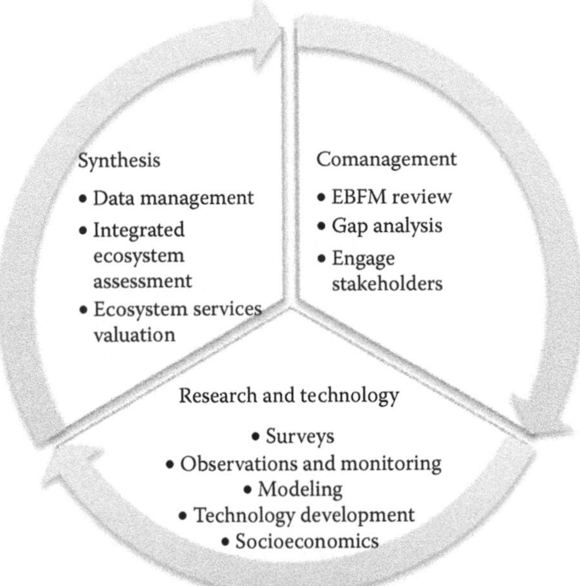

Figure 9.3 Recommended actions to enable EBM for gulf deep coral ecosystem goods and services represent a continuum including comanagement (Objective A), research and technology (Objective B), and synthesis (Objective C).

A actions provide the planning frameworks and foundational gap analyses required to prioritize Objective B actions, including data collection, research, and technology development. Objective C utilizes the resulting data and information to support analyses, syntheses, and products required by decision makers to assess risks and uncertainty, evaluate the success of a gulf regional recovery, and adapt management options and work plans (Garcia and Cochrane 2005, Granek et al. 2010). As envisaged, the gulf restoration programs will collaboratively sponsor relevant outputs for a gulfwide restoration plan (Table 9.3). Identification of restoration program sponsors is not meant to imply they will support these tasks alone, but must coordinate with and build on existing program and agency capabilities.

Objective A: Identify priority needs and gaps in knowledge for comanagement of Gulf of Mexico coral ecosystems

Action A1: Review ecosystem-based management (EBM) successes and lessons learned from other ecosystems

EBM may fail for many reasons including insufficient funding, political stalemates, lack of effective governance, and inadequate science (Granek et al. 2010). These challenges vary across the spectrum of management tools used to implement EBM (Garcia and Cochrane 2005, Curtin and Prellezo 2010). Murawski (2007) documented experiences of efforts to accomplish multistakeholder decision making to effect ecosystem objectives at regional scales, and recommended that an initial step towards EBM should be a systematic assessment ways managers communicated with and adapted to stakeholder concerns, and other collaborative methods to enhance communications and leveraging

Table 9.3 Outputs and Sponsors for Recommended Tasks

Objectives	Task	Suggested outputs	Proposed sponsor
A: Comanagement	1. Review EBM from other ecosystems	Review report of EBM successes and lessons learned to inform gulf regional coral ecosystem effort	RA1604 competitive contract/grant
	2. Engage stakeholders to identify gaps/priorities	EBM implementation plan for coral ecosystems developed through comanagement at appropriate temporal/spatial scales	RA1604 fund working group led by GMFMC and GSMFC to develop EBM implementation plan for corals; workshops to develop plans should be cofunded by NAS, NFWF, and RA1603
B: Data and technology	1. Expand surveys/research	Increased fisheries-independent and -dependent sampling and survey efforts to address gaps and needs identified by Action A2 frameworks	RA1603 council and states allocations support for permanent fund to increase and stabilize support for fisheries-independent monitoring and NOAA Marine Recreational Information Program; RA1604 and NFWF programs focus on research and technologies to improve sampling, landings data, assess fishing pressure, and understand fisheries ecology
		Coral habitat maps for the entire gulf exclusive economic zone (EEZ) supported through collaboration of existing data and information management partners in the gulf	RA1603 council and state funding to expand seafloor habitat mapping coverage of gulf EEZ–survey time and information management services; any related project funded by restoration programs should contribute mapping data/metadata for archival and public access
		New research discoveries applied to EBM science needs identified in Action A2	Joint RA1604 and 1605 competitive program

Table 9.3 (Continued) Outputs and Sponsors for Recommended Tasks

Objectives	Task	Suggested outputs	Proposed sponsor
	2. Develop observing and monitoring program	Promote and fund development of GCOOS (2011) build-out plan elements that support EBM data and information needs; new technologies to enhance gulf observing system	RA1603 council and state funding to support initial investment for backbone infrastructure, determined after implementation plan is finalized, and endowment or permanent fund ($1 billion minimum) for annual operating costs; RA1604, RA1605, NAS, and NFWF utilize program data for research and develop and utilize new technologies (e.g., platforms, sensors) that enhance applications
	3. Ecosystem models	Gulfwide conceptual models; coral ecosystem models for ecological forecasting; regional ecosystem modeling working group	RA1604 for support of current NOAA efforts and proposed outputs; NFWF for development of new products and applications via open competition
	4. Support technology developments	New technologies applied in support of EBM, resulting in reduced bycatch, and less destructive, more efficient fishing gear	Joint RA1604 and 1605 competitive program
	5. Socio economic values	Improved online public databases for all ecosystem values related to coral ecosystem services	NAS fund grants including deep coral ecosystem valuation studies via competition with RFP building on NRC (2012, 2013); NAS fund contracts to support existing efforts (e.g., Gecoserv 2013, NOEP 2012)
C: Synthesis	1. Integrate data, products, and results	Gulfwide restoration information system including data archives and policies; applications of new information technology tools for facilitating access to data and data derivative products to support EBM decision makers	RA1603 council establish information management (IM) policies, requirements, and best practices for all restoration funded programs; RA1604 support competitive grants/ contracts and internal support for continued improvement and use of existing IM programs; RA1604 sponsor grants/ contracts for new products; all restoration projects contribute related data to the proposed system-of-systems

(*continued*)

Table 9.3 (Continued) Outputs and Sponsors for Recommended Tasks

Objectives	Task	Suggested outputs	Proposed sponsor
	2. Support synthesis outputs	Gulfwide integrated ecosystem assessment (IEA) including coral ecosystem indicators, as paper and electronic (Web-based, interactive) report; coral ecosystem chapter of Gulf Report Card report, also developed as interactive Web-based products accessible on various platforms (e.g., mobile phones)	Joint RA1603 and NAS contract for up to 10 years; RA1604 mixed (internal and external) funding model support for NOAA IEA team efforts

Note: See Table 9.1 for sponsor program names and objectives.

of capabilities and resources. Regional examples of efforts to implement EBM for coral ecosystems include management plans for the Florida Keys and Flower Garden Banks National Marine Sanctuaries, both of which cover shallow and deep coral ecosystems (FKNMS 2007, FGBNMS 2012), and the South Atlantic Fishery Management Council's comprehensive ecosystem amendment related to designated coral HAPCs (SAFMC 2009). This review should be carried out prior to implementation of restoration projects for coral ecosystems.

Action A2: Engage stakeholders in developing EBM operational frameworks for coral ecosystem goods and services that identify science goals, gaps in science information, management priorities, and desired outcomes

The Food and Agriculture Organization of the United Nations (FAO) defined a process for operational implementation of regional EBM for ocean resources, which includes setting high-level policy goals and operational objectives; developing indicators, reference points, and decision rules for the application of measures; and monitoring and evaluating performance (FAO 2005). The Gulf Coast Ecosystem Restoration Council's initial comprehensive restoration plan (GCERC 2013a) includes strategic objectives but little detail on how it will solicit, select, and integrate sponsored projects. Prior to implementing regional restoration, the council and related state plan leadership should make use of established guidance to develop operational frameworks for prioritized ecosystem services, including coral habitat and associated fisheries (e.g., FAO 2005). Federal and state agencies should establish and coordinate a comanaged entity responsible for engaging coral ecosystem stakeholders (scientists, managers, public users) in the tasks of developing, integrating, and sustaining restoration plans at the appropriate temporal and spatial scales to support adaptive management (Wilson 2006, Ostrom 2009, Gutierrez et al. 2011).

Objective B: Conduct research, monitoring, modeling, and technology developments required to fill gaps and support adaptive management

Action B1: Expand surveys and research needed, as determined by gap analyses, to support management of activities that affect coral ecosystems

Stock assessments require data and information on both fish populations and fisheries. In 2002, the National Marine Fisheries Service and regional management councils in the gulf, Caribbean, and U.S. South Atlantic established the Southeast Data Assessment and Review (SEDAR) process to improve the rigor and transparency of stock assessments. "Benchmark assessments" entail a full suite of data, models, and related peer review and may take over a year to complete. In addition to benchmark assessments, more frequent updates may be done once an assessment is approved through SEDAR, requiring new input data and few, if any, model configuration improvements (SEDAR 2012). These assessments and related ecosystem models require inputs from historic and current data sources for gulf commercial and recreational fisheries, with continuous, long-term time series being particularly valuable for detecting population changes over time. Both fishery-dependent and fishery-independent data need to be enhanced in quality, time, and space. This is particularly true and challenging for deep reef (>300 m) fisheries, which cannot be sampled using traditional trawl surveys, most acoustic technologies, or scuba divers (e.g., Foster et al. 2013, McIntyre et al. 2013, Price et al. 2013).

Efforts to enhance and supplement fisheries and ecosystem data collection should build on and improve existing data collection programs, for example, the Southeast Area Monitoring and Assessment Program (SEAMAP) in the South Atlantic, Caribbean, and Gulf of Mexico regions, which received an inadequate total (all three regions) of $5 million in FY2012 for fisheries-independent data collection. SEAMAP annual reports (e.g., SEAMAP 2012) summarize the traditional activities employed, including surveys, data management and synthesis, and planning and reporting. For coral ecosystems, these must be supplemented with new nondestructive, innovative assessment approaches (see Action B4 later in this chapter). Priority activities should be guided by the planning activities recommended in Actions A1 and A2.

A related critical need is to enhance and improve recreational fisheries-dependent data collections (NRC 2006) including increase Access Point Angler Intercept Surveys (APAIS) in the National Oceanic and Atmospheric Administration's (NOAA) Marine Recreational Information Program; standardize sampling protocol and estimators for state and federal programs; and require charter boat, head boat, and other for-hire recreational fishing operations to maintain electronic logbooks of fish landed and kept, as well as fish caught and released.

Deep coral ecosystems in the Gulf of Mexico are, in general, poorly mapped. Seafloor habitat mapping information is required to support both the sustainability of exploited populations and the maintenance of biological diversity. Maps required should depict cumulative area impacted by all human activities (e.g., oil and gas development, fishing effort by gear types); distribution of habitats and diversity of key taxa (e.g., structure forming and bed corals); and linkages between habitat and the dynamics of exploited populations (Auster 2001). Although the gulf seafloor may be extensively mapped in support of oil and gas exploration-related activities, these data were not collected for the purpose of mapping habitats and associated biotic communities, and are not easily accessible. Resulting map products supported by restoration program efforts need to be quality controlled and easily accessible for all constituents. Priority targets need to include ultra-deep oil and gas lease blocks in pending lease sales, and known and predicted (based on habitat suitability models, e.g., Guinotte and Davies 2012, Kinlan et al. 2014) deep coral ecosystems in the gulf.

Exploration in the context of the gulf most often refers to oil and gas prospecting. The Gulf of Mexico Ecosystem Restoration Task Force (GCERTF 2011:50), however, called for "… investing in research and basic exploration to understand the ecosystems in the gulf and how they can be resilient to impacts from episodic events, such as hurricanes or oil spills, and long-term changes such as climate impacts." The GCERC comprehensive restoration plan (GCERC 2013b) list of proposed projects must include exploratory research projects that address gaps identified in the development of EBM operational frameworks (Action A2), utilize mapping products to target and explore new gulf frontiers, and invest in developing advanced technologies (i.e., Class I research ship for the gulf, and undersea technologies such as robotic and human-occupied vehicles). Exploration is expensive and should be coordinated and leveraged with programs that now support frontier science such as the Bureau of Ocean Energy Management (BOEM), the National Science Foundation (NSF), NOAA's cooperative institutes in the gulf region, and NOAA's Ocean Exploration Program.

Action B2: Develop and fund an expanded (applications, time and space scales) gulfwide observing and monitoring program

The Gulf Coast Ecosystem Restoration Council's comprehensive restoration plan calls for both project-level and regional ecosystem monitoring to support science-based decision making (GCERC 2013a). Monitoring requires integration of input from observations and research (Busch and Trexler 2003). As Weisberg (2011:38) observed, "fisheries science cannot advance by just studying fish. Instead, the entire context in which fish make their living must be understood." Several plans now exist for regional ocean observing systems and required priority observations, for example, the Gulf of Mexico Coastal Ocean Observing System (GCOOS) draft "build out" plan (GCOOS 2011). Development of a gulf-wide observing and monitoring program in support of gulf restoration and EBM should include the following critical elements related to coral ecosystems:

1. Interdisciplinary approaches to collecting data and conducting research on all key ecosystem components—physical, chemical, and biological. An acoustic (passive and active) network, similar to Canada's Acoustic Ocean Tracking Network (OTN) (Block et al. 2012, Welch et al. 2012) should be deployed at deep gulf stations. Observations from ships, buoys, moorings, remote sensing, and gliders are required to support high-resolution habitat maps, resource surveys (fisheries-dependent and -independent), and water quality monitoring.
2. A permanent endowment that provides on-going support from interest versus principal is recommended to sustain required observing network operations. Data collection must be sustained at operational levels for the long term. Cowan et al. (2012:506) observed that EBM may be hindered by the ability and willingness of society "…to understand and tolerate the tradeoffs needed to forego short-term economic benefits, and to abandon open-access utilization of natural resources, in favor of protecting or restoring ecosystem goods and service over much longer time periods." Most ocean ecosystem observation efforts in the gulf have been sustained by short-term grants and contracts and terminated at the end of the award period.
3. Data collection must be sustained at appropriate temporal and spatial scales. Appropriate scales in an EBM context for coral ecosystems vary from large-scale (decades in time, hectares in distance) oceanographic and climatic processes (e.g., North Atlantic Oscillation, Pacific Decadal Oscillation, location and paths of

boundary currents, upwelling zones) to intermediate and local-scale features in time and space (e.g., bathymetry, substrate types, localized upwelling, jets, and riverine plumes). Operational frameworks developed in Action A2 will help define the required density of observations.
4. Technology innovations to improve the pace, scope, and efficiency of observing and monitoring efforts. Annual operating costs must include funding for capitalization that maintains and repairs existing assets, and integrates new technologies into the gulfwide observing and monitoring network, resulting in improvements in the efficiency, effectiveness, and relevance of observations and resulting science.

Action B3: Develop and apply ecosystem models to support decision makers

Cowan et al. (2012) argued for a holistic ecosystem modeling approach to support EBM, while acknowledging the empirical challenge of ecosystem-scale applications, and deficiencies in providing an actual basis for tactical decision making in any particular fishery. Murawski (2007) suggested both conceptual and mathematical modeling efforts are warranted. Conceptual models of coral ecosystem functions (including basic food webs, keystone species and fisheries, and factors that impact on productivity and sustainability) are needed to help guide "…the establishment of plausible subsets of potential outcomes, particularly in an adaptive management scenario, where provisional management policies are considered 'experiments' for the purposes of gathering additional information" (Murawski 2007:686). These conceptual models should be associated with operational frameworks developed in Action A2.

Plagányi (2007) reviewed the strengths and limitations of multispecies and ecosystem models, and noted that their forecasting skill is constrained by limitations in fundamental knowledge of ecosystem dynamics and structure. Ecological forecasting is central to EBM, allowing managers to assess diverse policy options prior to implementation (Kaplan et al. 2012). More sophisticated mathematical modeling of ecosystem interactions and the bioeconomic impacts of stressors and policy choices are used to predict change and identify feasible subsets of management options. These are particularly needed for coral ecosystems that are sensitive to a range of stressors in time and space. Given the range and scope of possible modeling solutions, and their importance to effective EBM, the Gulf Ecosystem Restoration Council should convene a standing (sustained) working group for ecosystem modeling to develop an action plan as part of the comprehensive restoration plan (GCERC 2013a).

Action B4: Support and incentivize new technologies to improve fisheries efficiency and reduce the negative impacts of fishing activities on coral ecosystems

Research priority 4 of the National Ocean Policy Plan (SOST 2013) calls for both applied research and technological development to support marine natural resources. In addition to exploration, research and mapping technologies addressed in Action B1, and observing and monitoring technologies in Action B2, innovative technologies are needed to reduce bycatch, overfishing, and habitat degradation. Nondestructive fishing and assessment technologies are particularly needed for coral habitats that can be destroyed in a day and take decades or more to recover (Reed et al. 2007). Additionally, these technologies may offer less expensive, faster, and safer approaches for fish stock and habitat assessment (Foster et al. 2013). Priorities should again emerge from gap analyses in Action A2. Research and development programs sponsored by RESTORE Act Section 1604 and 1605 programs should have a cooperative element, whereby stakeholders (e.g., the fishing and diving communities) participate in the planning and implementation of projects by providing ships, gear, or expertise and by helping to ensure innovations achieved to support EBM science are also applied to improve fishing operations.

Action B5: Determine socioeconomic value of coral ecosystem services

The Gulf Task Force called for expanding ecosystem services and benefit analysis tools and capabilities to determine the socioeconomic benefits that ecosystems provide throughout the gulf region (GCERTF 2011). Several studies highlight the benefits and need to improve the assessment and valuation of ecosystem services, including both market and nonmarket values (e.g., Granek et al. 2010, Barbier et al. 2011). The National Ocean Economics Program (NOEP 2012) adequately estimates the market values of ocean and coastal services, although these are at least a year out of date, and the database does not include nonmarket values. Available inventories for gulf nonmarket valuation studies by ecosystem service and habitat type (Gecoserv 2013, NOEP 2013) do not currently include any studies for deep coral ecosystems in the gulf. Funding should be provided to facilitate and expand the delivery of NOEP data and derivatives, including the development of nonmarket values for major estuarine, coastal, and oceanic ecosystems that can be applied across the gulf landscape and seascape.

Objective C: Support information management and synthesis

Action C1: Integrate data, derivative products, and research results to support EBM for coral ecosystems

The Gulf Coast Ecosystem Restoration Task Force initiated by executive order right after the DWHOS in 2010 was charged with creating a comprehensive gulf restoration strategy (GCERTF 2011). This plan prioritized the development of decision-making visualization aids that were overlaid on the myriad uses of the gulf which can potentially interact with energy and mineral development. Actions recommended by Garcia and Cochrane (2005) to support EBM include the need to develop and build on existing information platforms to integrate information and data on indicators for different fisheries and ecosystems, and to create and distribute large-scale, multicriteria descriptions of ecosystems (functions, structure, and services) using accessible information technologies such as global information systems (GIS) and Web-based databases. Several gulf-based regional partners have advanced data and information management using open architecture software, cloud services, public repositories, and innovative visualization tools. These should be reviewed and scaled up for gulf-wide application. The challenge will be to integrate interoperable, quality-controlled, and accessible (with metadata records) ecological data sets into these archives, usually designed to hold geospatial data and metadata records. Programs such as NOAA's Deep Sea Coral Program is making strides in working with scientists in the grant stages to develop data management agreements that recognize intellectual property rights and promote the sharing and integration of ecological information into national databases. All restoration programs should strive to provide access via this system-of-systems through standardization of data collection protocols and policies that require sharing by funded projects.

Action C2: Support synthesis of useful products for managers and communication to all stakeholders

The Gulf Task Force strategy recommendations also mentioned the need to develop integrated decision-support tools and systems, including expansion and enhancement of predictive, simulation, and risk assessment models and ecological forecasting capabilities, and to establish indicators of success and monitoring assessments to evaluate how well program elements meet their stated goals (GCERTF 2011). These are priority objectives in IEAs, as defined by Levin et al. (2009). A gulf interagency IEA has been underway since 2010, intended to provide a framework for organizing and synthesizing science to inform multiscale, multisector EBM (Schirripa et al. 2012). This interagency effort should continue and expand to include new partners from science, management, and industry, and a joint education campaign should be launched that explains the assessment process to students, professionals, and the public.

The task force recommended development of a gulfwide progress report to provide "summary status information on ecosystem endpoints and communicate progress of management in improving ecosystem functions and services" (GCERTF 2011). This objective should be addressed in collaboration with the gulf IEA, which includes identification and tracking of ecosystem indicators. As McKinney et al. (2011) envisaged for their Report Card project, the goal is to develop a graphical representation of the environmental condition of the gulf that will be science based, widely accessible, and readily understandable by policy makers, stakeholders, scientists, and, most importantly, the American public. They use a drivers-pressures-stressors-state-impacts-response (DPSSIR) conceptual framework to guide and focus scientific research on identifying and addressing the most important risks to priority ecosystem services, including "a broad diversity of ecosystem types, from deepwater bottom communities, pelagic habitats, coral reefs, sea grass communities, salt and freshwater marshes, and riverine systems to barrier islands, coastal forests, and the larger watershed." A coral ecosystems chapter should be developed in collaboration with the gulf IEA project (Schirripa et al. 2012).

Conclusion

The recommended action plan will build the capacity required to develop an EBM approach for gulf deep (>50 m depth) coral ecosystems. Science programs required to support EBM have been grossly underfunded in comparison to programs seeking to address the economic impact of the DWHOS on the gulf's ecosystems, from estuaries to the deep sea (Shepard et al. in press). Coral ecosystems are among the most vulnerable and least understood ecosystems potentially affected by the DWHOS. Civil and criminal penalties for ecosystem restoration are an unprecedented opportunity to adequately support the science and EBM actions required to sustain gulf ecosystem goods and services for many generations. These funds are being distributed to a variety of state and federal programs with, in many cases, overlapping objectives and nonoverlapping management structures. The recommended EBM approach will help focus and coordinate these programs, address both primary (direct) and compensatory damages required to restore coral ecosystems, and mitigate regional ecosystem degradation that has been happening over decades due to a mix of stressors. Proposed sponsors and outputs provide a framework for performance evaluation. Success will not just compensate for damages by restoring the system to prespill conditions, a benchmark for natural resource damage assessment restoration (NOAA 1996), but may reverse decades of environmental degradation and sustain gulf coral ecosystem goods and services for generations.

Acknowledgments

The author is grateful for reviews and edits from N. Thompson, C. Friess, and C. Robbins. The author acknowledges the Walton Family Foundation, Ocean Conservancy, Florida

Institute of Oceanography, and University of South Florida for the support of workshops that provided input to this paper.

References

Al-Dahash LM, Mahmoud HM. 2012. Harboring oil-degrading bacteria: A potential mechanism adaptation and survival in corals inhabiting oil-contaminated reefs. Mar Poll Bull. 72(2):364–374. Available from: http://www.ncbi.nlm.nih.gov/pubmed/23014479 via the Internet. Accessed 22 January, 2014.

Auster PJ. 2001. Defining thresholds for precautionary habitat management actions in a fisheries context. N Am J Fish Manage. 21(1):1–9.

Barbier EB, Hacker SD, Kennedy C, Koch EW, Stier AC, Silliman BR. 2011. The value of estuarine and coastal ecosystem services. Ecol Monogr. 81(2):169–193.

Becker EL, Cordes EE, Macko SA, Fisher CR. 2009. Importance of seep primary production to *Lophelia pertusa* and associated fauna in the Gulf of Mexico. Deep-Sea Res Pt I. 56:786–800.

Bellwood DR, Hughes TP, Folke C, Nystrom M. 2004. Confronting the coral reef crisis. Nature. 429:827–833.

Block BA, Holland K, Costa D, Kocik J, Fox D, Mate B, Grimes C, Moustahfid H, Seitz A, Behzad M, et al. 2012. Toward a US animal telemetry observing network (US ATN) for our oceans, coasts and Great Lakes. Community white paper developed for the Interagency Ocean Observation Committee, IOOS Summit 2012, Herndon, VA. Available from: http://www.iooc.us/2012/list-of-community-white-papers-now-available/ via the Internet. Accessed 1 July, 2013.

Brooke S, Schroeder WW. 2007. State of deep coral ecosystems in the Gulf of Mexico region: Straits of Florida to Texas. *In:* Lumsden SE, Hourigan TF, Bruckner AW, Dorr G, editors. The state of deep coral ecosystems of the United States. NOAA Technical Memorandum CRCP-3. Silver Spring, MD: NOAA. p. 271–306.

Busch DE, Trexler JC. 2003. The importance of monitoring in regional ecosystem initiatives. *In:* Busch DE, Trexler JC, editors. Monitoring ecosystems: Interdisciplinary approaches for evaluating ecoregional initiatives. Washington, DC: Island Press. p. 1–23.

Coleman FC, Baker PB, Koenig CC. 2004. A review of Gulf of Mexico marine protected areas: Successes, failures, and lessons learned. Fisheries. 29(2):10–21.

Cowan JH, Rice JC, Walters CJ, Hilborn R, Essington TE, Day JW, Boswell KM. 2012. Challenges for implementing an ecosystem approach to fisheries. Mar Coast Fish Dynam Manag Ecosyst Sci. 4(1):496–510.

CSAI (Continental Shelf Associates International, Inc.). 2007. Characterization of Northern Gulf of Mexico deep-water hard bottom communities with emphasis on *Lophelia* coral. OCS Study MMS 2007-044. New Orleans, LA: U.S. Department of the Interior, Minerals Management Service, Gulf of Mexico OCS Region.

Curtin R, Prellezo R. 2010. Understanding marine ecosystem-based management: A literature review. Mar Policy. 34:821–830.

CWA (Clean Water Act). 2002. Federal Water Pollution Control Act, 33 U.S.C. 1251 et seq., as amended through P.L. 107–303, p. 234. 27 November, 2002.

Deepwater Horizon Natural Resource Trustees. 2012a. Natural resource damage assessment April 2012 status update for the Deepwater Horizon oil spill. Available from: http://www.gulfspill-restoration.noaa.gov/wp-content/uploads/FINAL_NRDA_StatusUpdate_April2012.pdf via the Internet. Accessed 1 July, 2013.

Deepwater Horizon Natural Resource Trustees. 2012b. Deepwater Horizon oil spill phase I early restoration plan and environmental assessment. Available from: http://www.gulfspillrestoration.noaa.gov/wp-content/uploads/Final-ERP-EA-041812.pdf via the Internet. Accessed 1 July, 2013.

Deepwater Horizon Natural Resource Trustees. 2013. Deepwater Horizon oil spill phase II early restoration plan and environmental assessment. Available from: http://www.gulfspillrestoration.noaa.gov/wp-content/uploads/DRAFT-Phase-II-DERP-ER-10-29-12.pdf via the Internet. Accessed 1 July, 2013.

Department of the Treasury. 2013. Proposed rule: Gulf Coast Restoration Trust Fund. Federal Register Notice 78 FR 54801. p. 54801–54813. Available from: https://federalregister.gov/a/2013-21595 via the Internet. Accessed 23 September, 2013.

EPA (Environmental Protection Agency). 2011. National coastal condition report IV, Chapter 5, Gulf Coast region. EPA-842-R-10-003. Washington, DC: United States Environmental Protection Agency, Office of Research and Development/Office of Water. p. 5-1–5-28. Available from: http://www.fws.gov/wetlands/Documents/National-Coastal-Condition-Report-IV-part-1-of-2.pdf via the Internet. Accessed 1 July, 2013.

Etnoyer PJ. 2009. NOAA Gulf of Mexico Digital Atlas: Gulf of Mexico Deep Sea Coral Database. Layer created by P. Etnoyer, August 2009, using ESRI ArcGIS 9.3. Available from: http://gulfatlas.noaa.gov. Accessed 4 March, 2014.

FAO (Food and Agriculture Organization of the United Nations). 2005. Putting into practice the ecosystem approach to fisheries. Rome, Italy: Food and Agriculture Organization of the United Nations. Available from: http://www.fao.org/docrep/009/a0191e/a0191e00.htm via the Internet. Accessed 25 September, 2013.

FGBNMS (Flower Garden Banks National Marine Sanctuary). 2012. Flower Garden Banks National Marine Sanctuary management plan. Galveston, TX: FGBNMS. Available from: http://flowergarden.noaa.gov/management/2012mgmtplan.html via the Internet. Accessed 1 July, 2013.

FKNMS (Florida Keys National Marine Sanctuary). 2007. Florida Keys National Marine Sanctuary management plan. Key West, FL: FKNMS. Available from: http://floridakeys.noaa.gov/mgmtplans/2007.html via the Internet. Accessed 1 July, 2013.

FOSC (Federal On-Scene Coordinator). 2011. On-Scene Coordinator report, Deepwater Horizon oil spill. Report to the National Response Team. Available from: http://www.uscg.mil/foia/docs/dwh/fosc_dwh_report.pdf via the Internet. Accessed 1 July, 2013.

Foster G, Gleason A, Costa B, Battista T, Taylor C. 2013. Acoustic applications. *In:* Goodman JA, Purkis SJ, Phinn ST, editors. Coral Reef Remote Sensing: A Guide for Mapping, Monitoring and Management, Chapter 9. New York, NY: Springer Science. p. 221–251.

Garcia SM, Cochrane KL. 2005. Ecosystem approach to fisheries: A review of implementation guidelines. ICES J Mar Sci. 62:11–318.

Gass SE, Roberts JM. 2006. The occurrence of the cold-water coral, *Lophelia pertusa* (Scleractinia) on oil and gas platforms in the North Sea: Colony growth, recruitment and environmental controls on distribution. Mar Pollut Bull. 52:549–559.

GCERC (Gulf Coast Ecosystem Restoration Council). 2013a. Initial comprehensive plan: Restoring the Gulf Coast's ecosystem and economy. Gulf Coast Ecosystem Restoration Council. Available from: http://www.restorethegulf.gov/sites/default/files/Final%20Initial%20Comprehensive%20Plan.pdf via the Internet. Accessed 1 July, 2013.

GCERC (Gulf Coast Ecosystem Restoration Council). 2013b. Initial comprehensive plan, Appendix A: Background information preliminary list of authorized but not yet commenced projects and programs. Gulf Coast Ecosystem Restoration Council. Available from: http://www.restorethegulf.gov/sites/default/files/Authorized%20But%20Not%20Yet%20Commenced%20List_8-6-13_FINAL.pdf via the Internet. Accessed 1 July, 2013.

GCERTF (Gulf Coast Ecosystem Research Task Force). 2011. Gulf of Mexico Regional Ecosystem Restoration Strategy. Stennis Space Center, MS: Gulf Coast Ecosystem Research Task Force. Available from: http://www.epa.gov/gcertf/pdfs/GulfCoastReport_Full_12-04_508-1.pdf via the Internet. Accessed 1 July, 2013.

GCOOS (Gulf of Mexico Coastal Ocean Observing System). 2011. Draft build-out plan for Gulf of Mexico Coastal Ocean Observing System. Texas A&M University, College Station, TX: Gulf of Mexico Coastal Ocean Observing System. Available from: http://gcoos.tamu.edu/BuildOut/GCOOS_BuildoutPlan_V1.pdf via the Internet. Accessed 1 July, 2013.

Gecoserv. 2013. Gulf of Mexico ecosystem services valuation database. Corpus Christi, TX: Harte Research Institute. Available from: http://www.gecoserv.org/index.html via the Internet. Accessed 1 July, 2013.

GMFMC (Gulf of Mexico Fishery Management Council). 2010. Final report: Gulf of Mexico Fishery Management Council 5-year review of the final generic amendment number 3 addressing essential fish habitat requirements, habitat areas of particular concern, and adverse effects of fishing in the fishery management plans of the Gulf of Mexico. Tampa, FL: Gulf of Mexico Fishery Management Council.

GMFMC (Gulf of Mexico Fishery Management Council). 2012. Species listed in the fishery management plans of the Gulf of Mexico Fishery Management Council. Tampa, FL: Gulf of Mexico Fishery Management Council. Available from: http://www.gulfcouncil.org/Beta/GMFMCWeb/downloads/species%20managed.pdf via the Internet. Accessed 25 September, 2013.

Goodbody-Gringley G, Wetzel DL, Gillon D, Pulster E, Miller A, Ritchie KB. 2013. Toxicity of Deepwater Horizon source oil and the chemical dispersant, Corexit® 9500, to coral larvae. PLoS ONE 8(1):e45574. Available from: http://www.plosone.org/article/info%3Adoi%2F10.1371%2Fjournal.pone.0045574 via the Internet. Accessed 22 January, 2014.

Granek EF, Polasky S, Kappel CV, Reed DJ, Stoms DM, Koch EW, Kennedy CJ, Cramer LA, Hacker SD, Barbier EB, et al. 2010. Ecosystem services as a common language for coastal ecosystem-based management. Conserv Biol. 24(1):207–216.

Guinotte JM, Davies AJ. 2012. Predicted deep-sea coral habitat suitability for the U.S. West Coast. Report to National Oceanic and Atmospheric Administration, National Marine Fisheries Service, Deep Sea Coral Research and Technology Program, Silver Spring, MD.

Gutierrez NL, Hilborn R, Defeo O. 2011. Leadership, social capital and incentives promote successful fisheries. Nature. 470:386–389.

Hoegh-Guldberg O, Mumby PJ, Hooten AJ, Steneck RS, Greenfield P, Gomez E, Harvell CD, Sale PF, Edwards AJ, Caldeira K, et al. 2007. Coral reefs under rapid climate change and ocean acidification. Science. 318:1737–1742.

IEM (Innovative Emergency Management). 2010. A study of the economic impact of the Deepwater Horizon oil spill, Part 1: Fisheries. Research Triangle Park, NC: Innovative Emergency Management, Inc. Available from: http://www.uflib.ufl.edu/docs/Economic%20Impact%20Study,%20Part%20I%20-%20Full%20Report.pdf via the Internet. Accessed 1 July, 2013.

Kaplan IC, Horne PJ, Levin PS. 2012. Screening California current fishery management scenarios using the Atlantis end-to-end ecosystem model. Prog Oceanogr. 102:5–18.

Kinlan B, Caldow C, Etnoyer PJ. 2014. Deep Coral Predictive Habitat Modeling in the U.S. Atlantic and Gulf of Mexico: Focusing on Uncharted Deep-Sea Corals. National Centers for Coastal Ocean Science Projects Explorer. Available from: http://www.coastalscience.noaa.gov/projects/detail?key=35. Accessed 15 March, 2014.

Kujawinski MC, Soule K, Valentine DL, Boysen AK, Longnecker K, Redmond MC. 2011. Fate of dispersants associated with the Deepwater Horizon oil spill. Environ Sci Technol. 45:1298–1306.

Levin PS, Fogarty MJ, Murawski SA, Fluharty D. 2009. Integrated ecosystem assessments: Developing the scientific basis for ecosystem-based management of the ocean. PLoS Biol. 7(1):23–28.

Loya Y, Rinkevich B. 1980. Effects of oil pollution on coral reef communities. Mar Ecol Prog Ser. 2:167–180.

MacDonald IR, Schroeder WW, Brooks JM. 1995. Chemosynthetic ecosystems studies, final report. OCS Study MMS 95-0023. Prepared by the Geochemical and Environmental Research Group for the U.S. Department of the Interior, Minerals Management Service, Gulf of Mexico OCS Region, New Orleans, LA.

McIntyre FD, Collie N, Stewart M, Scala L, Fernandes PG. 2013. A visual survey technique for deep-water fishes: Estimating anglerfish *Lophius* spp. abundance in closed areas. J Fish Biol. 83:739–753. Available from: http://onlinelibrary.wiley.com/doi/10.1111/jfb.12114/pdf via the Internet. Accessed 22 January, 2014.

McKinney L, Tunnell W, Harwell M, Kelsey H, Dennison B. 2011. Vision for the Gulf of Mexico report card. Corpus Christi, TX: Harte Research Institute. Available from: http://harteresearchinstitute.org/newsletter/docs/gulfofmexico_reportcard_brochure.pdf via the Internet. Accessed 1 July, 2013.

MMS (Minerals Management Service). 2009. Notice to lessees and operators of federal oil, gas, and sulphur leases and pipeline right-of-way holders in the outer Continental Shelf, Gulf of Mexico OCS Region. NTL No. 2009-G39. New Orleans, LA: Minerals Management Service [Bureau of Ocean Energy Management as of 2010], Gulf of Mexico Region.

Murawski SA. 2007. Ten myths concerning ecosystem approaches to marine resource management. Mar Policy. 31:681–690.

Murawski SA, Hogarth WT, Peebles EB, Stein JE, Ylitalo GM, Barbieri L. In press. Prevalence of fish diseases in the Gulf of Mexico, Post-*Deepwater Horizon*. Transactions of the American Fisheries Society.

NMFS (National Marine Fisheries Service). 2011. ERMA Deepwater Gulf Response: Cumulative Emergency Fishery Closure. Layer created by SERO GIS Coordinator, April 2011, using ESRI ArcGIS 9.3. Available from: http://gomex.erma.noaa.gov/. Accessed 4 March, 2014.

NMFS (National Marine Fisheries Service). 2012a. Fisheries economics of the United States, 2011. U.S. Department of Commerce, NOAA Technical Memorandum NMFS-F/SPO-118. Available from: https://www.st.nmfs.noaa.gov/st5/publication/index.html via the Internet. Accessed 25 September, 2013.

NMFS (National Marine Fisheries Service). 2012b. Annual report to Congress on the status of U.S. Fisheries-2011. Silver Spring, MD: U.S. Department of Commerce, NOAA, National, Marine Fisheries Service. Available from: http://www.nmfs.noaa.gov/sfa/statusoffisheries/2011/RTC/2011_RTC_Report.pdf via the Internet. Accessed 25 September, 2013.

NOAA (National Oceanic and Atmospheric Administration). 1996. Guidance document for natural resource damage assessment under the Oil Pollution Act of 1990. Silver Spring, MD: Damage Assessment and Restoration Program, National Oceanic and Atmospheric Administration.

NOAA (National Oceanic and Atmospheric Administration). 2001. Toxicity of oil to reef-building corals: A spill response perspective. NOAA Technical Memorandum NOS OR&R 8. Seattle, WA: Hazardous Materials Response Division, Office of Response and Restoration, National Oceanic and Atmospheric Administration.

NOAA (National Oceanic and Atmospheric Administration). 2012. Biennial report to Congress on the Deep Sea Coral Research and Technology Program. Silver Spring, MD: National Oceanic and Atmospheric Administration, National Marine Fisheries Service.

NOAA (National Oceanic and Atmospheric Administration). 2013. EFH areas protected from fishing in the Gulf of Mexico. National Oceanic and Atmospheric Administration Habitat Program. Available from: http://www.habitat.noaa.gov/protection/efh/newInv/EFHI/docs/gfmc_datasheet.pdf via the Internet. Accessed 20 September, 2013.

NOAA Subsurface Monitoring Unit. 2013. BP Deepwater Horizon Oil Spill Cumulative NESDIS SAR Composite. Layer created by NESDIS Satellite Analysis Branch, 9/28/11 using ESRI ArcGIS 9.3. Available from: http://gomex.erma.noaa.gov/. Accessed 4 March, 2014.

NOEP (National Ocean Economics Program). 2012. National Ocean Economics Program. Available from: http://www.oceaneconomics.org via the Internet. Accessed 1 July, 2013.

NOEP (National Ocean Economics Program). 2013. Environmental & Recreational (Non-Market) Values: Valuation Studies Search. Available from: http://www.oceaneconomics.org/nonmarket/NMsearch2.asp via the Internet. Accessed 1 September, 2013.

NRC (National Research Council). 2006. Review of recreational fisheries survey methods. National Research Council, Committee on the Review of Recreational Fisheries Survey Methods. Washington, DC: The National Academies Press.

NRC (National Research Council). 2012. Approaches for ecosystem services valuation for the Gulf of Mexico after the Deepwater Horizon oil spill: Interim report. National Research Council, Committee on the Effects of the Deepwater Horizon Mississippi Canyon-252 Oil Spill on Ecosystem Services in the Gulf of Mexico. Washington DC: National Academies Press. Available from: http://www.nap.edu/catalog.php?record_id=13141 via the Internet. Accessed 25 September, 2013.

NRC (National Research Council). 2013. An ecosystem services approach to assessing the impacts of the Deepwater Horizon oil spill in the Gulf of Mexico. National Research Council, Committee on the Effects of the Deepwater Horizon Mississippi Canyon-252 Oil Spill on Ecosystem Services in the Gulf of Mexico. Washington DC: National Academies Press. Available from: http://www.nap.edu/catalog.php?record_id=18387 via the Internet. Accessed 25 September, 2013.

Oil Spill Commission. 2011. Deep water: The Gulf oil disaster and the future of offshore drilling. Washington, DC: National Commission on the Deepwater Horizon Oil Spill and Offshore Drilling. Available from: http://www.oilspillcommission.gov/final-report via the Internet. Accessed 25 September, 2013.

Osgood KE. 2008. Climate impacts on U.S. living marine resources: National Marine Fisheries Service concerns, activities and needs. U.S. Department of Commerce, NOAA Technical Memorandum NMFSF/SPO-89.

Ostrom E. 2009. A general framework for analyzing sustainability of social-ecological systems. Science. 325:419–422.

Paul JH, Hollander D, Coble PG, Daly K, Murasko S, English D, Basso J, Delaney J, McDaniel L, Kovach CW, et al. 2013. Toxicity and mutagenicity of Gulf of Mexico waters during and after the Deepwater Horizon oil spill. Environ Sci Technol. 47:9651–9659. Available from: http://pubs.acs.org/doi/pdf/10.1021/es401761h via the Internet. Accessed 22 January, 2014.

Peterson CH, Rice SD, Short JW, Esler D, Bodkin JL, Ballachey BE, Irons DB. 2003. Long-term ecosystem response to the Exxon Valdez oil spill. Science. 302:2082–2086.

Plagányi ÉE. 2007. Models for an ecosystem approach to fisheries. FAO Fisheries Technical Paper No. 477. Rome: FAO.

Price VE, Auster PJ, Kracker L. 2013. Use of high-resolution DIDSON sonar to quantify attributes of predation at ecologically relevant space and time scales. Mar Tech Soc J. 47(1):34–46.

Public Law 112-141. 2012. Public Law 112-141: Moving Ahead for Progress in the 21st Century Act (MAP–21), 126 Stat. 405, enacted HR4348 on 7/6/12. Available from: http://thomas.loc.gov/cgi-bin/query/z?c112:H.R.4348 via the Internet. Accessed 26 November, 2012.

Reddy CM, Arey JS, Seewald JS, Sylva SP, Lemkau KL, Nelson RK, Carmichael CA, McIntyre CP, Fenwick J, Ventura GT, et al. 2012. Composition and fate of gas and oil released to the water column during the Deepwater Horizon oil spill. Proc Natl Acad Sci U S A. 109(50):20229–20234.

Reed JK, Koenig CC, Shepard AN. 2007. Impacts of bottom trawling on deep-water Oculina coral ecosystem off Florida. Bull Mar Sci. 81(3):481–496.

SAFMC (South Atlantic Fishery Management Council). 2009. Comprehensive ecosystem-based amendment 1 for the South Atlantic Region. Charleston, SC: South Atlantic Fishery Management Council.

Schirripa MJ, Allee B, Cross S, Kelble C, Rost-Parsons A. 2012. Progress towards an integrated ecosystem assessment for the Gulf of Mexico. Collective Volume of Scientific Papers ICCAT, SCRS/2012/082. Available from: http://www.noaa.gov/iea/Assets/iea/gulf/SCRS-12-082_Schirripa.pdf via the Internet. Accessed 5 September, 2013.

Schofield P. 2010. Update on geographic spread of invasive lionfishes (*Pterois volitans* [Linnaeus, 1758] and *P. miles* [Bennett, 1828]) in the western North Atlantic Ocean, Caribbean Sea and Gulf of Mexico. Aquat Invasions. 5(1):S117–S122.

Schroeder WW, Brooke SD, Olson JB, Phaeuf B, McDonough J, Etnoyer P. 2005. Occurrence of deep-water *Lophelia pertusa* and *Madrepora oculata* in the Gulf of Mexico. *In:* Freiwald A, Roberts JM, editors. Cold-water corals and ecosystems. Berlin: Springer-Verlag. p. 297–307.

SEAMAP (Southeast Area Monitoring and Assessment Program). 2012. FY2011 annual report for the Southeast Area Monitoring and Assessment Program. Ocean Springs, MS: Gulf States Marine Fisheries Commission.

SEDAR (Southeast Data Assessment and Review). 2012. SEDAR background and description. North Charleston, SC: Southeast Data Assessment and Review. Available from: http://www.sefsc.noaa.gov/sedar/ via the Internet. Accessed 1 July, 2013.

Shepard AN, Valentine JF, D'Elia CF, Yoskowitz DW, Dismukes DE. In press. Economic impact of Gulf of Mexico ecosystem goods and services and integration into restoration decision-making. Gulf Science (2014).

Shepard AN, Valentine JF, D'Elia CF, Yoskowitz DW, and Dismukes DE. In press. Economic impact of Gulf of Mexico ecosystem goods and services and integration into restoration decision-making. Gulf Science (2014)

SOST (Subcommittee on Ocean Science and Technology). 2013. Science for an ocean nation: Update of the Ocean Research Priorities Plan. Washington, DC: Subcommittee on Ocean Science and Technology, National Science and Technology Council.

Weisberg R. 2011. Coastal ocean pollution, water quality, and ecology. Mar Tech Soc J. 45(2):35–42.

Welch DW, Brosnan I, Rechisky EL, Porter AD, Challenger W. 2012. Extension of large-scale Continental margin observing systems from fisheries science applications to a complete ocean observing system. Community white paper developed for the Interagency Ocean Observation Committee, IOOS Summit 2012, Herndon, VA. Available from: http://www.iooc.us/2012/list-of-community-white-papers-now-available/ via the Internet. Accessed 1 July, 2013.

White HK, Hsing P, Choc W, Shank TM, Cordes EE, Quattrini AM, Nelson RK, Camilli R, Demopoulos AWJ, German CR, et al. 2012. Impact of the Deepwater Horizon oil spill on a deep-water coral community in the Gulf of Mexico. Proc Natl Acad Sci U S A. 109(50):20303–20308. Available from http://www.pnas.org/cgi/doi/10.1073/pnas.1118029109 via the Internet. Accessed 22 January, 2014.

Wilson JA. 2006. Matching social and ecological systems in complex ocean fisheries. Ecol Soc. 11(1):1–9.

Yender RA, Michel J, Shigenaka G, Mearns A, Hunter CL. 2010. Oil spills in coral reefs: Planning and response considerations. Silver Spring, MD: National Oceanic and Atmospheric Administration, National Ocean Service, Office of Response and Restoration.

chapter ten

Mangrove connectivity helps sustain coral reef fisheries under global climate change

Janet A. Ley

Contents

Introduction

Severe impacts of global climate change on marine fisheries resources are seemingly inevitable. Political and social pressures are not likely to bring about the necessary economic, societal, and resource use transformations soon enough to reverse damage to fisheries (Hoegh-Guldberg and Bruno 2010). Resource managers seek to implement conservation measures that may allow at least some exploited fish populations on coral reefs to survive under altered ecosystem conditions. Conservation of coral reef fisheries centers on the tropics, where people have traditionally relied on food from the sea. Thus, incentives to implement steps for offsetting losses due to climate change may be particularly valuable in those regions. This chapter presents an overview of climate change–induced impacts on two of the most important habitats in tropical fisheries conservation: mangroves and coral reefs. After the potential benefits of protecting mangrove habitats as an adaptive measure to conserve coral reef–associated fisheries under global climate change are highlighted, two emerging areas of research are addressed: (1) quantifying the degree to which coral reef finfish populations of fishery and conservation value depend on mangrove habitats as nurseries and (2) determining whether initiatives for protecting mangroves can mitigate the loss of coral reef habitat, thus helping to protect valuable fisheries.

While this chapter emphasizes the value of mangroves as finfish nurseries, other back reef habitats (e.g., seagrass beds, coral patches, and macroalgae) are also valuable for the juveniles of a variety of fishery-associated species, including crustaceans (Heck et al. 2003,

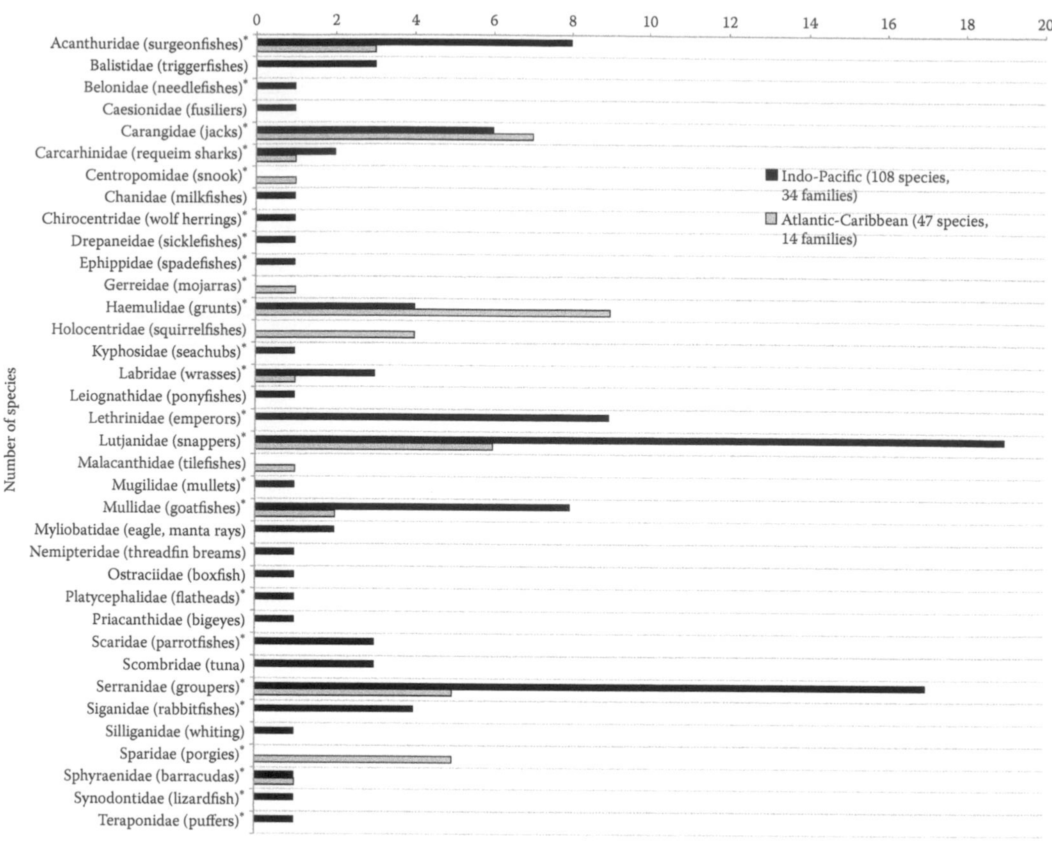

Figure 10.1 Coral reef finfish families (n = 37) of fishery importance by global region. Families with mangrove nursery species (n = 24) are indicated by an asterisk (*). A list of these nursery species is given in Table 10.2. (From Nagelkerken et al. 2000, Pet-Soede et al. 2001, Green et al. 2006, NOAA/Western Pacific Regional Fishery Management Council 2010, State of Queensland 2010, NOAA/Caribbean Fishery Management Council 2011, Smith et al. 2011.)

Adams et al. 2006). This paper addresses "finfish species" that are caught by coral reef fisheries and likely to be consumed as food by humans. Many of the fisheries discussed in this paper are managed by governmental bodies such as the National Marine Fisheries Service (USA) and the Queensland Government (Australia), or have been assessed under the auspices of recognized conservation agencies (e.g., Nature Conservancy; Green et al. 2006) (see references in Figure 10.1).

Trends in global climate change: Coral reefs

The concentration of carbon dioxide (CO_2) in the atmosphere has increased from a global average of 280 ppm in preindustrial times (i.e., prior to 1750) to 390 ppm in 2007, and is predicted to increase to 560 ppm by 2100; consequently, the average global temperature is expected to increase by between 1.1°C and 6.4°C (Solomon et al. 2007). Given these atmospheric changes, a global rise in average sea level of up to 2 m is predicted by 2100 (Vermeer and Rahmstorf 2009). Changes in these fundamental physical conditions will generate further changes in other conditions and processes, such as storm intensity, rainfall distribution, and perhaps most crucially for coral reefs, chemical cycling (Record et al. 2013). The world's coral reefs have developed and grown for thousands of years, supported by a stable carbonate cycle and consistent surface water temperatures (Silverman et al. 2009). Changing global climate conditions have destabilized this balance, resulting in ocean acidification and episodes of coral bleaching. Calcium carbonate, which provides the structure of the coral reef, is dissolving, and new growth is either decreasing in linear extension rate or carbonate skeletal density (Hoegh-Guldberg et al. 2007). Ocean acidification is now occurring faster than at any time in the past, and based on projected trends, most of the world's coral reefs will be destroyed by 2100 (Silverman et al. 2009, Hoegh-Guldberg 2012, Kennedy et al. 2013). Coral reefs are subject to severe impacts from these global phenomena, as well as local and regional anthropogenic stressors.

Australian monitoring programs reveal that live coral cover on the Great Barrier Reef declined from 28% in 1985 to 14% in 2012 (De'ath et al. 2013). For the Caribbean Sea, a meta-analysis of 65 separate studies found that live coral cover declined from 55% in 1977 to 10% in 2001 (Gardner et al. 2003). Attributing the reasons for these declines to specific causes can be difficult. However, a recent Caribbean study modeled the effects of stressors due to climate change (e.g., elevated temperatures, ocean acidification) and biological perturbations (e.g., disease, sea urchin population surge and dieoff), as well as local anthropogenic stressors (e.g., overfishing, eutrophication) (Kennedy et al. 2013). This model was used to explore future scenarios for coral reef resilience through 2080, considering the effects of: (1) alternative local management actions (e.g., no-take marine reserves, reduced eutrophication), and (2) alternative global actions that reduced emissions of greenhouse gases. The model showed that implementing a variety of local management actions, without global actions to reduce the rate of emissions, led to a delay in reef losses of about 10 years, but by 2080 all coral reefs were destroyed. Alternatively, model scenarios in which a variety of local management actions were implemented, and in addition global actions were implemented to substantially reduce emissions, resulted in the survival of at least some areas with live coral cover through 2080. Thus, this comprehensive model for the Caribbean suggested that both local management actions protecting coral reefs and strong global action reducing greenhouse gas emissions are needed to prevent catastrophic degradation of reef structures and loss of their benefits to humans.

Coral reefs and associated fisheries

Global fisheries yield from coral reefs has an estimated annual value of at least $5 billion. This figure does not fully account for the value of coral reef fisheries to the millions of people in tropical countries, particularly those in developing nations, who are directly dependent on food harvested from local reefs (Munday et al. 2008). Officially managed and locally valued coral reef fisheries catch at least 153 species from 37 finfish families (Figure 10.1). Thus, part of the concern over coral reef conservation involves the impacts of climate-induced changes on the food security and economic well-being of fishery-dependent communities (Bell et al. 2013). But because so many species are involved, the mechanisms and processes linking habitat changes to fisheries sustainability are often unknown.

To help predict the impact of climate-induced coral reef decline on fisheries, resource managers must rely on evidence from previous large-scale coral die-back events and on the results of focused experiments. The most immediate and predictable effect of climate-induced coral bleaching is a decline in live coral cover caused by the mortality of entire coral colonies as well as partial loss of live tissue in many surviving corals (Pratchett et al.

2008). Live coral cover is important for the recruitment and survival of coral-dependent fish that feed on coral itself (e.g., coralivores such as butterflyfishes [Chaetodontidae]). Following a devastating (90%) decline in live coral cover in Papua New Guinea, however, fishery-independent surveys revealed that 50% of the reef fish species had declined to less than half of their original numbers, even though most of these species were not coralivores (Jones et al. 2004). Attempting to explain these observations, Coker et al. (2012) working on experimental patch reefs, demonstrated that live coral attracts a broad spectrum of finfish species as recruits. Furthermore, finfish larvae apparently sense the presence of live coral through olfaction and select living coral reef habitats in preference to macroalgae-dominated habitats (Lecchini et al. 2013). Thus, live coral cover is important for the recruitment of both coral-dependents and other resident fishes (Pratchett et al. 2008, Coker et al. 2012). As climate-induced changes eliminate live coral, reduced recruitment of finfish on reefs can be expected.

Climate-induced coral bleaching kills corals but leaves the skeleton intact, at least for a time (Hoegh-Guldberg 1999). Branching corals eventually break down into rubble, while more massive corals erode in situ or are dislodged by storms. Dead branching and plate corals eventually collapse from bioerosion, reducing rugosity. These processes contribute to long-term declines in structural complexity, reducing reef carrying capacity for fish due to the loss of substrate for their prey (Pratchett et al. 2008, Fenner 2012). In addition, small fish living among dead coral colonies are more vulnerable to predation and, those fish that do survive, migrate to healthy stands if available, which makes them vulnerable to predation and other processes (Coker 2012).

Despite the trends revealed in these fishery-independent surveys and experiments, studies based on fishery-dependent data from reef-based fisheries have not reported declines immediately following episodes of climate-induced coral bleaching (McClanahan et al. 2002, Grandcourt and Cesar 2003). Many fisheries associated with coral reefs target herbivorous or planktivorous groups such as surgeonfishes (Acanthuridae), rabbitfishes (Siganidae), and parrotfishes (Scaridae). Degraded coral reefs become colonized by macroalgae, which provides a forage base for herbivorous fishes, and there appears to be little impact on the planktonic forage base on degraded reefs. So far, then, climate-induced coral bleaching has not led to notable declines in these fisheries (Pratchett et al. 2008).

Clearly, however, the more dependent a species is on living coral substrate, the greater its expected decline in abundance if coral should die (Jones et al. 2004). The effects of climate-induced coral bleaching may take several years to influence the catch rates in local fisheries for species that rely on living coral substrate for recruitment (Graham et al. 2007). Resultant delays in effects may make the impacts of climate change difficult to detect and to separate from the effects of fishing and other impacts (Munday et al. 2008). But the impact of a decrease in live coral cover may be less extreme for reef fish that do not exclusively use coral for settlement and nursery sites. For example, emperors (Lethrinidae) and snappers (Lutjanidae) may be more resilient to loss of live coral substrate because they settle also in noncoral habitats, including mangroves. Mangroves, however, have also been subject to severe impacts by stressors, as discussed in the next section.

Trends in the status of mangroves

Global climate change is likely to influence the large-scale distributional patterns of mangroves. As air temperatures increase, the latitudinal ranges of certain mangrove species are expected to extend poleward (McKee 2012). But macroecological statistical models suggest that mangrove areas will be reduced in the parts of the western hemisphere where they now occur, due to decreased precipitation and runoff (Record et al. 2013). In contrast, the distribution of mangrove areas will be stable in parts of the eastern hemisphere where mangroves now occur, due to increased precipitation and runoff. The macroecological model used to predict the distributional patterns resulting from these changes provides a first approximation of how several of the prominent species of mangroves will respond to climate change on the global scale (Figure 10.2) (Record et al. 2013).

Given these trends in the global-scale impacts of climate change, for any specific location natural processes operating at smaller scales will influence mangrove distributions (Alongi 2008). For example, as sea levels rise under global climate change, in areas with significant tides and riverine sediment input, mangrove substrate will accrete as peat accumulates in the forest and the soil surface expands upward (Valiela et al. 2001, Feller et al. 2010). Thus, based on their adaptability to a variety of conditions, mangroves may inherently have the capacity to survive despite predicted climate change-induced impacts (Alongi 2008, Feller et al. 2010). Unfortunately, in recent decades mangroves have suffered large-scale anthropogenic displacement throughout their distributional range, eliminating

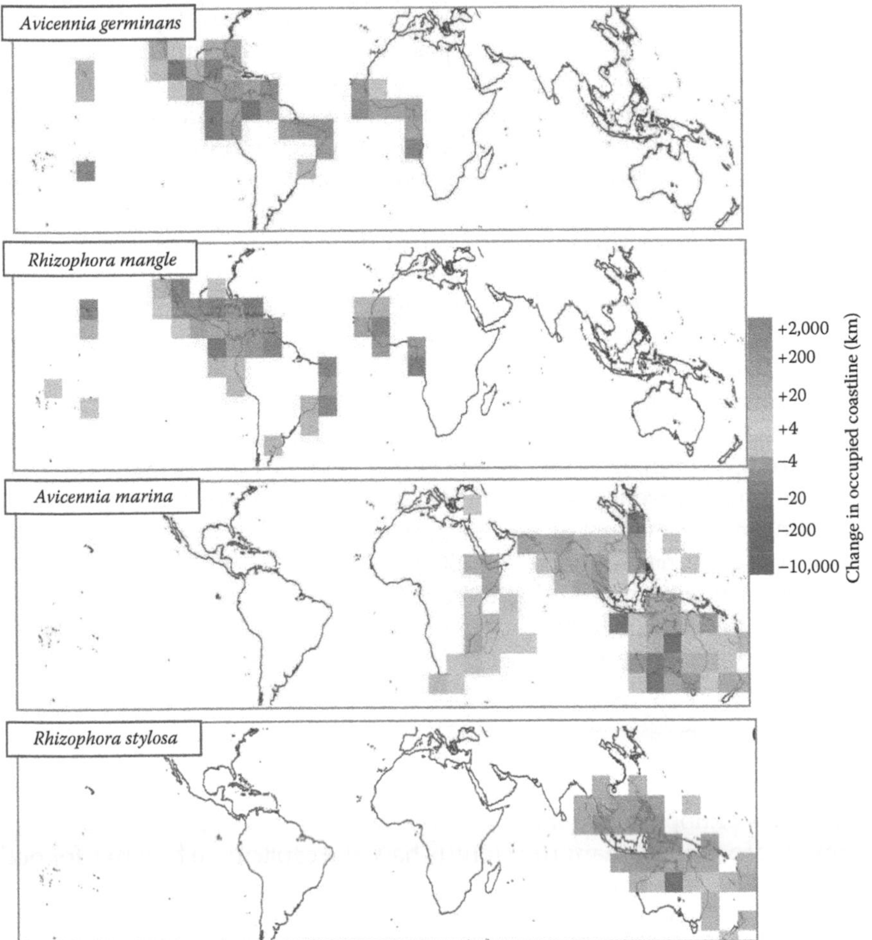

Figure 10.2 **(See color insert)** Change in predicted occupancy for four prominent mangrove species projected for 2080 relative to current occupancy. (From Record, S., Charney, N.D., Zakaria, R.M., Ellison, A.M., *Ecosphere*, 4, art34, 2013. With permission.)

their ability to protect against shoreline erosion as well as their value as habitat for fish and wildlife.

Mangrove forests once occupied more than 200,000 km^2 of the world's tropical coasts (Duke et al. 2007). Though many national and international organizations have recognized that mangroves are valuable, direct elimination continues unabated almost worldwide (Polidoro et al. 2010). Of the 70 species of mangroves, 11 are at an elevated threat of extinction (Polidoro et al. 2010). A fine-scale analysis using more than 1000 recent Landsat images showed that the spatial area historically covered by mangroves has been reduced by one-third, to 137,760 km^2 (Giri et al. 2011). Between 1980 and 2000 alone, about 35% of the area of the world's mangroves was destroyed through a variety of human activities (Valiela et al. 2001, Polidoro et al. 2010). Hamilton (2013) studied historical and recent records from several countries with large mangrove areas (i.e., Ecuador, Bangladesh, Brazil, India, Indonesia, Thailand, and Vietnam) to determine the fate of former mangrove areas. Between 1972 and 2010, 37% of the area historically covered by mangroves in those countries was destroyed; aquaculture facilities were constructed in 63% of the former mangrove areas, while most of the remaining mangroves were uprooted and removed so that the former low-lying wetlands could be filled and developed (Hamilton 2013). Destruction of mangroves to construct aquaculture facilities proliferated in the 1980s and 1990s with financing from agencies such as the World Bank (Valiela et al. 2001). The lifespan of aquaculture ponds established in areas formerly occupied by mangroves is <10 years, after which growers abandon the facility, moving on to destroy a new area of mangroves (Valiela et al. 2001). Abandoned ponds become wastelands, with soils unusable for other purposes due to high levels of salinity, acidity, and toxic chemicals. Most of the aquaculture ponds produce commercially traded shrimp, now comprising the most valuable exported fishery in the world (FAO 2012). Between 1981 and 2005, the proportion of commercial shrimp raised in aquaculture facilities worldwide increased from 3% to 43% (FAO 2012). Thus, commercial aquaculture is a growing industry that continues to decimate mangrove habitats.

Drainage, uprooting, and removal of mangrove forests to create filled land for development has been the second-greatest cause of mangrove destruction (37%) (Hamilton 2013). Decisions to dredge, drain, and fill mangrove wetlands are often made on a piecemeal basis. Preventing mangrove loss at the community level may be at least partly addressed by incorporating mangrove protection as part of planning through zoning codes and incentives for conservation (Gilman et al. 2006).

The market for mangrove wood and wood chips further threatens these habitats. To create ponds and developable dry land, mangroves may be initially harvested and sold to wood chip, charcoal, and other forestry-related industries. In Indonesia, for example, factories to create paper and chipboard products from mangrove wood have proliferated (Kathiresan 2008). Although mangrove forestry can be a sustainable industry, clearfelling on a large scale without replanting remains one of the greatest threats to the world's mangroves (Alongi 2002).

There are rays of hope in this story, however. Because they have extraordinarily high rates of primary production, mangrove forests have the capacity to be sinks for both atmospheric CO_2 and for carbon stored in biomass and sediments (Duke et al. 2007). Carbon captured by the world's oceans ("blue" waters) is stored in mangroves, seagrasses, and salt marshes and is often referred to as "blue carbon" (Siikamaki et al. 2012). Under the United Nations Framework Convention on Climate Change, institutions such as Resources for the Future (Washington, DC) and the United Nations Environmental Programme (Norway) are promoting the concept of trading credits earned for protecting mangroves (primarily in developing countries) in exchange for leniency in reducing carbon emissions (primarily in developed countries). With about 90% of global mangroves growing in developing countries, such carbon payment programs may be highly effective. Daunting challenges are associated with establishing carbon trading programs, such as creating legal and governance instruments (Kathiresan 2008, Donato et al. 2012). Alternatively, conservation and governmental agencies could simply allocate funds for the protection and sustainable management of mangrove ecosystems, justified by the multitude of benefits they provide (Alongi 2011, Caldeira 2012).

Mangroves appear to be more naturally resilient to the impacts of climate-induced changes than coral reefs. However, mangrove wetlands are being lost due to the cumulative effects of decisions at local and national levels, replacing mangroves with aquaculture ponds and more permanent land developments. Like coral reefs, the continued existence of mangroves is vitally important due to their intrinsic value and the services they provide, including habitat benefits to fishery resources.

Mangroves as nursery habitats for coral reef fish

Little scientific information was available on the community of fishes occupying mangrove habitats until the early 1990s. This habitat is so structurally complex that it is difficult to sample with traditional fisheries research gear, and thus few surveys were available until underwater visual census (UVC) methodologies were developed and validated (i.e., initially for coral reefs, Bohnsack and Bannerot 1986). Although UVC has been impracticable in many mangrove settings due to crocodiles or low water clarity, early survey results indicated that well-developed mangroves in marine salinities often harbored later-stage juveniles of fish species usually associated with coral reefs (Rooker and Dennis 1991, Ley 1992). In one of the most comprehensive studies of its kind, long-term (1999–2007) UVC surveys in the Biscayne Bay area of Florida have been conducted in mangrove and coral reef habitats (Bohnsack et al. 1999, Serafy et al. 2003). An analysis of combined data collected under these surveys revealed that 68 finfish species occurred in both mangrove and coral reef habitats (Jones et al. 2010). For 10 of those species that were consistently present and abundant in both habitat types, mean fish length was less in mangroves than on reefs (Figure 10.3). Three of these species are targeted in recreational fisheries (*Lutjanus apodus*, *Lutjanus griseus*, and *Sphyraena barracuda*) and were present in relatively high densities in both mangrove and coral reef habitats. For all three species, their young-of-the-year were highly abundant in the mangrove habitats of Biscayne Bay in 2000; the following year, because fish in this strong cohort had presumably migrated out of the mangrove nursery habitat to reefs as they approached 1 year of age, densities of these species surged in the nearby coral reef habitats. These findings strongly support the hypothesis that mangrove systems in large reef–bay ecosystems provide nursery habitats for coral reef species of importance to fisheries.

Several other studies have addressed hypotheses about ontogenetic connectivity for fishes between mangrove and coral reef habitats at the large ecosystem scale (>500 km^2) (Table 10.1). In these studies, investigators have drawn upon large-scale data compilations such as FishBase (Froese and Pauly 2013), or have combined data from mangrove and coral reef studies conducted in a region (Berkstrom et al. 2012). In these extensive systems, juvenile fish may move long distances (>20 km) as they make ontogenetic habitat shifts along complex gradients. They may temporarily live in fresh, brackish, and fully marine aquatic conditions in a variety of structural settings including streams, marshes, and seagrass beds as well as mangroves.

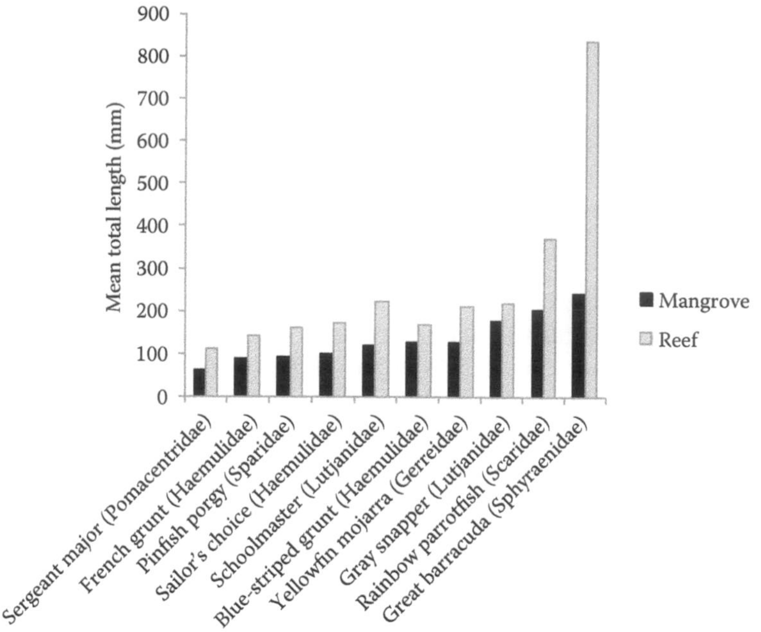

Figure 10.3 Mean total length (mm) of fish in mangrove and coral reef habitats based on underwater visual census surveys in Biscayne Bay, Florida. (Graph derived from Jones, D.L., Walter, J.F., Brooks, E.N., Serafy, J.E., *Mar Ecol-Prog Ser*, 401, 245–258, 2010.)

Studies at a smaller scale (<10 km^2) have focused on mangroves in clear water marine lagoons or bays, having nearby or directly adjacent coral reefs. Surveys have been conducted in singular lagoons and in several lagoons along island shorelines. Often referred to as back reef habitats, mangroves in small lagoons or bays may be a component of a mosaic of habitats that also includes seagrass beds, channels, rocks and rubble, mudflats, patches of coral, and macroalgae (Adams et al. 2006; Ogden et al., Chapter 14, this volume) (Figure 10.4). Detailed surveys within such habitat mosaics have provided valuable information about the processes involved in ontogenetic habitat connectivity. Intensive surveys in single and multiple small mangrove habitats have shown that many species of coral reef finfish begin life by settling in shallow seagrass or rubble habitats, move to mangroves and patch reefs as larger juveniles, and eventually migrate to coral reef habitats as they reach maturity (Mumby et al. 2004).

Regardless of the scale of the ecosystem, an increasing number of studies on tropical and subtropical ecosystems worldwide have revealed that mangrove and associated shallow-water habitats provide nurseries for many coral reef fishery species. Apparent ontogenetic connectivity for finfishes between mangrove and coral reef systems has been documented in the Atlantic/Caribbean region for at least 31 fishery-associated species from 16 families, and in the Indo-Pacific for at least 52 fishery-associated species from 23 families (Table 10.2). These species are usually called mangrove nursery fish (Nagelkerken et al. 2002). Throughout the world's tropical regions, snappers (Lutjanidae) stand out as the most speciose fishery-targeted family occurring in mangroves as juveniles and migrating to coral reefs (and other habitats) as adults, accounting for 13 of the 81 species listed in Table 10.2. Coral reef fisheries for snappers are attractive for anglers and consumers, making the fishery economically valuable as

Table 10.1 Survey-Based Studies of Fish Assemblages in Mangrove Habitats, by Scale (Small <5 km^2, Large >100 km^2), Location, and Migration Distance between Mangrove (Juvenile) and Reef (Adult) Habitats

Scale	Author (year published)	Ecosystem study location	Methods	Distance from mangrove to reef
Small	Rooker and Dennis (1991)	Caribbean Sea Cayo Collado Southwestern Puerto Rico	UVC	Adjacent
Small	Nagelkerken et al. (2000)	Caribbean Sea Lac Bay Bonaire (island)	UVC	<5 km
Small	Nagelkerken and van der Velde (2002)	Caribbean Sea Spanish Waters lagoon Curacao (island)	UVC	<5 km
Small	Aguilar-Perera and Appeldoorn (2007)	Caribbean Sea Montalva Bay Southwestern Puerto Rico	UVC	<5 km
Small	Shibuno et al. (2008)	East China Sea Fukido River mangrove basin Ishigaki Island, Japan	UVC	<5 km
Small	Barnes et al. (2012)	Great Barrier Reef lagoon Pioneer and Hazard Bays Orpheus Island	UVC	Adjacent
Multiple Small	Eggleston et al. (2004)	Florida Keys Reef Key West National Wildlife Refuge Lakes and Marquesas Keys	UVC	<5 km
Multiple Small	Dorenbosch et al. (2004)	Caribbean Sea 11 Reefs with/without neighboring mangrove/seagrass bays Curacao	UVC	<5 km
Multiple Small	Unsworth et al. (2008)	Banda Sea Seagrass and mangrove sites along five islands in Wakatobi Marine National Park Southeastern Sulawesi, Indonesia	Seine nets UVC	<6 km
Multiple Small	Nagelkerken et al. (2002)	Caribbean Sea Coral reefs on six islands with/without mangroves	UVC	<5 km
Large	Sheaves (1995) Ley et al. (2002) Baker and Sheaves (2005)	Great Barrier Reef lagoon Estuarine mangrove-dominated tributaries Northeastern Australia	Fish traps Gill nets	>20 km
Large	Ley (1992) Ley et al. (1999) Ley and McIvor (2002) Faunce et al. (2004)	Florida Bay/Keys Reef Mangrove islands and tributary shorelines Key Largo/Everglades National Park	Block nets with rotenone UVC	>20 km

(*continued*)

Table 10.1 (Continued) Survey-Based Studies of Fish Assemblages in Mangrove Habitats, by Scale (Small <5 km², Large >100 km²), Location, and Migration Distance between Mangrove (Juvenile) and Reef (Adult) Habitats

Scale	Author (year published)	Ecosystem study location	Methods	Distance from mangrove to reef
Large	Bohnsack et al. (1999) Serafy et al. (2003) Jones et al. (2010)	Biscayne Bay/Florida Keys Reef Mangrove shorelines and coral reefs Biscayne Bay National Park, Miami	UVC	>20 km
Large	Mumby et al. (2004)	Caribbean Sea and Mesoamerican Barrier Reef 6 Large atolls and reefs with/without mangroves Belize and Mexico	UVC	
Large	Ley (2005)	Great Barrier Reef lagoon 11 Estuarine mangrove-dominated tributaries Townsville to Cape York, Australia	Gill nets	>20 km
Large	Berkstrom et al. (2012)	Western Indian Ocean Mangroves, seagrass beds, and coral reefs Zanzibar, Tanzania, Africa	UVC	Various

Note: All the studies listed here documented mangrove habitat use by juveniles of species that are known to live on coral reefs as adults.

a commercial enterprise. In Florida, for example, the commercial fishery for yellowtail snapper (*Ocyurus chrysurus*) in Monroe County (Florida Keys) is worth $4 million annually (Florida Fish and Wildlife Conservation Commission 2013). In Figure 10.1, 37 families included species of importance to coral reef fisheries worldwide and among those, 24 families are also listed in Table 10.2, indicating that these families include species that use mangroves and associated habitats as nurseries. Thus, at least 65% of the coral reef fish families that support global fisheries include species that use mangrove habitats as a nursery.

The list of mangrove nursery species in Table 10.2 includes nine species recognized as threatened with extinction by the International Union for the Conservation of Nature (IUCN 2013). Such coral reef species have long been desirable food resources for humans because of their large size and ease of capture (McClenachan 2009, McClenachan and Kittinger 2013). Five species of large serranids (groupers) were popularly exploited for sport and food on both coral reefs as adults and in mangroves where they occur as large juveniles (Table 10.2). Among the parrotfishes (Scaridae), the largest parrotfish species in the world (*Bolbometopon muricatum*) is also designated as a threatened species. Known as the bumphead parrotfish, it occurs as adults on coral reefs throughout the Indo-Pacific and Red Sea, but its juveniles are found mainly in mangrove lagoons (Hamilton et al. 2008, Bellwood and Choat 2011). Likewise, the largest parrotfish on Caribbean and Florida reefs (*Scarus guacamaia*) is listed as vulnerable to extinction. Known as the rainbow parrotfish, this species appears to require access to mangrove habitats as a juvenile and has suffered

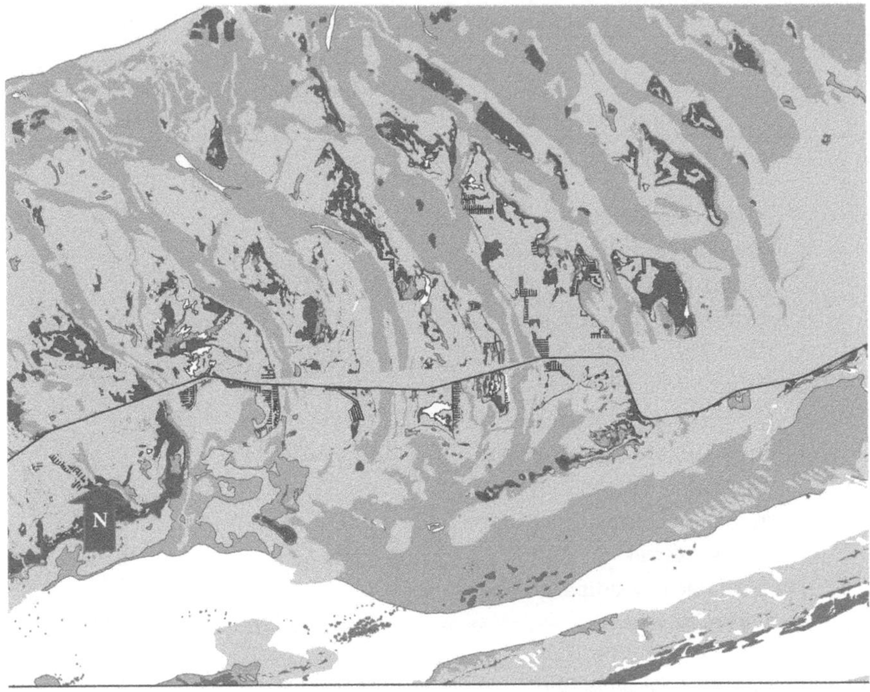

Figure 10.4 **(See color insert)** Mosaic of habitats for the area surrounding Big Pine Key, Florida Keys. Terrestrial and shoreline habitats are uplands (gray) and mangrove (brown). Aquatic habitats are aligned north to south between the islands, with continuous seagrass beds (dark green) and patchy seagrass beds (light green). Coral patch reefs and the continuous offshore reef tract are shown in red. The Overseas Highway is shown as a black line running east to west.

local extinction when mangroves have been removed (Mumby et al. 2004, Machemer et al. 2012). These seven species of large-bodied groupers and parrotfish have historically been subject to heavy exploitation leading to their current status as threatened. Two threatened requiem sharks (*Carcarhinus leucas* [bull shark] and *Negaprion acutidens* [lemon shark]) also use mangrove habitats as nurseries. These sharks are exploited primarily for their fins, but also die when they become inadvertently caught as bycatch in a variety of coastal fisheries (Simpfendorfer and Burgess 2009).

The IUCN has recognized that even if fishing pressure on these world-valued species is reduced, the loss of critical juvenile habitat—mangroves—may impede the recovery of populations. The recovery of these fish species depends on both their protection from exploitation and the conservation of their habitat, including protection of the mangrove habitats they rely on for nurseries.

Connectivity between mangrove and coral reef habitats may be important in sustaining fish populations and habitat value, but the degree of habitat interdependency varies among species. Surveys such as those listed in Table 10.2 can be used to quantify the number or density of juveniles that occur in a purported nursery area, as well as the number of adults that occur on a coral reef. It cannot be assumed, however, that juveniles recruit to adult habitats in direct proportion to the number of juveniles found in a given nursery

Table 10.2 Finfish Found on Coral Reefs as Adults That Use Mangrove Habitats as Nurseries

Region	Coral reef adults	Mangrove nursery	Mangrove and seagrass nursery	Mosaic nursery
Atlantic/ Caribbean	Acanthuridae (surgeonfishes)			
	Acanthurus coeruleus			(4)
	Carangidae (trevallies, jacks)			
	Caranx hippos	(1) (19)		
	Carcarhinidae (requiem sharks)			
	Carcarhinus leucas[a]	(2)		
	Centropomidae (snook)			
	Centropomus undecimalis	(2) (3) (19)		
	Ephippidae (spadefish, batfish)			
	Chaetodipterus faber	(2)		
	Gerreidae (mojarras)			
	Gerres cinereus	(3) (2)	(22)	(8)
	Haemulidae (sweetlips, grunts)			
	Haemulon sciurus	(1) (2) (9) (3)	(4) (6) (7) (21) (22)	(8)
	Haemulon flavolineatum	(1) (3) (21)		(4) (7) (9)
	Haemulon parra	(2) (7) (3)		
	Haemulon plumierii	(1)		(9)
	Kyphosidae (sea chubs)			
	Kyphosus sectatrix	(3)		
	Lutjanidae (snappers)			
	Ocyurus chrysurus	(1) (7)	(6)	(9)
	Lutjanus apodus	(1) (2) (7) (3) (21)	(4) (6) (9)	
	Lutjanus mahogoni	(1) (7)		
	Lutjanus jocu	(2)	(22)	
	Lutjanus synagris			(8)
	Lutjanus analis		(7)	
	Lutjanus griseus	(1) (2) (7) (3) (19)	(22)	(8)
	Mugillidae (mullets)			
	Mugil curema	(2)		
	Pomacentridae (damselfishes)			
	Abudefduf saxatilis	(3)		
	Pomacanthidae (angelfishes)			
	Pomacanthus arcuatus			(8)
	Scaridae (parrotfish)			
	Scarus guacamaia[a]	(9) (3)		
	Scarus iserti		(6)	(7)
	Scarus taeniopterus	(21)		
	Sparisoma chrysoterum			(7)

Table 10.2 (Continued) Finfish Found on Coral Reefs as Adults That Use Back Reef Habitats as Nurseries

Region	Coral reef adults	Mangrove nursery	Mangrove and seagrass nursery	Mosaic nursery
	Serranidae (groupers)			
	Epinephelus striatus[a]			(5)
	Epinephelus itajara[a]	(10) (11)		
	Sparidae (porgies)			
	Archosargus probatocephalus	(2)		
	Archosargus rhomboidalis	(22)		
	Lagodon rhomboides	(3)		
	Sphyraenidae (barracudas)			
	Sphyraena barracuda	(1) (2) (7) (3)	(4) (8)	
Indo-Pacific	Acanthuridae (surgeonfishes)			
	Naso vlamingii			(23)
	Belonidae (needlefishes)			
	Tylosurus crocodilus	(13) (18)		
	Carangidae (trevallies, jacks)			
	Atule mate	(18)		
	Caranx ignobils	(13) (18)		(23)
	Caranx sexfasciatus	(18)		
	Gnathonodon speciosus	(13)		
	Scomberoides commersonnianus	(13) (18)		
	Scomberoides lysan	(13) (18)		
	Scomberoides tala	(13) (18)		
	Scomberoides tol	(13)		
	Trachinotus blochii	(13)		
	Carcarhinidae (requiem sharks)			
	Carcharhinus leucas[a]	(13)		
	Negaprion acutidens[a]	(13)		
	Chirocentridae (wolf herring)			
	Chirocentrus dorab	(13)		
	Drepaneidae (sicklefishes)			
	Drepane punctata	(13)		
	Gerreidae (mojarras)			
	Gerres oyena	(13)		(23)
	Haemulidae (sweetlips, grunts)			
	Plectorhinchus gibbosus	(13)		
	Plectorhinchus gaterinus	(16)		
	Hemiramphidae (halfbeaks)			
	Hyporhamphus dussumieri		(23)	
	Labridae (wrasses)			
	Choerodon anchorago			(23)

(continued)

Table 10.2 (Continued) Finfish Found on Coral Reefs as Adults That Use Back Reef Habitats as Nurseries

Region	Coral reef adults	Mangrove nursery	Mangrove and seagrass nursery	Mosaic nursery
	Lethrinidae (emperors)			
	Lethrinus harak	(23)	(16)	
	Lethrinus lentjan		(16)	
	Lethrinus obsoletus			(23)
	Lethrinus ornatus			(23)
	Lutjanidae (snappers)			
	Lutjanus fulviflamma		(12) (16)	(17)
	Lutjanus fulvus		(12)	
	Lutjanus monostigma	(16)	(12)	
	Lutjanus gibbus		(12)	
	Lutjanus russelli	(3) (14) (18)		
	Lutjanus argentimaculatus	(3) (14) (13) (16)		
	Mullidae (goatfishes)			
	Mulloidichthys flavolineatus		(16)	
	Parupeneus barberinus		(12) (16)	
	Parupeneus macronemus		(16)	(23)
	Platycephalidae (flatheads)			
	Platycephalus indicus	(13)		
	Plotosidae (eeltail catfishes)			
	Paraplotosus albilabris	(13)		
	Pomacentridae (damselfishes)			
	Abudefduf bengalensis			(17)
	Abudefduf sexfasciatus			(17)
	Scaridae (parrotfishes)			
	Bolbometopon muricatum[a]			(20)
	Scarus ghobban		(16)	
	Scarus psittacus		(16)	
	Hipposcarus harid		(16)	
	Serranidae (groupers)			
	Epinephelus coioides[a]	(13) (15)		
	Epinephelus lanceolata[a]			
	Epinephelus malabaricus[a]	(13) (15) (16)		
	Siganidae (rabbitfishes)			
	Siganus sutor		(16)	
	Siganus canaliculatus		(16) (23)	(17)
	Siganus lineatus			
	Silliganidae (whiting)			
	Sillago sihama	(13) (18)		
	Sphyraenidae (barracudas)			
	Sphyraena barracuda	(13)	(16)	

Table 10.2 (Continued) Finfish Found on Coral Reefs as Adults That Use Back Reef Habitats as Nurseries

Region	Coral reef adults	Mangrove nursery	Mangrove and seagrass nursery	Mosaic nursery
	Sphyraena jello	(13)	(16)	
	Synodontidae (lizardfishes)			
	Saurida gracilis	(18)		
	Tetraodontidae (puffers)			
	Arothron hispidus	(13)		

Sources: Atlantic/Caribbean: (1) Rooker and Dennis (1991) (Cayo Collado, southwestern Puerto Rico); (2) Ley (1992), Ley et al. (1999, 2002) (northeastern Florida Bay); (3) Serafy et al. (2003), Jones et al. (2010) (Biscayne Bay); (4) Nagelkerken et al. (2000) (Bonaire); (5) Dahlgren and Eggleston (2001) (Great Exuma, Bahamas); (6) Nagelkerken et al. (2002) (six Caribbean islands); (7) Nagelkerken and van der Velde (2002), Dorenbosch et al. (2004) (Curacao); (8) Eggleston et al. (2004) (lower Florida Keys); (9) Mumby et al. (2004) (Belize); (10) Frias-Torres (2007) (lower Florida Keys); (11) Koenig et al. (2007) (Ten Thousand Islands, Everglades National Park, southwestern Florida); (19) Faunce et al. (2004) (three tributaries to northeastern Florida Bay); (21) Aguilar-Perera and Appeldoorn (2007) (southwestern Puerto Rico); (22) Clark et al. (2009) (southwestern Puerto Rico, gill nets).

Indo-Pacific: (12) Shibuno et al. (2008) (Ryukyu Islands, southern Japan); (13) Ley et al. (2002), Ley (2005) (11 large estuaries, Great Barrier Reef lagoon); (14) Thollot and Kulbicki (1988) (St. Vincent Bay, southwestern New Caledonia); (15) Sheaves (1995) (estuaries and reefs, Great Barrier Reef lagoon); (16) Berkstrom et al. (2012) (Island of Zanzibar, Tanzania, western Africa); (17) Barnes et al. (2012) (Orpheus Island, Great Barrier Reef lagoon); (18) Baker and Sheaves (2005) (17 estuaries, Great Barrier Reef lagoon); (20) Bellwood and Choat (2011); (23) Unsworth et al. (2008) (Indonesia, Kaledupa Island near Sulawesi Island).

Notes: Species in these surveys were assigned to mangrove habitats as reported in the source and defined as: (1) mangrove, (2) mangrove and seagrass, or (3) mosaic (mangrove, seagrass, channel, rubble, and patch reef habitats). In each of these studies, mangroves were used as a nursery habitat, but not all the studies included data on the use of nonmangrove habitats (Mangrove and seagrass, Mosaic). Species have been included in this list if: (a) reported abundance was >10; (b) it was listed as reef associated in FishBase (Froese and Pauly 2013); and (c) the maximum size collected was less than two-thirds the maximum total length given for the species in FishBase (i.e., assumed to be small or large juveniles; see Nagelkerken and van der Velde 2002).

[a] IUCN (2013).

habitat. Many juveniles may not survive to the migration stage, or they may become prey as they migrate to adult habitats. Studies that go beyond quantification of densities by size classes in particular habitats are needed to answer the question: Do mangroves and associated habitats contribute substantial numbers of successful recruits to coral reefs?

Quantifying mangrove dependence: Concepts

Habitat protection has only recently become institutionally recognized as essential in managing recreational and commercial fisheries. For example, in 1996, the Magnuson-Stevens Fishery Conservation and Management Act (U.S. PL 94-265) was amended, mandating that fisheries management agencies in the United States and its territories identify and protect essential fish habitat, defined as "...those waters and substrates necessary to fish for spawning, breeding, feeding or growth to maturity" (16 USC 1853, Section 303). Thus, conserving essential fish habitat encompasses measures that protect areas important during adult stages and also younger life-history stages.

Beck et al. (2001:635) refined the definition of essential fish habitat as applied to fish "nurseries," as areas whose "...contribution per unit area to the production of individuals that recruit to adult populations is greater, on average, than production from other habitats in which juveniles occur." This definition of fish nursery habitat has been widely accepted by the scientific and resource management communities. In many ways, it has fundamentally changed the science of fisheries biology and ecology, particularly with respect to the need to quantify the "major contribution of recruits" to adult populations on a spatially explicit basis (Figure 10.5). The questions presented in Figure 10.5 outline this new approach to fisheries research, which expands upon traditional surveys of fish densities.

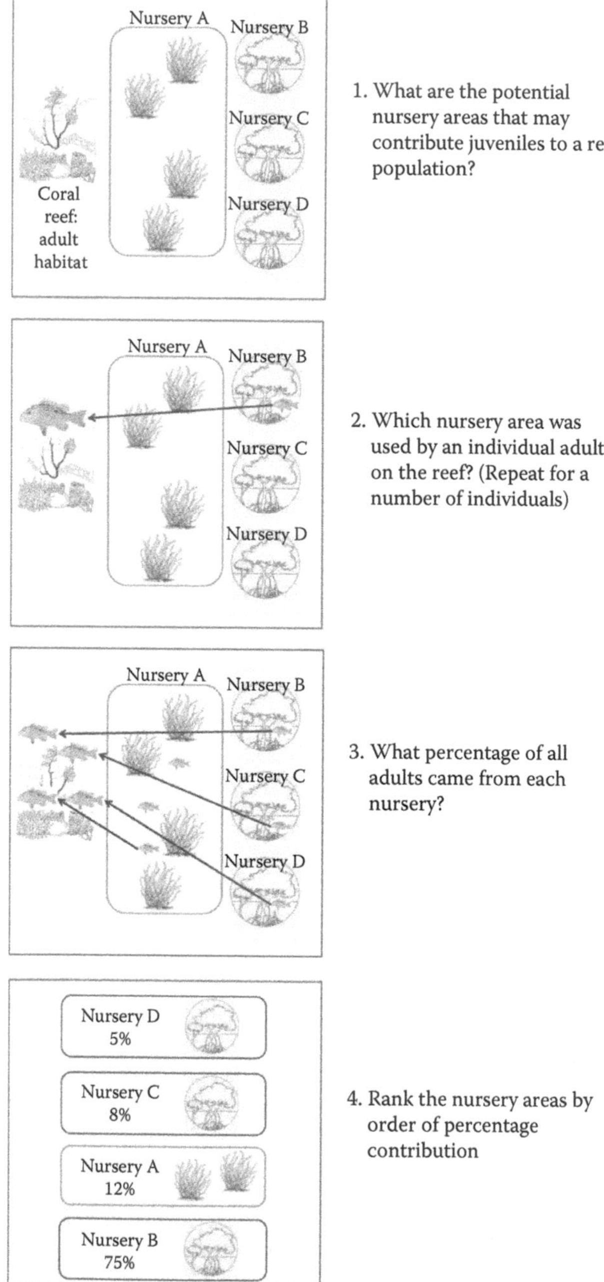

Figure 10.5 Overview of an approach to retrospectively identifying and quantifying the contribution of recruits from four purported nursery habitats to a coral reef fish population.

One new aspect is that the basic sampling unit in the new approach is an individual fish and its characteristics, rather than abundances or densities of fish sampled along a visual census transect or obtained via a net. Recently, the development of new techniques to track movement from nursery to adult habitats has paralleled the growing demand for these types of data.

Quantifying mangrove dependence: Methods

Studies using newer methodologies that facilitate tracing nursery habitat use are now being published. They use four general approaches: (1) acoustic telemetry, (2) stable isotope analysis, (3) otolith (earbone) microchemistry analysis, and (4) genetic parentage analysis (Table 10.3). In Biscayne Bay, Florida, over a few months, 50% of later-stage juvenile gray snapper (*L. griseus*) that had been acoustically tagged migrated from mangrove habitats toward offshore reefs at least 8 km away (Luo et al. 2009). In Japan, stable-isotopic analysis of fish muscle tissue revealed that for 88% of the blacktail snapper (*Lutjanus fulvus*) analyzed, individuals foraged in the mangrove habitat as juveniles for several months and then made a complete habitat transition as they grew and began to consume coral reef prey (Nakamura et al. 2008).

Three recent studies applied the concept that fish otoliths grow continuously throughout the life of a fish and remain chemically inert once formed. These bony structures in the ear can thus provide a detailed history of a fish's environment if it resided in a particular habitat for a sufficient period of time. In Puerto Rico and St. Croix, analysis of otolith chemistry revealed that while French grunt (*Haemulon flavolineatum*) recruited to the coral reef from more than one type of nursery habitat, young schoolmaster snapper (*L. apodus*) had only occupied mangrove nurseries (Mateo et al. 2010) (Table 10.3). In the Red Sea, otolith analysis revealed that 21% of blackspot snapper (*Lutjanus ehrenbergii*) collected at inshore reefs had migrated from mangrove and seagrass nurseries, but 79% on mid- and offshore habitats apparently developed in coral reef nursery habitats (McMahon et al. 2012). In studies conducted on coral reefs off Tanzania, otolith analysis revealed that 70% of adult pink ear emperor (*Lethrinus lentjan*) had coral reef signatures, indicating that its populations may be largely replenished by juveniles that have settled and developed on the coral reef itself. In contrast, adult thumbprint emperor (*Lethrinus harak*) collected on coral reefs had mainly recruited from seagrass (53%) and mangrove (29%) nurseries. Finally, 53% of adult dory snapper (*Lutjanus fulviflamma*) on coral reefs had grown up in mangrove nurseries and 10% in seagrass beds, as determined by otolith chemistry (Kimirei et al. 2013).

The fourth new approach to the study of nursery–adult habitat connectivity is known as genetic parentage analysis. This approach determines where offspring from a spawning coral reef–associated population went as they dispersed to nursery habitats (Christie et al. 2010). In Table 10.3, an example of the application of genetic parentage analysis is summarized for yellow tang (*Zebrasoma flavescens*). While not including mangroves, this study analyzed connectivity between parents and offspring in several coral-dominated bays along the island of Hawaii. Similarly, if adults spawning on a coral reef can be genotyped, genetically matching juveniles in purported mangrove nurseries may reveal linkages between the mangrove nursery and reef adult habitats.

The examples described above and summarized in Tables 10.3 and 10.4 provide the type of quantitative information that can be applied in defining nursery habitats for the early life stages of coral reef fishery species. Dahlgren et al. (2006:292) modified the nursery definition of Beck et al. (2001), defining effective juvenile habitat as "…a habitat

Table 10.3 Methods Used for Directly Tracing Connectivity of Individuals between Nursery and Coral Reef Adult Finfish Populations

General approach	Author (year published), location	Mangrove–reef connectivity research questions	Methods	Results
Acoustic telemetry	Luo et al. (2009), Biscayne Bay National Park, Florida	How many *Lutjanus griseus* in mangrove habitats migrated to nearby coral reefs?	Fish were surgically implanted with individually coded acoustic transmitters and tracked using an array of automated receivers.	Of 14 *L. griseus* implanted with transmitters in bay mangrove habitats and tracked for up to 336 days, seven moved through bay-to-ocean passes and three of these were detected on the coral reef (10–15 km away).
Stable isotope muscle tissue analysis	Nakamura et al. (2008), Ishigaki Island, East China Sea, Japan	What are the nursery origins of subadult *Lutjanus fulvus* on coral reefs?	Tracing the food web-derived stable isotopic signature of muscle tissue indicates if *L. fulvus* recently migrated to coral reefs from mangroves.	Of 41 subadult *L. fulvus* collected on the coral reef, 36 had a carbon isotopic signature indicating previous growth in the mangrove habitat. The other five had always lived on the coral reef.
Otolith elemental and isotopic analysis	Mateo et al. (2010), Puerto Rico and St. Croix, Caribbean Sea	What are the nursery origins of subadult *Lutjanus apodus* and *Haemulon flavolineatum* on coral reefs?	The elemental and stable isotopic signatures of juvenile ($n = 378$) otoliths were matched to the juvenile (core) portion of subadult ($n = 189$) otoliths	Of 96 subadult *H. flavolineatum* collected on the coral reef, 60 had migrated from mangroves while 36 had migrated from seagrass habitats. Of 90 subadult *L. apodus* collected on the coral reef, 89 had migrated from mangroves while one had migrated from seagrass habitats.

Otolith and muscle tissue stable isotopic analysis	McMahon et al. (2012), Saudi Arabia, Red Sea	What is the nursery origin of adult *Lutjanus ehrenbergii* collected along an inshore–offshore gradient?	The essential amino acid d^{13}C signature in fish (n = 89) muscle was determined for five purported nursery areas. Then the d^{13}C signatures of otolith cores for each adult collected across a 50 km inshore–offshore gradient were matched to nursery values.	Of 125 adult *L. ehrenbergii* collected from reefs, 30 had migrated from the seagrass/mangrove habitats (all within 32 km from shore).
Otolith stable isotopic analysis	Kimirei et al. (2013), Tanzania, East Indian Ocean	What are the nursery origins of adult *Lethrinus harak*, *Lethrinus lentjen*, and *Lutjanus fulviflamma* on coral reefs?	The d^{13}C and d^{18}O signatures of juvenile (n = 265) otoliths were matched to the juvenile (core) portion of adult otoliths. Adults and juveniles were collected in seagrass, mangrove, and coral reef habitats.	Of 25 adult *L. harak* collected on the reef, 20 had migrated from seagrass/mangrove nurseries. Of 74 adult *L. lentjen* collected on the reef, 22 had migrated from seagrass/mangrove nurseries. Of 40 *L. fulviflamma* collected on the reef, 25 had migrated from seagrass/mangrove nurseries.
Genetic parentage analysis	Christie et al. (2010), Hawaii,Pacific Ocean	What bays did *Zebrasoma flavescens* (yellow tang) offspring use as nurseries after dispersing from the bay in which their parents spawned?	DNA extraction was used to genotype 506 adults and 566 juveniles on 20 microsatellite loci. Fish were collected from coral reef habitats in nine bays around the island of Hawai'i.	Four parent–offspring pairs were found. Parents ranged from bays 15 to 184 km distant from offspring

Table 10.4 Applications of Two Juvenile Habitat Definitions

Author (year published)	Species	Habitat types	Percent contribution (%)	Effective juvenile habitat (Dahlgren et al. 2006)	Percent total area (%)	Percent contribution per percent total area (%)	Nursery (Beck et al. 2001)
Nakamura et al. (2008)	*Lutjanus fulvus* (blacktail snapper)	Coral reef	12	No	96	0.13	No
		Mangrove	88	Yes	4	22.00	Yes
		Mean	*50*				
Mateo et al. (2010)	*Haemulon flavolineatum* (French grunt)	Seagrass	37	No	31	1.20	Yes
		Mangrove	73	Yes	69	0.91	Yes
		Mean	*50*				
	Lutjanus apodus (schoolmaster snapper)	Seagrass	1	No	31	0.30	No
		Mangrove	99	Yes	69	1.43	Yes
		Mean	*50*				

McMahon et al. (2012)	*Lutjanus ehrenbergii* (blackspot snapper)	Mangrove/seagrass	21	Yes
		Coastal reefs	12	No
		Shelf reefs	13	No
		Island reefs	29	Yes
		Oceanic reefs	25	Yes
		Mean	*20*	
Kimirei et al. (2013)	*Lethrinus harak* (thumbprint emperor)	Seagrass	53	Yes
		Mangrove	29	No
		Coral reef	18	No
		Mean	*33*	
	Lethrinus lentjan (pink ear emperor)	Mangrove/seagrass	70	Yes
		Coral reef	30	No
		Mean	*50*	
	Lutjanus fulviflamma (dory snapper)	Seagrass	10	No
		Mangrove	53	Yes
		Coral reef	38	Yes
		Mean	*33*	

Note: Habitat definitions are offered by Beck et al. (2001) ("Nursery") and Dahlgren et al. (2006) ("Effective juvenile habitat") derived for the studies presented in Table 10.3.

thatvcontributes a greater proportion of individuals to the adult population than the mean level contributed by all habitats used by juveniles, regardless of area coverage." In the current context, applying either the density-based nursery (Beck et al. 2001) or non-density limited (Dahlgren et al. 2006) approach, "contribution" would be identified as the percentage of juveniles from mangrove habitats entering the adult population on coral reef habitats. For the four studies included in Table 10.4, the percentage contribution of juveniles from mangroves to adult reef populations ranged from 21% (*L. ehrenbergii*) to 99% (*L. apodus*). Factoring in contribution per unit area revealed that even though mangrove areas only occupied 4% of the study area, mangroves contributed 88% of the juvenile *L. fulvus* on Ishigaki Island in Japan (Nakamura et al. 2008) (Table 10.4).

Information on the percentage contribution from various nursery habitats to coral reefs not only provides guidance for prioritizing habitat conservation efforts but also indicates the relative vulnerability of a species to climate change. The greater the percentage nursery contribution from coral reefs, the more likely a species will be strongly impacted by the loss of coral cover. Those species that use mangroves as nurseries are potentially less vulnerable to coral habitat loss. However, though more naturally resilient to climate change, mangroves have been severely impacted by widespread removal, threatening any benefits they may offer reef fishes under duress because of the effects of global climate change.

Conservation implications

Coral reef fish populations are frequently limited by the availability of juvenile habitat (Schmitt and Holbrook 2000). As coral reefs die due to climate-induced degradation, the amount of living coral and associated food and shelter for all fish is reduced. Recruitment of most juvenile fishes is expected to decline accordingly (Adam et al. 2011). But if a fish species is a recruitment generalist (e.g., *H. flavolineatum*, Table 10.4) that settles in a variety of habitats in back reef systems (e.g., mangroves, seagrass beds) and successfully migrates to coral reefs, its population may be somewhat buffered against collapse (Graham et al. 2007).

The resilience of an entire coral reef ecosystem may be strengthened by the presence of healthy populations of fishes with the capacity to settle in a broad range of habitats, as was recently observed in Moorea, French Polynesia. Adam et al. (2011) found that at least two abundant species of herbivorous parrotfish (*Chlorurus sordidus* and *Scarus psittacus*) thrived even though corals in fore reef areas were destroyed by a surge in the population of coralivorous crown-of-thorns starfish. The investigators hypothesized that adult parrotfish populations survived the loss of coral on the fore reef because they settled and grew in back reef lagoonal areas as juveniles where crown-of-thorns starfish were not as effective at destroying coral cover. These observations suggest that connectivity between offshore reefs and inshore nursery areas (including mangroves and associated habitats) can buffer populations against climate change and the related decline of coral reefs (Adam et al. 2011). As an additional benefit, the herbivorous parrotfish grew and migrated to the fore reef where they consumed algae that were beginning to colonize coral patches laid bare by the crown-of-thorns starfish. Removing the algal growth, in turn, allowed coral to recruit successfully, eventually leading to reef recovery (Figure 10.6).

A similar synergy apparently enhanced the ecosystem resilience of reefs in Moreton Bay, in eastern Australia. Mangroves there serve as both nurseries and attractive feeding habitats for herbivorous fishes (Olds et al. 2012). Removal of macroalgae was greater

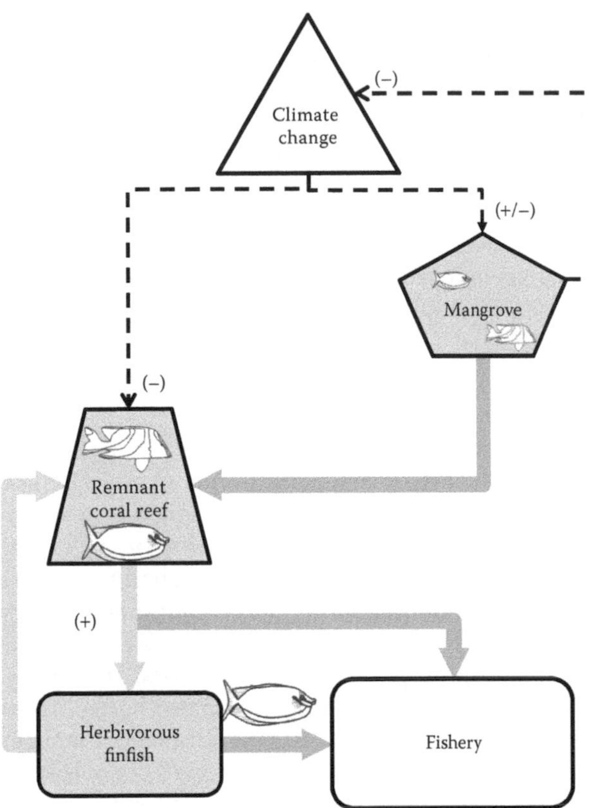

Figure 10.6 Relationships between mangrove and coral reef habitats, linking juvenile and adult life-history phases of numerous finfish species. As climate-induced degradation of coral reefs intensifies (broken arrow), and recruitment associated with live coral cover diminishes, locally protected mangroves provide increasingly important larval settlement and nursery sites for finfish. For herbivorous fishes that settle in mangroves and recruit to coral reefs as adults, consumption of macroalgae clears substrate for coral recruits (shaded arrow). Mangroves sequester carbon, reducing atmospheric greenhouse gases (broken arrow).

on coral reefs that had abundant mangrove habitats nearby, indicating that herbivory was enhanced by fish foraging trips from mangroves to reefs. Establishing no-take zones that protected the herbivore populations from commercial harvest and spearfishing also improved grazing. These results support the idea that reefs may recover from climate-induced bleaching more readily if mangrove-associated herbivores are abundant enough to help control algal recruitment (Figure 10.6).

While herbivorous species may find sufficient food and cover on remnant coral reefs as described above, questions remain concerning the survival of carnivorous species. Most reef fish species are carnivores and their biomass can be three times that of herbivorous fishes on a given coral reef (Heck et al. 2008). Food linkages are more complex for these upper trophic level species. For example, serranids (groupers) tend to be reef-attached, sedentary predators, consuming reef-associated prey fish and invertebrates (Jory and Iversen 1989). Thus, even if some grouper species use a variety of nursery habitats, they may be highly reliant on healthy coral reef systems as adults, and therefore vulnerable to climate change–induced impacts. In contrast, many lutjanids (snappers) and haemulids (grunts) are less reliant on coral reefs for all their needs. Not only are they recruitment generalists, but as adults some species of snappers and grunts make foraging migrations into surrounding seagrass beds and other habitats, supplementing the limited prey base on the reef itself (Ogden and Ehrlich 1977, Luo et al. 2009). Thus, herbivorous acanthurids and ecologically euryphagic lutjanids and haemulids may continue to support exploitable population levels. As global climate change impacts intensify, ontogenetic connectivity between remnant coral reef and protected mangrove habitats may at least partly buffer impacts on associated fisheries (Figure 10.6). Mangrove habitats may enhance reef ecosystem resilience and help sustain valuable reef fisheries. While no panacea for climate change–induced impacts, the protection of mangrove nursery habitats has clear fisheries benefits and should be a top management priority at all institutional levels.

Conclusions

Two broad types of conservation strategies have been referenced in the current review: global activities aimed at reducing or reversing trends in emissions of greenhouse gases, and local/regional activities aimed at reducing impacts due to pollution, deforestation, overfishing, and other stressors. Mangrove-dominated systems have the inherent ecological and biological capacity to adjust to global climate change, but are highly vulnerable to anthropogenic destruction for aquaculture and development. If effective management measures are implemented to protect them, natural processes may allow mangrove systems to survive climate change impacts. In contrast, coral reefs may not have the natural characteristics required to withstand global climate change impacts such as oceanic acidification and increased temperatures, even if localized management measures are implemented. For coral reef fisheries that target the adults of species reliant on mangrove nurseries, the integrity of the two habitats has provided food security for people in tropical regions for thousands of years. These fisheries will be severely impacted as coral reefs continue to decline. Fishery species that use a broad range of nursery habitats including mangroves, and that require three-dimensional cover during later stages, but not necessarily live coral cover, will be more resilient. Perhaps surprisingly, two-thirds of families fished on coral reefs include species that use mangroves and associated back reef habitats as nurseries. While this review represents a first attempt to bring together a diverse range of concepts concerning the impacts of climate change on coral reef fisheries with connectivity to mangroves, it seems clear that protecting mangrove habitats is a rational strategy in regions where coral reef fisheries are of paramount importance.

References

Adam TC, Schmitt RJ, Holbrook SJ, Brooks AJ, Edmunds PJ, Carpenter RC, Bernardi G. 2011. Herbivory, connectivity, and ecosystem resilience: Response of a coral reef to a large-scale perturbation. PLoS ONE. 6(8):e23717.

Adams AJ, Dahlgren CP, Kellison GT, Kendall MS, Layman CA, Ley JA, Nagelkerken I, Serafy JE. 2006. Nursery function of tropical back-reef systems. Mar Ecol-Prog Ser. 318:287–301.

Aguilar-Perera A, Appeldoorn RS. 2007. Variation in juvenile fish density along the mangrove-seagrass-coral reef continuum in SW Puerto Rico. Mar Ecol-Prog Ser. 348:139–148.

Alongi DM. 2002. Present state and future of the world's mangrove forests. Environ Conserv. 29(3):331–349.

Alongi DM. 2008. Mangrove forests: Resilience, protection from tsunamis, and responses to global climate change. Estuar Coast Shelf Sci. 76(1):1–13.

Alongi DM. 2011. Carbon payments for mangrove conservation: Ecosystem constraints and uncertainties of sequestration potential. Environ Sci Policy. 14(4):462–470.

Baker R, Sheaves M. 2005. Redefining the piscivore assemblage of shallow estuarine nursery habitats. Mar Ecol-Prog Ser. 291:197–213.

Barnes L, Bellwood DR, Sheaves M, Tanner JK. 2012. The use of clear-water non-estuarine mangroves by reef fishes on the Great Barrier Reef. Mar Biol. 159(1):211–220.

Beck MW, Heck KL Jr, Able KW, Childers DL, Eggleston DB, Gillanders BM, Halpern B, Hays CG, Hoshino K, Minello TJ, et al. 2001. The identification, conservation, and management of estuarine and marine nurseries for fish and invertebrates. BioScience. 51(8):633–641.

Bell JD, Ganachaud A, Gehrke PC, Griffiths SP, Hobday AJ, Hoegh-Guldberg O, Johnson JE, Le Borgne R, Lehodey P, Lough JM, et al. 2013. Mixed responses of tropical Pacific fisheries and aquaculture to climate change. Nat Clim Change. 3(6):591–599.

Bellwood DR, Choat JH. 2011. Dangerous demographics: The lack of juvenile humphead parrotfishes *Bolbometopon muricatum* on the Great Barrier Reef. Coral Reefs. 30(2):549–554.

Berkstrom C, Gullstrom M, Lindborg R, Mwandya AW, Yahya SAS, Kautsky N, Nystrom M. 2012. Exploring knowns and unknowns in tropical seascape connectivity with insights from East African coral reefs. Estuar Coast Shelf Sci. 107:1–21.

Bohnsack JA, Bannerot SP. 1986. A stationary visual census technique for quantitatively assessing community structure of coral reef fishes. NOAA Technical Report NMFS. 41:1–15.

Bohnsack JA, McClellan DB, Harper DE, Davenport GS, Konoval GJ, Eklund A-M, Contillo JP, Bolden SK, Fischel PC, Sandorf GS, et al. 1999. Baseline data for evaluating reef fish populations in the Florida Keys, 1979–1998. Miami: National Oceanic and Atmospheric Administration.

Caldeira K. 2012. Avoiding mangrove destruction by avoiding carbon dioxide emissions. Proc Natl Acad Sci U S A. 109(36):14287–14288.

Christie MR, Tissot BN, Albins MA, Beets JP, Jia Y, Ortiz DM, Thompson SE, Hixon MA. 2010. Larval connectivity in an effective network of marine protected areas. PLoS ONE. 5(12):e15715.

Clark RD, Pittman S, Caldow C, Christensen J, Roque B, Appeldoorn RS, Monaco ME. 2009. Nocturnal fish movement and trophic flow across habitat boundaries in a coral reef ecosystem (SW Puerto Rico). Caribb J Sci. 45:282–303.

Coker DJ. 2012. The importance of live coral habitat for reef fishes and its role in key ecological processes. PhD Thesis. James Cook University, Townsville, Queensland.

Coker DJ, Graham NAJ, Pratchett MS. 2012. Interactive effects of live coral and structural complexity on the recruitment of reef fishes. Coral Reefs. 31(4):919–927.

Dahlgren CP, Eggleston DB. 2001. Spatio-temporal variability in abundance, size and microhabitat associations of early juvenile Nassau grouper *Epinephelus striatus* in an off-reef nursery system. Mar Ecol-Prog Ser. 217:145–156.

Dahlgren CP, Kellison GT, Adams AJ, Gillanders BM, Kendall MS, Layman CA, Ley JA, Nagelkerken I, Serafy JE. 2006. Marine nurseries and effective juvenile habitats: Concepts and applications. Mar Ecol-Prog Ser. 312:291–295.

De'ath G, Fabricius KE, Sweatman H, Puotinen M. 2013. The 27-year decline of coral cover on the Great Barrier Reef and its causes. Proc Natl Acad Sci U S A. 109(44):17995–17999.

Donato DC, Kauffman JB, Mackenzie RA, Ainsworth A, Pfleeger AZ. 2012. Whole-island carbon stocks in the tropical Pacific: Implications for mangrove conservation and upland restoration. J Environ Manage. 97:89–96.

Dorenbosch M, Van Riel MC, Nagelkerken I, Van der Velde G. 2004. The relationship of reef fish densities to the proximity of mangrove and seagrass nurseries. Estuar Coast Shelf Sci. 60(1):37–48.

Duke NC, Meynecke JO, Dittmann S, Ellison AM, Anger K, Berger U, Cannicci S, Diele K, Ewel KC, Field CD, et al. 2007. A world without mangroves? Science. 317(5834):41–42.

Eggleston DB, Dahlgren CP, Johnson EG. 2004. Fish density, diversity, and size-structure within multiple back reef habitats of Key West National Wildlife Refuge. Bull Mar Sci. 75:175–204.

FAO (Food and Agriculture Organization of the United Nations). 2012. The state of world fisheries and aquaculture 2012. Rome: FAO.

Faunce CH, Serafy JE, Lorenz JJ. 2004. Density-habitat relationships of mangrove creek fishes within the southeastern saline Everglades (USA), with reference to managed freshwater releases. Wetl Ecol Manag. 12(5):377–394.

Feller IC, Lovelock CE, Berger U, McKee KL, Joye SB, Ball MC. 2010. Biocomplexity in mangrove ecosystems. Annu Rev Mar Sci. 2:395–417.

Fenner D. 2012. Challenges to managing fisheries on diverse coral reefs. Diversity. 4(1):105–160.

Florida Fish and Wildlife Conservation Commission. 2013. http://myfwc.com//research. Accessed 20 June 2013.

Frias-Torres S. 2007. Habitat use of juvenile goliath grouper *Epinephelus itajara* in the Florida Keys, USA. Endanger Species Res. 3(3):1–6.

Froese R, Pauly D, editors. FishBase [Internet]. Penang, Malaysia: WorldFish Office (HQ); 2013, Revised June, 2013. Available from: http://www.fishbase.org.

Gardner TA, Cote IM, Gill JA, Grant A, Watkinson AR. 2003. Long-term region-wide declines in Caribbean corals. Science. 301(5635):958–960.

Gilman E, Ellison JC, Jungblut V, Van Lavieren H, Wilson L, Areki F, Brighouse G, Bungitak J, Dus E, Henry M, et al. 2006. Adapting to Pacific Island mangrove responses to sea level rise and climate change. Clim Res. 32(3):161–176.

Giri C, Ochieng E, Tieszen LL, Zhu Z, Singh A, Loveland T, Masek J, Duke N. 2011. Status and distribution of mangrove forests of the world using earth observation satellite data. Global Ecol Biogeogr. 20(1):154–159.

Graham NAJ, Wilson SK, Jennings S, Polunin NVC, Robinson JAN, Bijoux JP, Daw TM. 2007. Lag effects in the impacts of mass coral bleaching on coral reef fish, fisheries, and ecosystems. Conserv Biol. 21(5):1291–1300.

Grandcourt EM, Cesar HSJ. 2003. The bio-economic impact of mass coral mortality on the coastal reef fisheries of the Seychelles. Fish Res. 60:539–550.

Green A, Ramohia P, Ginigele M, Leve T. 2006. Fisheries resources: Coral reef fishes. Solomon Islands Marine Assessment: Technical report of survey conducted May 13 to June 17, 2004. TNC Pacific Island Countries Report No. 1/06. Brisbane, Australia: The Nature Conservancy, Indo-Pacific Resource Centre.

Hamilton S. 2013. Assessing the role of commercial aquaculture in displacing mangrove forest. Bull Mar Sci. 89:585–601.

Hamilton RJ, Adams S, Choat JH. 2008. Sexual development and reproductive demography of the green humphead parrotfish (*Bolbometopon muricatum*) in the Solomon Islands. Coral Reefs. 27(1):153–163.

Heck KL Jr, Carruthers TJB, Duarte CM, Hughes AR, Kendrick G, Orth RJ, Williams SW. 2008. Trophic transfers from seagrass meadows subsidize diverse marine and terrestrial consumers. Ecosystems. 11(7):1198–1210.

Heck KL, Hays G, Orth RJ. 2003. Critical evaluation of the nursery role hypothesis for seagrass meadows. Mar Ecol-Prog Ser. 253:123–136.

Hoegh-Guldberg O. 1999. Climate change, coral bleaching and the future of the world's coral reefs. Mar Freshwater Res. 50(8):839–866.

Hoegh-Guldberg O. 2012. The adaptation of coral reefs to climate change: Is the Red Queen being outpaced? Sci Mar. 76(2):403–408.

Hoegh-Guldberg O, Bruno JF. 2010. The impact of climate change on the world's marine ecosystems. Science. 328(5985):1523–1528.

Hoegh-Guldberg O, Mumby PJ, Hooten AJ, Steneck RS, Greenfield P, Gomez E, Harvell CD, Sale PF, Edwards AJ, Caldeira K, et al. 2007. Coral reefs under rapid climate change and ocean acidification. Science. 318(5857):1737–1742.

IUCN (International Union for Conservation of Nature). 2013. The IUCN Red List of Threatened Species. Version 2013.2. http://www.iucnredlist.org. Accessed 20 June 2013.

Jones GP, McCormick MI, Srinivasan M, Eagle JV. 2004. Coral decline threatens fish biodiversity in marine reserves. Proc Natl Acad Sci U S A. 101(21):8251–8253.

Jones DL, Walter JF, Brooks EN, Serafy JE. 2010. Connectivity through ontogeny: Fish population linkages among mangrove and coral reef habitats. Mar Ecol-Prog Ser. 401:245–258.

Jory DE, Iversen ES. 1989. Species profiles: Life histories and environmental requirements of coastal fishes and invertebrates (South Florida): Black, red and Nassau groupers. U.S. Fish and Wildlife Service Biological Report 82 (11.110). Washington, DC: U.S. Fish and Wildlife Service.

Kathiresan K. 2008. Threats to mangroves: Degradation and destruction of mangroves. *In:* Kathiresan K, Ajmal-Kahn S, editors. Training course on mangroves and biodiversity. Tamil Nadu, India: Annamalai University Centre of Advanced Study in Marine Biology.

Kennedy EV, Perry CT, Halloran PR, Iglesias-Prieto R, Schoenberg CHL, Wisshak M, Form AU, Carricart-Ganivet JP, Fine M, Eakin CM, et al. 2013. Avoiding coral reef functional collapse requires local and global action. Curr Biol. 23:912–918.

Kimirei IA, Nagelkerken I, Mgaya YD, Huijbers CM. 2013. The mangrove nursery paradigm revisited: Otolith stable isotopes support nursery-to-reef movements by Indo-Pacific fishes. PLoS ONE. 8:e66320.

Koenig CC, Coleman FC, Eklund AM, Schull J, Ueland J. 2007. Mangroves as essential nursery habitat for goliath grouper (*Epinephelus itajara*). Bull Mar Sci. 80(3):567–585.

Lecchini D, Waqalevu VP, Parmentier E, Radford CA, Banaigs B. 2013. Fish larvae prefer coral over algal water cues: Implications of coral reef degradation. Mar Ecol-Prog Ser. 475:303–307.

Ley JA. 1992. Influence of changes in freshwater flow on the use of mangrove prop root habitat by fishes. PhD Thesis, University of Florida, Gainesville.

Ley JA. 2005. Linking fish assemblages and attributes of mangrove estuaries in tropical Australia: Criteria for regional marine reserves. Mar Ecol-Prog Ser. 305:41–57.

Ley JA, Halliday IA, Tobin AJ, Garrett RN, Gribble NA. 2002. Ecosystem effects of fishing closures in mangrove estuaries of tropical Australia. Mar Ecol-Prog Ser. 245:223–238.

Ley JA, McIvor CC. 2002. Linkages between estuarine and reef fish assemblages: Enhancement by the presence of well-developed mangrove shorelines. The Everglades, Florida Bay and coral reefs of the Florida Keys: An ecosystem sourcebook. Boca Raton, FL: CRC Press. p. 539–562.

Ley JA, McIvor CC, Montague CL. 1999. Fishes in mangrove prop-root habitats of northeastern Florida Bay: Distinct assemblages across an estuarine gradient. Estuar Coast Shelf Sci. 48(6):701–723.

Luo J, Serafy JE, Sponaugle S, Teare PB, Kieckbusch D. 2009. Movement of gray snapper *Lutjanus griseus* among subtropical seagrass, mangrove, and coral reef habitats. Mar Ecol-Prog Ser. 380:255–269.

Machemer EGP, Walter III JFW, Serafy JE, Kerstetter DW. 2012. Importance of mangrove shorelines for rainbow parrotfish *Scarus guacamaia*: Habitat suitability modeling in a subtropical bay. Aquat Biol. 15:87–98.

Mateo I, Durbin EG, Appeldoorn RS, Adams AJ, Juanes F, Kingsley R, Swart P, Durant D. 2010. Role of mangroves as nurseries for French grunt *Haemulon flavolineatum* and schoolmaster *Lutjanus apodus* assessed by otolith elemental fingerprints. Mar Ecol-Prog Sci. 402:197–212.

McClanahan TR, Maina J, Pet-Soede L. 2002. Effects of the 1998 coral mortality event on Kenyan coral reefs and fisheries. Ambio. 31:543–550.

McClenachan L. 2009. Documenting loss of large trophy fish from the Florida Keys with historical photographs. Conserv Biol. 23(3):636–643.

McClenachan L, Kittinger JN. 2013. Multicentury trends and the sustainability of coral reef fisheries in Hawai'i and Florida. Fish Fish. 14(3):239–255.

McKee KL. 2012. Ecological functioning of mangroves under changing climatic conditions. *In:* Dahdouh-Guebas F, Satyanarayana B, editors. Proceedings of the international conference: Meeting on mangrove ecology, functioning and management. Vlaams Instituut voor de Zee (VLIZ) Special Publication 57. Galle, Sri Lanka: VLIZ.

McMahon KW, Berumen ML, Thorrold SR. 2012. Linking habitat mosaics and connectivity in a coral reef seascape. Proc Natl Acad Sci U S A. 109(38):15372–15376.

Mumby PJ, Edwards AJ, Arias-Gonzalez JE, Lindeman KC, Blackwell PG, Gall A, Gorczynska MI, Harborne AR, Pescod CL, Renken H, et al. 2004. Mangroves enhance the biomass of coral reef fish communities in the Caribbean. Nature. 427(6974):533–536.

Munday PL, Jones GP, Pratchett MS, Williams AJ. 2008. Climate change and the future for coral reef fishes. Fish Fish. 9(3):261–285.

Nagelkerken I, Dorenbosch M, Verberk W, de la Morinière EC, van der Velde G. 2000. Importance of shallow-water biotopes of a Caribbean Bay for juvenile coral reef fishes: Patterns in biotope association, community structure and spatial distribution. Mar Ecol-Prog Ser. 202:175–192.

Nagelkerken I, Roberts CM, van der Velde G, Dorenbosch M, Van Riel MC, de la Morinière C, Nienhuis PH. 2002. How important are mangroves and seagrass beds for coral-reef fish? The nursery hypothesis tested on an island scale. Mar Ecol-Prog Ser. 244:299–305.

Nagelkerken I, van der Velde G. 2002. Do non-estuarine mangroves harbour higher densities of juvenile fish than adjacent shallow-water and coral reef habitats in Curaçao (Netherlands Antilles)? Mar Ecol-Prog Ser. 245:191–204.

Nakamura Y, Horinouchi M, Shibuno T, Tanaka Y, Miyajima T, Koike I, Kurokura H, Sano M. 2008. Evidence of ontogenetic migration from mangroves to coral reefs by black-tail snapper *Lutjanus fulvus*: Stable isotope approach. Mar Ecol-Prog Ser. 355:257–266.

NOAA/Caribbean Fisheries Management Council. 2011. Reef fish fishery management plan of Puerto Rico and the U.S. Virgin Islands. San Juan, PR: Caribbean Fisheries Management Council.

NOAA/Western Pacific Regional Fishery Management Council. 2010. Western Pacific region reef fish trends. Honolulu, HI: Western Pacific Regional Fishery Management Council.

Ogden JC, Ehrlich PR. 1977. Behavior of heterotypic resting schools of juvenile grunts (Pomasyidae). Mar Biol. 42:273–280.

Olds AD, Connolly RM, Pitt KA, Maxwell PS. 2012. Primacy of seascape connectivity effects in structuring coral reef fish assemblages. Mar Ecol-Prog Ser. 462:191–203.

Pet-Soede C, van Densen WLT, Pet JS, Machiels MAM. 2001. Impact of Indonesian coral reef fisheries on fish community structure and the resultant catch composition. Fish Res. 51(1):35–51.

Polidoro BA, Carpenter KE, Collins L, Duke NC, Ellison AM, Ellison JC, Farnsworth EJ, Fernando ES, Kathiresan K, Koedam NE, et al. 2010. The loss of species: Mangrove extinction risk and geographic areas of global concern. PLoS ONE. 5:e10095.

Pratchett MS, Munday PL, Wilson SK. 2008. Effects of climate-induced coral bleaching on coral-reef fishes: Ecological and economic consequences. Oceanogr Mar Biol. 46:251–296.

Record S, Charney ND, Zakaria RM, Ellison AM. 2013. Projecting global mangrove species and community distributions under climate change. Ecosphere. 4(3):art34.

Rooker JR, Dennis GD. 1991. Diel, lunar and seasonal changes in a mangrove fish assemblage off southwestern Puerto Rico. Bull Mar Sci. 49(3):684–698.

Schmitt RJ, Holbrook SJ. 2000. Habitat-limited recruitment of coral reef damselfish. Ecology. 81(12):3479–3494.

Serafy JE, Faunce CH, Lorenz JJ. 2003. Mangrove shoreline fishes of Biscayne Bay, Florida. Bull Mar Sci. 72(1):161–180.

Sheaves M. 1995. Large lutjanid and serranid fishes in tropical estuaries: Are they adults or juveniles? Mar Ecol-Prog Ser. 129(1):31–40.

Shibuno T, Nakamura Y, Horinouchi M, Sano M. 2008. Habitat use patterns of fishes across the mangrove-seagrass-coral reef seascape at Ishigaki Island, Southern Japan. Ichthyol Res. 55(3):218–237.

Siikamaki J, Sanchirico JN, Jardine SL. 2012. Global economic potential for reducing carbon dioxide emissions from mangrove loss. Proc Natl Acad Sci U S A. 109(36):14369–14374.

Silverman J, Lazar B, Cao L, Caldeira K, Erez J. 2009. Coral reefs may start dissolving when atmospheric CO_2 doubles. Geophys Res Lett. 36(5):L05606.

Simpfendorfer C, Burgess GH; *Carcharhinus leucas*. *In:* IUCN 2013. IUCN red list of threatened species [Internet]. Gland, Switzerland: International Union for the Conservation of Nature; 2009, Version 2013.1. Available from: www.iucnredlist.org.

Smith SG, Ault JS, Bohnsack JA, Harper DE, Luo J, McClellan DB. 2011. Multispecies survey design for assessing reef-fish stocks, spatially explicit management performance, and ecosystem condition. Fish Res. 109(1):25–41.

Solomon S, Qin D, Manning M, Chen Z, Marquis M, Averyt KB, Tignor M, Miller HL. 2007. IPCC Climate Change 2007: The physical science basis. Contribution of Working Group I to the Fourth Assessment Report of the Intergovernmental Panel on Climate Change. Cambridge: Cambridge University Press.

State of Queensland. 2010. Annual status report, coral reef fin fish fishery. Brisbane: Department of Employment, Economic Development and Innovation.

Thollot P, Kulbicki M. 1988. Overlap between the fish fauna inventories of coral reefs, soft bottoms and mangroves in Saint-Vincent Bay (New Caledonia). Proc 6th Int Coral Reef Symp. 6:613–618.

Unsworth RKF, De Leon PS, Garrard SL, Jompa J, Smith DJ, Bell JJ. 2008. High connectivity of Indo-Pacific seagrass fish assemblages with mangrove and coral reef habitats. Mar Ecol-Prog Ser. 353:213–224.

Valiela I, Bowen JL, York JK. 2001. Mangrove forests: One of the world's threatened major tropical environments. BioScience. 51(10):807–815.

Vermeer M, Rahmstorf S. 2009. Global sea level linked to global temperature. Proc Natl Acad Sci U S A. 106(51):21527–21532.

chapter eleven

Augmented catch-MSY approach to fishery management in coral-associated fisheries

Marlowe Sabater and Pierre Kleiber

Contents

Introduction

Fishing on coral reefs in the Western Pacific region has been practiced by the indigenous people of American Samoa, Guam, the Commonwealth of the Northern Mariana Islands (CNMI), and Hawaii for more than 3 millennia (Dye and Graham 2004). This practice is embedded in the fabric of the local cultures and traditions despite changes in socioeconomic and sociocultural conditions brought about by urbanization and Western influence (Allen and Amesbury 2012, Levine and Sauafea-Leau 2013). In an age of globalization and modernization, coral reefs and associated fisheries are being threatened from multiple fronts and on multiple scales: land-based pollution resulting in phase shifts (Pastorok and Bilyard 1985, Hughes 1994, Edinger et al. 1998), global warming coupled with climate change (Brander 2007, Munday et al. 2008), and destructive fishing combined with overexploitation (e.g., Edinger et al. 1998, Jackson et al. 2001, Newton et al. 2007, McClanahan et al. 2008). The multidimensionality of the coral reef fisheries poses a significant challenge to management, hence multiple tools have been developed to address the various impacts affecting the fisheries. These management tools range from spatial–temporal management-like rotational closures or permanent no-take marine protected areas (Roberts and Polunin 1993) to traditional fishery tools such as input controls (e.g., gear restriction, limited entry program, and effort limits) and output controls (e.g., size limits, bag limits, seasonal closures, and catch limits). All these tools are geared toward conserving and managing stocks that are regarded as being in decline on a regional and global scale (Pandolfi et al. 2003, Newton et al. 2007, Zeller et al. 2007, Worm et al. 2009).

Application of these diverse fishery management tools depends on the long-term goal for the stocks. The reauthorization of the Magnuson-Stevens Fishery Conservation and Management Act in 2007 required the implementation of annual catch limits for the different fisheries in the United States and its territories, with an overall goal of preventing overfishing and at the same time developing underutilized or unutilized fisheries to ensure that the citizens benefit from the employment, food supply, and revenue thereby generated. It is therefore inherent that, in order to provide sustainable economic benefit for the nation, fishery stocks should be sustainable in the long term.

However, what makes sense on a national level may not necessarily apply on a regional scale given the diversity of fishing community culture, fishing practices, and the fish stocks being managed. The U.S. federal waters in the Western Pacific Ocean are managed by the Western Pacific Regional Fishery Management Council. This region is comprised of the Pacific Remote Island Areas consisting of the small islands and atolls of Palmyra, Jarvis, Johnston, Wake, Howland and Baker, the State of Hawaii, CNMI, and Guam, and, in the southern hemisphere, American Samoa (Figure 11.1). The council manages hundreds of marine species through its fishery ecosystem plans (FEPs), including corals and coral reef fishes. The scientific information for each stock and fishery varies. To comply with the requirements of the Magnuson-Stevens Act to end overfishing and National Standard 1 (implementing guidelines on annual catch limit specifications), the council amended the Pacific Remote Island Areas, American Samoa, the Marianas, and Hawaii FEPs to include a tier system of control rules in specifying acceptable biological catches and a set of options

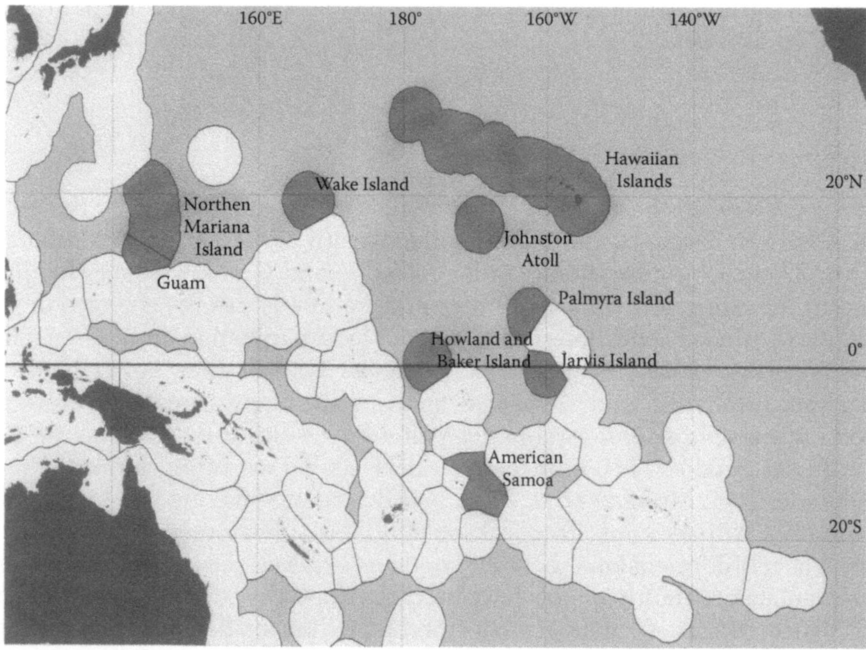

Figure 11.1 **(See color insert)** Map of the Western Pacific region showing the exclusive economic zones (EEZs). The red EEZs (3–200 nmi) are under the fishery management jurisdiction of the Western Pacific Regional Fishery Management Council.

for specifying annual catch limits below the acceptable biological catches (WPRFMC 2011). The tiers ranged from Tier 1 stocks with the best quality information (i.e., typically with a stock assessment and an estimated risk of overfishing) to Tier 5 stocks (i.e., stocks with only catch information available). The majority of the coral reef fish stocks have been categorized as Tier 5. Conventional stock assessment is impractical for many coral reef stocks due to the number of species, limited life history information, and multiple gears used to harvest various subsets of stocks simultaneously. Therefore, not only are coral reef fish stocks data poor, but also managing coral reef fisheries on a stock basis poses a management challenge.

Problems with annual catch limit specification for data-poor stocks

The most critical requirement for a successful catch limit–based management is knowing the status of the stock. This information is usually generated from stock assessments, but stock assessments are not available for the Western Pacific region. The starting point for the annual catch limit specification process is to estimate the maximum fishing mortality threshold. This is the level of fishing mortality, on an annual basis, above which overfishing occurs. The annual catch associated with this fishing mortality corresponds to the overfishing limit. These parameters cannot be estimated if an assessment is not done for the species subject to this type of management measure. In the absence of a stock assessment, the specification process becomes subjective and precautionary principle dictates that management should err on the side of caution, therefore forcing the manager to be extremely conservative. For the coral reef fish stocks, there were no estimated overfishing limits and the acceptable biological catches and annual catch limits were based purely on catch data.

The initial Tier 5 acceptable biological catch limit specifications for the coral reef ecosystem management unit species in the Western Pacific region were based on the guideline suggested by Restrepo et al. (1998) where the catch limit is set equal to, or a fraction of, the long-term average of reliable annual catch from a period in the fishery when there was no quantitative or qualitative evidence of declining abundance. However, the catch trends in the coral reef fishery did not exhibit any periods with little or no decline for most of the reef fish families (Figure 11.2). Coral reef fish species were categorized to the family level because species level catch information was not available for most of the areas given the way the data collection had been designed. In light of the large fluctuations in catch, the council utilized the entire catch time series (American Samoa: 1990–2008; Guam: 1985–2008; CNMI: 2000–2008; Hawaii: 1948–2007) from creel surveys in the territories and fishers' trip reports from the State of Hawaii. Catch data from creel surveys are not entirely reliable because they do not provide an estimate of total catch. The fisher trip reporting system also does not provide an estimate of total catch because it only focuses on commercial landings. Despite the underestimation of total catch, these were readily available sources of catch information that could be used for management.

The council chose the acceptable biological catches equal to the 75th percentile of historic catches rather than the long-term median. This would provide a nonparametric approach and a 75% chance of catches being below the potential limit in any given year. The annual catch limit was set equal to the acceptable biological catch since there were indications from the biomass estimates that catches were a relatively small proportion of the corresponding biomass (Luck and Dalzell 2010, Sabater and Tulafono 2011). There are indications that coral reef fisheries in some parts of the Western Pacific region

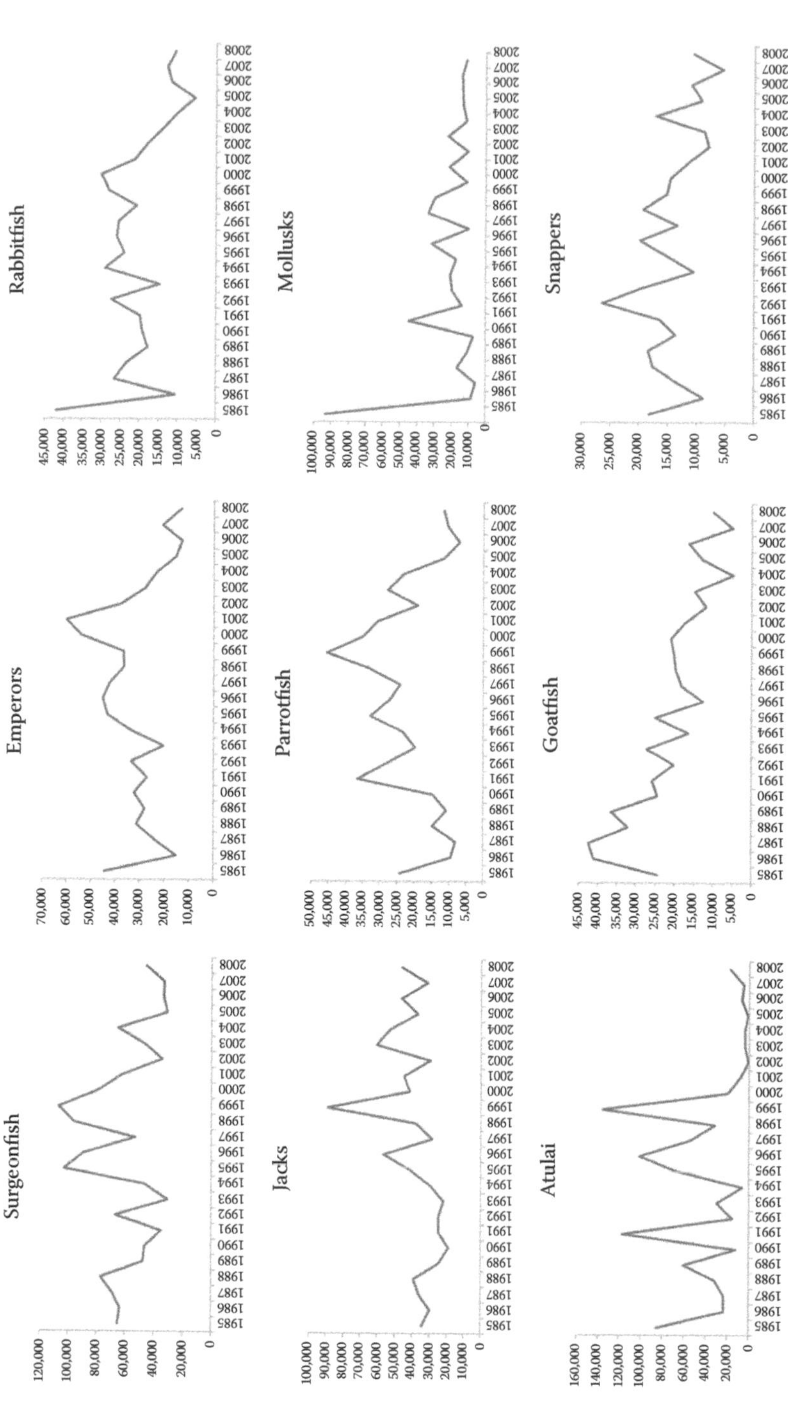

Figure 11.2 Sample coral reef fish catch time series from Guam indicating interannual fluctuations in catch from 1985 to 2008. Catch landings in pounds are indicated on the *y* axes, while the *x* axes are years.

might be sustainable based on comprehensive analysis using fishery-dependent, fishery-independent, archaeological, and socioeconomic information (Sabater and Carroll 2009). This is contrary to the general notion of a major decline in productivity based solely on either biomass or catch information (Newton et al. 2007, Zeller et al. 2007, Williams et al. 2011, Houk et al. 2012).

The initial annual catch limit specifications were solely based on catch information and do not directly incorporate biomass information or other relevant data in the calculations. Some of the initial harvest limits were also very restrictive since they were based only on creel surveys, which, even when expanded, represented only a portion of total catch. Some of the fisheries, particularly night-time spearfishing, are inadequately documented in creel surveys and are better represented by commercial receipt-book data. It was calculated that the reported catches were underestimated by 2.2-, 2.5- and 7-fold for CNMI, Guam, and American Samoa, respectively (Zeller et al. 2007). Given the severe underestimation resulting in overly restrictive annual catch limits, the council is shifting to a model-based approach in specifying acceptable biological catches, which would incorporate biomass information and other life history traits to overcome the limitations of catch-only information.

In addition, the way in which the Tier 5 control rule was implemented generated an unforeseen and unintentional "ratchet-down" effect. Utilizing the entire catch time series in calculating the annual catch limits, required the council to incorporate the most recent data once they became available as mandated by the U.S. National Marine Fisheries Service (NMFS) National Standard 2 (i.e., use of the best scientific information available). Over the long term, implementation of the catch limits and with the fishery in compliance to keep the catch below the specified limits, will result in a "ratchet-down" effect once new data are added in the time series when new annual catch limits are calculated. Conversely, if the fishery is not compliant and the catches are consistently above the 75th percentile, application of the control rule will cause the acceptable biological catch to increase over time, creating a disincentive to comply with the limits.

Moving from the data-poor tier to the model-based tier

Biomass, abundance, species composition, average length, coral reef habitat, qualitative estimates of natural mortality and limited fishing mortality, and life history information are available for the coral reef species in the Western Pacific region. All this information needs to be utilized to move the coral reef stocks from the data-poor tier (Tier 5), which applies only catch information, to Tier 3 to generate an estimate of sustainable harvest levels through model-based approaches. Four models were explored to enhance the annual catch limit specification process. These were (1) a bulk estimator of maximum sustainable yield (MSY) using a modified Schaefer and Fox model (Garcia et al. 1989); (2) depletion-corrected average catches (DCAC) (MacCall 2009); (3) depletion-based stock reduction analysis (DB-SRA) (Dick and MacCall 2011); and (4) a catch-MSY estimator (Martell and Froese 2012).

The Garcia et al. (1989) bulk estimator generates a point estimate of MSY from straightforward derivations of two well-known surplus production models (Schaefer 1954, Fox 1970). These equations are suitable for those cases where there are no available long time series of catch or effort and where the only estimates available are the total catch, average total biomass, and an expert guess of fishing mortality needed to obtain the MSY of the fish stocks in question. These equations have similar limitations and constraints as the models from which they were derived. The main assumptions were that the biological processes involved are deterministic, the fishery is on a single stock with stable age/size characteristics, the catchability is not density dependent, and there are no time lags between catch and productivity. Using this model would be challenging due to its poor applicability to a complex fishery like the coral reef fishery and the oversimplification of the assumptions, particularly with the use of mortality estimates applied equally across a broad range of species within each reef fish family.

The depletion-based models like DCAC and DB-SRA (MacCall 2009, Dick and MacCall 2011) provide an estimate of potential yield from an equation that originated from Gulland (1970) where sustainable yield is half of the virgin biomass once the natural mortality is accounted for. The unsustainable windfall effect of depletion of the stock biomass and the potential yield dictate the level of sustainable annual harvest. This method requires a catch time series, an estimate of natural mortality, and nominal information on stock depletion (change in abundance from the first to the last year of the catch time series). Monte Carlo simulation allows for determination of probability distribution around the sustainable yield value, biomass at MSY, and catch at fishing mortality at MSY. Merging stock reduction analysis with DCAC provides a production function derived from a standard stock recruitment relationship (Dick and MacCall 2011). It also incorporates uncertainties in the natural mortality, stock dynamics, optimal harvest rates, and stock status via the Monte Carlo simulation. The depletion models were not chosen as they assume catch trends are directly associated with the abundance of fish. In reality, the fluctuations in coral reef catches in the Western Pacific region were driven mostly by changes in the amount of effort over time and possibly changes in the data collection system. Moreover, recent work by Vert et al. (2013) shows catch is not a good predictor of stock abundance for most stocks. The fluctuations in abundance are not directly correlated with increases in catches. Depletion-based models were also shown to be highly sensitive to assumed distribution for the ratio of starting to current biomass (stock depletion levels), which typically results in overestimation of the sustainable harvest levels when this parameter is set at optimistic levels (Wetzel and Punt 2011).

The catch-MSY estimator (Martell and Froese 2012) utilizes a time series of removals (catch time series), an estimate of r (rate of population increase) and k (carrying capacity), and some assumptions about biomass at the start and end of the time series. The range of r can be taken from FishBase (Froese and Pauly 2013) in the form of resilience. The Schaefer production model then creates annual biomass projections from the combination of r and k that would not result in biomass exceeding the carrying capacity and the stock being depleted. The assumption behind the biomass can be informed by augmenting the model with an independent source of biomass information. To maximize the potential and reliability of the model, fishery-independent information from underwater visual census surveys using stationary point counts by the Coral Reef Ecosystem Division (CRED) of the U.S. National Oceanographic and Atmospheric Agency (NOAA) was incorporated into the model to enhance the biomass projection. The augmented catch-MSY model will be the basis for moving the current Tier 5 reef fish stock to Tier 3, that is, stock that has a model-based estimate of MSY.

This chapter provides an overview of the modified catch-MSY approach to estimate a reference point for the coral reef fish stocks to improve specification of acceptable biological catches in the Western Pacific region. This is the first attempt to generate MSY estimates for the reef fish stocks, which are the starting point of the annual catch limit–based management framework.

Model-based approach to estimating MSY

Data preparation: Catch time series

Catch time series were generated from the boat-based and shore-based creel surveys conducted in American Samoa, Guam, and CNMI. The creel survey program generates an expanded catch from the participation counts that generate effort estimates and catch per unit effort from the catch intercept interview phase. The expanded catch covers only areas that are surveyed, and adjustment factors (when available and updated) are used to estimate total catches (limited). The catch data are summarized to family level. The data summaries were provided by the Western Pacific Fishery Information Network (WPacFIN), which is a program of the Pacific Island Fisheries Science Center of NMFS.

In addition, the commercial catch data for the island jurisdictions (American Samoa, Guam, and CNMI) (Figure 11.1) were extracted from the WPacFIN website (http://www.pifsc.noaa.gov/wpacfin). These data were generated from the commercial landings receipt book. The commercial landings from the website were summarized to family level so as to be compatible with the creel survey summaries. The commercial and creel survey catches were then summed to generate a more holistic total catch estimate. As mentioned earlier, some fisheries are better captured in one data collection system than another. It is noteworthy, however, that dealer reports and creel survey estimates likely underestimate the true total catches, hence the issue of double counting may not be significant.

The Hawaii catch time series was generated from the state's Division of Aquatic Resources commercial catch reporting system, which includes monthly catch reports from commercial marine license holders and vendors. This time series was summarized by coral reef fish families. Unfortunately, the reestimated recreational catch information (S Pooley, Pacific Island Fisheries Science Center, personal communication) was incomplete and could not be incorporated here. This would have improved the catch time series to facilitate evaluation of the noncommercial aspect of the fishery. The reestimation effort will be conducted for all U.S. Western Pacific states and territories, but this chapter will only focus on the preliminary results from American Samoa.

Data preparation: Biomass information

Standing biomass estimates were generated from the NOAA-CRED Rapid Assessment and Monitoring Program using stationary point count data (CRED-PIFSC 2013). Biomass estimates were summarized to family level. They were derived from two to four stationary point count surveys following a randomly stratified statistical design from approximately 1294 random sites in American Samoa, the Mariana Islands, and the main Hawaiian island. The mean biomass was then expanded to hard bottom areas (0–30 m deep) of different habitat types (treated as strata) from the mapping division of NOAA-CRED (Williams 2010) (Figure 11.3). This generated a standing stock biomass at the family level for each island in the Western Pacific region. These data included only species that were captured in the fishery and were greater than 15 cm in total length (the minimum fish size). There were biomass estimates for 3 years for American Samoa, 2 years for the Mariana Island archipelago, and 1 year for Hawaii. The dispersion of points around the mean biomass value per strata, also known as the coefficient of variation (CV), was estimated for each reef fish family for the most recent year of the survey and was weighted by sample size to determine the CVs for other years.

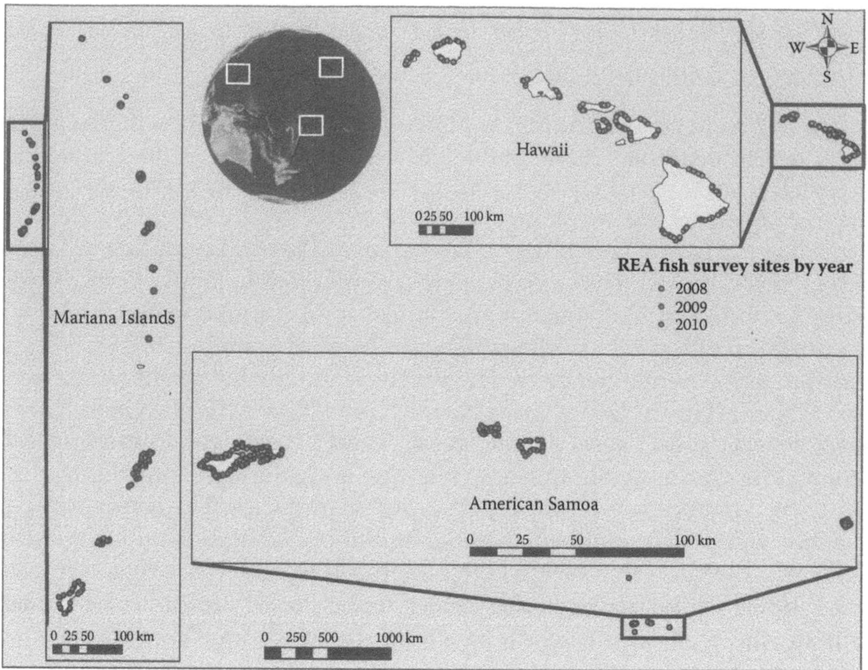

Figure 11.3 **(See color insert)** Map of the randomly selected sampling locations of the NOAA-CRED Rapid Assessment and Monitoring Program that generated the standing biomass estimates from 2008 to 2010. The sampling sites shown are only for biomass data used in the analysis. Biomass data exist for the Northwestern Hawaiian Islands and Pacific Remote Island Areas. (Map provided by Kaylyn McCoy of NOAA-CRED. With permission.)

Augmented catch-MSY method

MSY estimates were based on modification of a method for estimating MSY that relies only on a catch time series, assumptions about the approximate level of resilience of the stock, and the assumed ranges of depletion at the start and end of the time series (Martell and Froese 2012). It is assumed that the stock follows Schaefer model dynamics with variables r and k, from which MSY is given by

$$\text{MSY} = rk/4, \tag{11.1}$$

where r is the maximum population growth rate and k is the carrying capacity. Carrying capacity is the maximum equilibrium population biomass that the population will approach in the absence of interference (Gulland 1985). The maximum population growth rate is how fast the population grows to attain carrying capacity as affected by combined environment and biological factors. The rate of population growth varies depending on the population abundance relative to the carrying capacity. A positive population growth rate is expected when the population is below the carrying capacity and the reverse is true when it is above the carrying capacity. The population is controlled by a range of regulating factors such as space, food availability, predation rate by other populations, and so on.

The difference equation form of the Schaefer model is

$$b_{t+1} = \left[b_t + rb_t \left(1 - \frac{b_t}{k} \right) - c_t \right] \exp(\varepsilon_t), \tag{11.2}$$

where b_t is biomass at time t, c_t is catch at time t, and where the exponential term (the process error) allows for inaccuracies in the model predictions. The error, ε_t, is given by random draws from the normal distribution, $N(0, \sigma)$, where σ is an assumed measure of confidence in the applicability of the Schaefer model. The catch-MSY fits the Schaefer model to the known catch data by searching for combinations of r, k that produce plausible outcomes, that is, the Schaefer model output must pass a series of tests that are detailed below. The procedure is summarized in the flow chart in Figure 11.4.

Initiating the model (Step 1 in the chart), a time series of annual observed catch is read along with observed biomass and its CV for any year's biomass was measured. Then, a text item is read indicating "resilience," which describes stock productivity and resistance

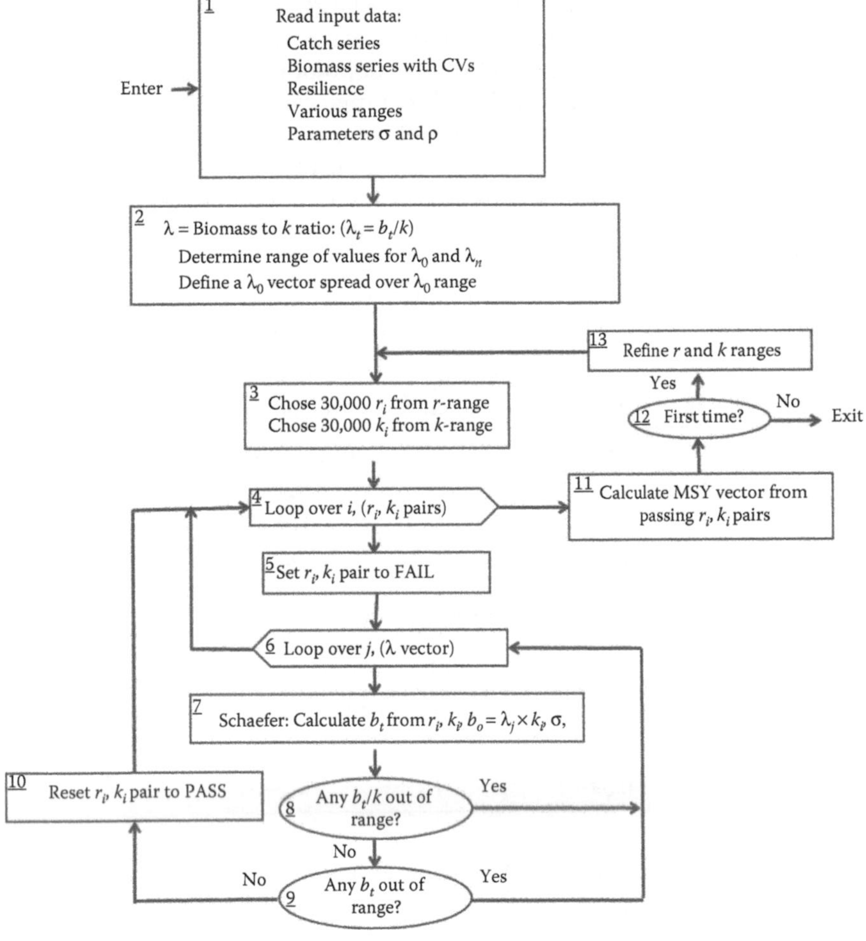

Figure 11.4 Flow diagram describing the model structure of the biomass augmented catch-MSY approach.

to fishing pressure. Resilience determines the range of r values to search (Table 11.1). Resilience descriptors are available for all stocks on FishBase (Froese and Pauly 2013). The next item, or items, in Step 1 provides optional overrides for ranges, including the r range given by resilience and other ranges described below. The variable k is generally more of an unknown than r, so its default is a very broad range from maximum observed annual catch up to a large value with a default of 100 times the maximum annual catch. The multiplier of 100 for the upper end of the k range can be overridden in Step 1 by input of a different multiplier. The process error, σ, can also be input in Step 1; it is zero by default. Finally the default value of 1.0 for the CV multiplier, which is used in Step 9, can be overridden by input of ρ in Step 1.

Step 2 in the flow chart deals with the variable λ, defined as the ratio of biomass to carrying capacity. The ratio λ_0 is at time zero, and λ_n is that ratio at time n where n is the last year in the catch time series. Step 2 determines an appropriate range for λ_0 and defines a vector of λ values spread over that range. By default, the range depends on whether the catch at the start of the time series is greater or less than half the maximum catch, as in Table 11.2. The assumption is that if the catch is small at the start, then it is likely that the population will have experienced minimal depletion from the fishery and the biomass will be close to the carrying capacity. Conversely, a large catch at the start implies the population will have been somewhat depleted at the start. Note that this assumption depends on the fishery having a significant effect on the population, at least at some time during the life of the fish being captured. The default λ_0 range can be overridden by a range entry in Step 1 of the flow chart. Once the range is determined, a vector of λ_0 values spread over that range is defined for later use in setting b_0 from the formula $b_0 = \lambda_0 / k$. A range of values is also determined by default for λ_n in a similar way to λ_0 (Table 11.3), with similar justification. This range can likewise be overridden by an entry in Step 1. It is used in testing r, k pairs (Step 11).

In Step 3, a large number (30,000 by default) of values for r and k are sampled from uniform distributions over the respective ranges. The r–k pairing is shown in Figure 11.5.

Table 11.1 Default Range of Rate of Population Increase for Each "Resilience" Level from Fish Base

Resilience	Range of r (year–1)
Very low	0.015–0.1
Low	0.05–0.5
Medium	0.2–1.0
High	0.6–1.5

Table 11.2 Default Range for ($\lambda_0 = b_0 / k$)

(Catch at time 1)/max(catch)	λ_0 range
<0.5	{0.5–0.9}
>0.5	{0.3–0.6}

Table 11.3 Default Range for ($\lambda_n = b_n / k$)

(Catch at time n)/max(catch)	λ_n range
<0.5	{0.01–0.4}
>0.5	{0.3–0.7}

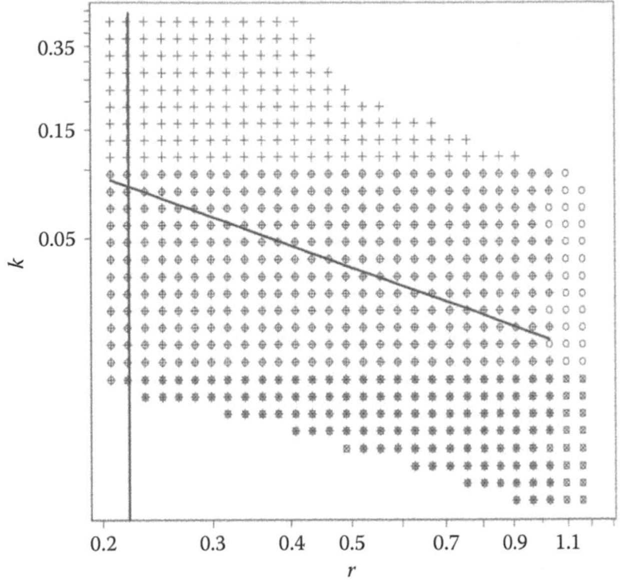

Figure 11.5 Examples for the upper end of the *k* range. + signs indicate k^* values from the first phase of testing. The vertical line indicates 1.1 times the low end of the first *r* range, and × signs indicate the values chosen for x_a. The diagonal line indicates the locus of r^* and k^* values corresponding to the geometric mean of all Y^* values, and O signs indicate values chosen for x_b. Overlapping signs (+ and X) indicate the *r*–*k* combinations that pass the tests and comprise the acceptable values to estimate MSY.

Step 4 runs a loop (indexed by *i*) testing each r_i, k_i pair for plausibility. The first step in the loop (Step 5) sets a test flag to FAIL for the *r*, *k* pair, after which Step 6 runs another loop over elements of the λ vector indexed by *j* and defined in Step 2.

Step 7 implements the Schaefer model (Equation 11.2) based on the current *r*, *k* pair and on b_0 calculated from the current *k*, as well as σ if it is nonzero. The output of the Schaefer model is a time trajectory of biomass values, b_t, starting from $t = 0$.

Following Step 7, the b_t/k ratio is first tested for all time steps. This includes testing whether in the final step the ratio b_n/k is outside the λ_n range, and for all other time steps, whether the ratio is outside the range 0–1 (i.e., biomass is <0 or $>k$). Note that this test depends on the current value of *k*, which is itself unknown.

The next test (Step 9) constitutes the augmented part of the catch-MSY technique. It is more stringent than the test in Step 8 in that the absolute value of b_t is tested against measured values of biomass for those times where such measurements were made. Failure is indicated if b_t is outside an allowable biomass range determined by the measured biomass and its CV times a multiplier, ρ, which defaults to 1.0.

Failure in either test ("yes" to the question in the ellipse) leads to the top of the inner loop (Step 6), and the tests are repeated with the next element of the λ vector and therefore with a new value of b_0. If the loop governed by Step 6 is exhausted, control will pass out the pointed end of the box to the top of the outer loop at Step 4 with the test flag still set to FAIL. Another *r*, *k* pair will then be tested. Otherwise, if both tests pass for any iteration of the inner loop, then control will pass to Step 10, where the test flag will be set to PASS. Controls will then break out of the inner loop and proceed to the top of the outer loop at Step 4 to test another *r*, *k* pair.

Once the outer loop is exhausted, control passes to Step 11, where MSY is calculated for all the r, k pairs with the test flag set to PASS. Then if Step 12 is entered for the first time, control will pass to Step 13, where refined ranges for r and k are established before proceeding back to Step 3 for a second phase of the procedure. The ranges are refined to focus the search in the second phase to areas of r and k space that are more likely to yield r, k pairs that pass the test.

To refine the r and k ranges, r^*, k^*, and Y^* are defined, respectively, as vectors of r and k values that passed the test and a vector of MSY values calculated from those r and k values, that is, from Equation 11.1, $Y_i^* = r_i^* \times k_i^*/4$. The new r range is then set by

$$\text{new-}r\text{-range} = \{\min(r^*) \ldots 1.2 \times \max(r^*)\}.$$

For the refined k range, a tentative maximum, x_a, is set to the minimum of the k^* values associated with r^* values within the lowest 10% of the original r range during the first phase of the procedure, that is,

$$x_a = \min\left[k^* \mid r^* < 1.1 \times \min(\text{first } r \text{ range})\right],$$

and a second tentative maximum, x_b, is set to the Y^* values that are less than the geometric mean of all the Y_m values, that is,

$$x_b = \max\left[k^* \mid Y^*(r^*, k^*) < \exp\left(\overline{\log(Y^*)}\right)\right].$$

The new k range is then set by

$$\text{new-}k\text{-range} = \{0.9 \times \min(k^*) \ldots \min(x_a, x_b)\}.$$

Ground-truthing the augmented catch-MSY approach using simulated data

In finding the plausible combination of r, k pair, the model appears to assume an inverse relationship between the priors (Figure 11.5). Carrying capacity is the asymptotic limit in the population controlled by environmental factors as well as biological factors such as density-dependent predation and resource availability. The lower the carrying capacity, the faster the population reaches the asymptotic maximum assuming that the population grows exponentially. The higher the carrying capacity, the slower the population reaches the asymptotic maximum. This population growth rate also depends on the stock in question, that is, whether the species that comprise the stock are r-select or k-select species. This relationship is true for most of the range of coral reef species, from slower growing groupers, parrotfish and wrasses to species with high turnover rates such as siganids (rabbitfish) and carangids (scads and jacks). The combinations of r and k that fall within the bounds of this inverse relationship are those accepted by the model to generate a distribution around the MSY estimate.

A simulation study of the catch-MSY approach examined the performance and sensitivity of the technique and the effect of incorporating observed biomass data. The simulation utilized data sets with and without biomass estimates, and the simulation results were compared to a known quantity of MSY. The model was tested for sensitivity to biomass information at varying degrees of fishing mortality (Figure 11.6). The simulation without biomass data showed the model generating a lower MSY estimate at $F=0.01$–0.05. The model generated an MSY estimate close to the true/known MSY at $F=0.10$ and remained close to the true value thereafter. When biomass information is included, the MSYs generated were consistently above the true value across a wide range of F, but not nearly as biased as the results with no biomass input and low F. Plots of good r, k pairs (Figure 11.7) show that more accurate results are obtained when the field of good r, k pairs spans true MSY, but biased results are obtained when this is not the case, as in Figure 11.7a, which corresponds to the lowest box in Figure 11.6.

The simulation results show that if catch is low relative to true MSY, and observed biomass data are not utilized, the method will consistently underestimate MSY.

Preliminary results using real data

For the analyses presented here, the resilience for all cases was assumed to be "medium," indicating a range in r of {0.2 … 1.0} year–1, depletion ranges at the start and end were set to the broad range of {0.01 … 0.99}, and process error, σ, was set to 0.05.

Table 11.4 offers the preliminary model results for management unit species in the American Samoa coral reef fish families comparing various model scenarios. The column labeled "Analysis 1" had no constraints on r and no biomass information. This run was not intended to represent any real estimate but merely to determine the sensitivity and effect of the r constraints on the estimated MSY value for model evaluation and comparative purposes. Analyses 2 and 3 had constraints applied to the priors and controlled for inclusion of biomass data (Analysis 2 had no biomass data, while Analysis 3 included biomass). In all cases, the MSY generated by Analysis 1 generated a higher estimated MSY.

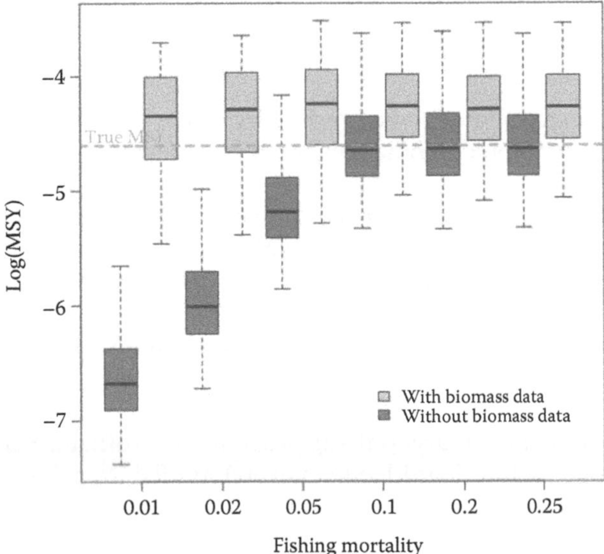

Figure 11.6 Paired MSY results across a broad range of fishing mortality values from model simulation using data with and without biomass information. Error bars (i.e., vertical dashed lines) indicate the coefficient of variation.

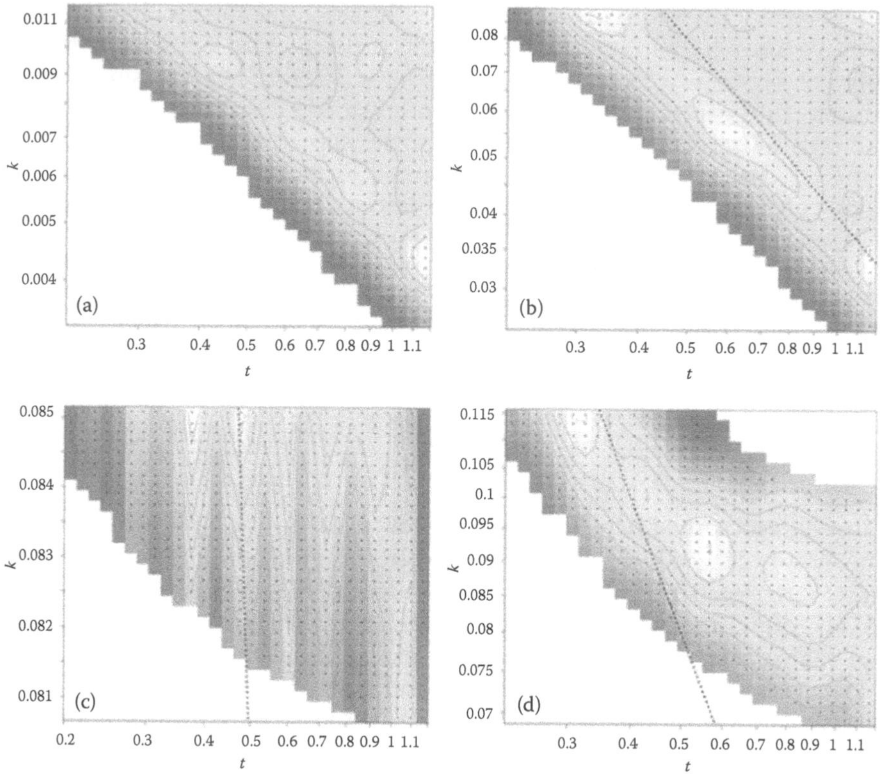

***Figure 11.7* (See color insert)** Pairs of good *r*, *k* values from the second phase of four simulations: (a) $F=0.01$, biomass ignored; (b) $F=0.10$, biomass ignored; (c) $F=0.01$, biomass included; (d) $F=0.10$, biomass included. The dotted lines show the locus of r^* and k^* values corresponding to the real MSY. In panel (a), the line is off the scale to the upper right. Single dots show grid squares containing at least one good *r*, *k* pair. Contour lines and color indicate density of good *r*, *k* pairs (white: high; red: low).

Analysis 2 simulations generated lower variability in the MSY estimates (CV) and lower MSY estimates. Incorporation of biomass estimates increased the MSY, in most cases with similar CVs.

The MSY estimates for the model run that has no constraints in *r* were consistently higher across all families tested. The range of values generated was typically larger if the *r* was not constrained. Constraining the prior and not incorporating biomass information resulted in a lower MSY, narrower distribution range, and smaller CV. Incorporating biomass information and constraining *r* generally resulted in a higher MSY estimate on a narrow range of values and CV. The level of enhancement from the model run depends on the amount of biomass information available and the CV around the biomass estimate. Acanthuridae (surgeonfish), Scaridae (parrotfish), Serranidae (grouper), Lutjanidae (snappers), and Lethrinidae (emperors) are fish families common in underwater visual census surveys in American Samoa (Page 1998, Sabater and Tofaeono 2007, PIFSC 2011, Williams et al. 2011). The differences in MSY between model runs with and without biomass values were small for Holocentridae (squirrelfish), Carangidae (jacks), and Carcharhinidae (requiem sharks) because fish in these families are either nocturnal and/or highly mobile and are not readily captured using stationary point count surveys (Williams 2010).

Table 11.4 Preliminary Model Results for Management Unit Species within American Samoa Coral Reef Fish Families Simulating Various Scenarios

	Analysis 1: No constraints applied				Analysis 2: Constraints applied but no biomass estimates				Analysis 3: Constraints applied with biomass estimates			
Management unit species	MSY	Low bound	High bound	CV	MSY	Low bound	High bound	CV	MSY	Low bound	High bound	CV
Acanthuridae	148	89	247	0.05	49	26	89	0.08	145	80	258	0.06
Scaridae	358	246	521	0.03	27	15	49	0.09	341	233	401	0.02
Serranidae	30	16	56	0.09	14	7	25	0.12	32	16	62	0.10
Lutjanidae	172	36	830	0.15	19	10	38	0.12	54	43	79	0.04
Lethrinidae	28	7	108	0.20	17	9	31	0.11	26	14	47	0.10
Holocentridae	10	5	20	0.16	6	3	12	0.17	13	5	21	0.17
Carangidae	44	5	384	0.29	13	7	24	0.12	19	11	31	0.08
Carcharhinidae	9	3	25	0.24	1	1	2	2.88	1	3	9	0.18

Notes: (1) No constraints on r priors; (2) priors are constrained with no independent input for biomass; (3) priors are constrained with biomass incorporated as input parameters. The numbers for MSY and bounds are expressed in 1000 pounds.

Implications of the augmented catch-MSY approach to fishery management in coral-associated fisheries

The augmented catch-MSY approach generates an estimate of MSY for the different coral reef fish stocks that can be used as a proxy for the overfishing limit under annual catch limit–based management. This elevates the coral reef fish stocks from the catch-only tier to a model-based tier utilizing the simple Schaefer production model and an independent estimate of biomass from fishery-independent surveys. In order to quantify the scientific uncertainty and determine the acceptable biological catch levels, the control rules (Figure 11.8) require the council to conduct a risk of overfishing analysis (denoted by P^*, henceforth referred to as P^* analysis) (WPRFMC 2011). The P^* analysis is a score-based system to semiquantitatively account for sources of scientific uncertainty based on four dimensions: (1) assessment information; (2) uncertainty characterization; (3) stock status; and (4) productivity susceptibility of the stock. The total uncertainty score will be deducted from the 50% risk of overfishing, which is equivalent to the proxy overfishing limit or the MSY estimate from the augmented catch-MSY approach.

The augmented catch-MSY approach generates a probability distribution around the mean MSY estimate. This can be used to generate a risk of overfishing (P^*) table using quantiles of the one-tail distribution of the MSY at 5% increment. The catch associated with each level of risk is the acceptable biological catch. Preliminary results of the 2013 P^* analysis indicate that the total uncertainty score ranges from 23.3 to 23.6, generating a P^* value range of 26.7%–26.3% (M Sabater, Western Pacific Regional Fishery Management Council,

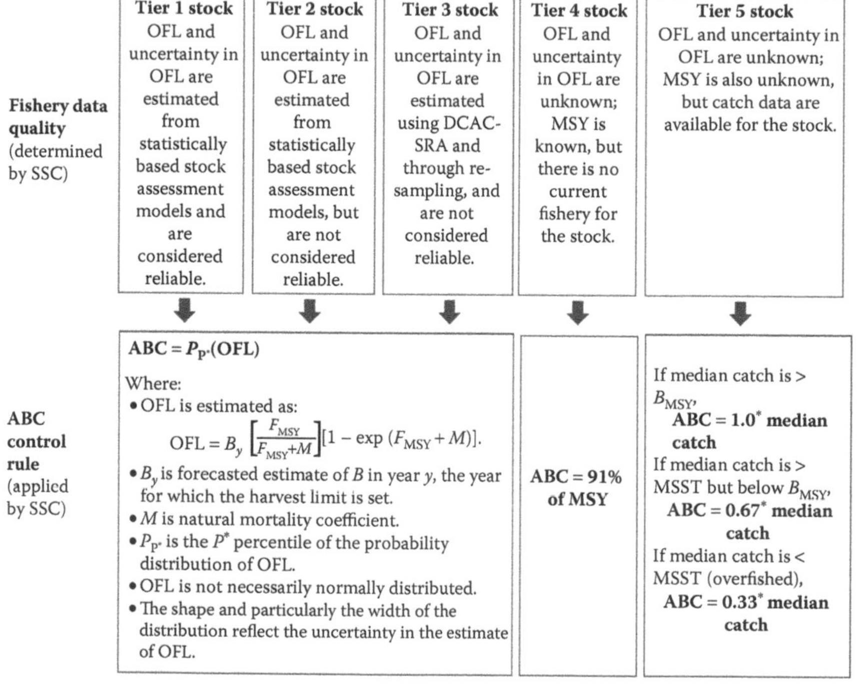

Figure 11.8 The tier-system approach of the Western Pacific Regional Fishery Management Council to determine the acceptable biological catch. DCAC-SRA, depletion-corrected average catch stock reduction analysis; MSST, minimum stock size threshold; OFL, overfishing limit; SSC, Scientific and Statistical Committee.

Table 11.5 Comparison of Recent Catches, Established Acceptable Biological Catch (ABC), and Estimated MSYs Using the Modified Catch-MSY Model in American Samoa

Family	MSY estimate	New ABCs	2012 ABCs	2012 Catch
Acanthuridae (surgeonfish)	145,500	116,000	19,516	6,394
Lutjanidae (snappers)	54,000	46,000	18,839	2,240
Carangidae (jacks)	18,400	15,400	9,460	2,374
Lethrinidae (emperors)	25,700	20,400	7,350	1,889
Scaridae (parrotfish)	341,300	299,000	8,145	2,807
Serranidae (groupers)	31,500	24,500	5,600	1,325
Holocentridae (squirrelfish)	13,700	11,900	2,585	905
Mugilidae (mullets)	3,100	2,500	2,857	1,252

Note: Values are expressed in pounds.

unpublished data). Table 11.5 compares the MSY estimates generated by the model, the acceptable biological catch associated with a P^* of 26.3, the 2012 acceptable biological catch based on the 75th percentile, and the 2012 coral reef fish catches for American Samoa. The new acceptable biological catch estimates from the model-based approach are generally higher than the estimates using the catch-only approach based on creel survey data. The new acceptable biological catch estimate for mullets (Mugilidae) is lower than the 2012 acceptable catch levels due to lack of biomass information to feed into the model. Mullets are not captured in the underwater visual census surveys since they are mostly found on reef flat and sandy bottom areas where surveys are not conducted. Striving for a better estimate of biomass as well as complete catch information improves the MSY estimation. The current Tier 5 control rule does not utilize any currently available biomass and generic life history information. This is the first regionwide attempt to estimate MSY for a broad range of coral reef stocks using a simple modeling approach.

The recent catches in American Samoa are small relative to the existing annual catch limits and annual catch limits were set equal to the acceptable biological catches (Table 11.5). In this particular case, the current acceptable biological catch appears to be adequate to protect the fisheries from overexploitation. However, lack of knowledge about the sustainable harvest limit obtained from a more scientifically robust method denies the fishing community the opportunity of exploring the development of their fishery. Based on the preliminary results, it appears that there is still a sufficient buffer for the fishery to develop. The coral reef fishery in the Western Pacific region is a low-value commercial fishery compared to the pelagic and bottomfish fishery (Gillett 2009). The current value of the coral reef fishery based on data from 2011 was estimated to be about $111,416 (WPacFIN, Pacific Island Fisheries Science Center, unpublished data). Fishing effort in American Samoa has declined over the past decades due to changes in socioeconomic conditions in this U.S. territory (Sabater and Carroll 2009). Reliance on fishing, although it still has cultural significance (Kilarski et al. 2006), has declined due to a limited market as well as dietary changes and higher economic status allowing American Samoans to purchase food instead of fishing for their protein source (Ponwith 1991, Craig et al. 1993, Saucerman 1995a,b, Coutures 2003). The cultural and aesthetic importance of coral reef and associated fishery resources, despite their low economic value, is invaluable to the indigenous people of the Western Pacific.

There is significant potential for maximizing the economic yield of the coral reef fishery. The information provided in Table 11.5, which refers to a subset of the coral reef fishery, can be translated into economic values. For this set of reef fish families, the 2012 catches can be valued at approximately $53,721, at an average of $2.80 per pound of reef fish. If the catches were to be maximized close to or at the annual catch limit (or the acceptable biological catch because both were set equal to each other) levels, then the potential economic value is estimated to be $208,186, quadruple the value of that caught in 2012. If the fishery operated close to the estimated new acceptable biological catches, the potential economic gain is estimated to be $1,499,960. This is eight times that allowed under the current annual catch limits and a 31-fold increase in 2012 landings. Although the potential for maximizing the economic yield is there, the coral reef fishery in the Western Pacific region is small and diverse and would require a significant increase in fishery participants and an investment upgrade of the current fishery operation. Due to low participation by fishers, local markets met the demand for fish by importing reef fish from neighboring island nations (Sabater and Carroll 2009). The current level of fishing in the small island U.S. territories and commonwealths may already be at its optimum. Coral reef fishery is important not only for commercial but also for traditional and cultural purposes (Levine and Allen 2009, Allen and Amesbury 2012). The coral reef fishery began as a means to feed the community and has communal importance. The cultural exchange of reef fishes is still practiced in the Western Pacific region through barter, trade, and customary exchange (Severance et al. 2013). The cultural and traditional significance of this fishery cannot be translated to an economic value but is deemed important in order to maintain cultural identity and relationships between communities.

Managing fisheries to attain optimum yields is one of the goals of the Magnuson-Stevens Fishery Conservation and Management Act. However, on a national level, this aim is overshadowed by the need to address overfishing. The model-based approach provides a reference point for the overfishing limit to be estimated based on more than just catch data. Optimum yield may have already been achieved for the Western Pacific coral reef fisheries, since it would entail significant investment in developing the fisheries to maximize the potential economic yield. Given the current modest commercial value of the coral reef fishery, a simple assessment method such as the augmented catch-MSY approach provides a practical way to provide scientific advice for managing the fishery compared to the cost of conducting a formal stock assessment for a large high-value commercial fishery.

References

Allen SD, Amesbury JR. 2012. Commonwealth of the Northern Mariana Islands as a fishing community. US Department of Commerce, NOAA Technical Memorandum NOAA-TM-NMFS-PIFSC-36. Available from: http://www.pifsc.noaa.gov/pubs/techpub_alpha.php via the Internet. Accessed 25 October, 2013.

Brander KM. 2007. Global fish production and climate change. Proc Natl Acad Sci U S A. 104(50):19709–19714.

CRED-PIFSC (Coral Reef Ecosystem Division, Pacific Islands Fisheries Science Center). 2013. Pacific Reef Assessment and Monitoring Program. Fish monitoring brief: American Samoa 2012. Pacific Islands Fisheries Science Center, PIFSC Data Report, DR-13-008. Available from: http://www.pifsc.noaa.gov/pubs/datapub_alpha.php via the Internet. Accessed 25 October, 2013.

Coutures E. 2003. The shoreline fishery of American Samoa: Analysis of 1-year data and implementation of a new sampling protocol. Biological Report Series No. 102. American Samoa: Department of Marine and Wildlife Resources.

Craig PC, Ponwith B, Aitaoto F, Hamm D. 1993. The commercial, subsistence, and recreational fisheries of American Samoa. Mar Fish Rev. 55:109–116.

Dick EJ, MacCall AD. 2011. Depletion-based stock reduction analysis: A catch-based method for determining sustainable yields for data-poor fish stocks. Fish Res. 110:331–341.

Dye TS, Graham TR. 2004. Review of archaeological and historical data concerning reef fishing in Hawaii and American Samoa. Honolulu: T. S. Dye and Colleagues, Archaeologists, Inc.

Edinger EN, Jompa J, Limmon GV, Widjatmoko W, Risk MJ. 1998. Reef degradation and coral biodiversity in Indonesia: Effects of land-based pollution, destructive fishing practices and changes over time. Mar Poll Bull. 36(8):617–630.

Fox WW. 1970. An exponential surplus-yield model for optimizing exploited fish populations. Trans Am Fish Soc. 99(1):80–88.

Froese R, Pauly D. 2013. FishBase [Internet]. June 5–6, 2013. Available from: http://www.fishbase.org.

Garcia S, Sparre P, Csirke J. 1989. Estimating surplus production and maximum sustainable yield from biomass data when catch and effort time series are not available. Fish Res. 8:13–23.

Gillett R. 2009. Fisheries in the economies of the Pacific island countries and territories. Mandaluyong City, Philippines: Asian Development Bank.

Gulland JA. 1970. The fish resources of the oceans. FAO Fisheries Technical Paper 97. Rome: Food and Agriculture Organization. p. 1–4.

Gulland JA. 1985. Fish stock assessment. Rome: Food and Agriculture Organization.

Houk P, Rhodes K, Cuetos-Bueno J, Lindfield S, Fread V, McIlwain JL. 2012. Commercial coral reef fisheries across Micronesia: A need for improving management. Coral Reefs. 31(1):13–26. Available from: http://dx.doi.org/10.1007/s00338-011-0826-3 via the Internet.

Hughes TP. 1994. Catastrophes, phase shifts, and large-scale degradation of a Caribbean coral reef. Science. 265(5178):1547–1551.

Jackson JBC, Kirby MX, Berger WH, Bjorndal KA, Botsford LW, Bourque BJ, Bradbury RH, Cooke R, Erlandson J, Estes JA, et al. 2001. Historical overfishing and the recent collapse of coastal ecosystems. Science. 293(5530):629–637.

Kilarski S, Klaus D, Lipscomb J, Matsoukas K, Newton R, Nugent A. 2006. Decision support for coral reef fisheries management: Community input as a means of informing policy in American Samoa. M.Sc. Thesis, The Donald Bren School of Environmental Science and Management. Santa Barbara: University of California.

Levine A, Allen S. 2009. American Samoa as a fishing community. US Department of Commerce, NOAA Technical Memorandum NOAA-TM-NMFS-PIFSC-19. Available from: http://www.pifsc.noaa.gov/pubs/techpub_alpha.php via the Internet. Accessed 25 October, 2013.

Levine A, Sauafea-Leau F. 2013. Traditional knowledge, use, and management of living marine resources in American Samoa: Documenting changes over time through interviews with elder fishers. Pac Sci. 67(3):395–407.

Luck D, Dalzell P. 2010. Western Pacific Region reef fish trends. A compendium of ecological and fishery statistics for reef fishes in American Samoa, Hawai'i and the Mariana Archipelago, in support of annual catch limit (ACL) implementation. Honolulu: Western Pacific Regional Fishery Management Council.

MacCall AD. 2009. Depletion-corrected average catch: A simple formula for estimating sustainable yields in data poor situations. ICES J Mar Sci. 66:2267–2271.

Martell S, Froese R. 2012. A simple method for estimating MSY from catch and resilience. Fish Fish. 14(4):504–514. Available from: http://dx.doi.org/10.1111/j.1467-2979.2012.00485.x via the Internet.

McClanahan TR, Hicks CC, Darling ES. 2008. Malthusian overfishing and efforts to overcome it on Kenyan coral reefs. Ecol Appl. 18(6):1516–1529.

Munday PL, Jones GP, Pratchett MS, Williams AJ. 2008. Climate change and future of coral reef fishes. Fish Fish. 9:261–285.

Newton K, Cote IM, Pilling GM, Jennings S, Dulvy NK. 2007. Current and future sustainability of island coral reef fisheries. Curr Biol. 17(7):655–658.

PIFSC (Pacific Islands Fisheries Science Center). 2011. Coral reef ecosystems of American Samoa: A 2002–2010 overview. Pacific Islands Fisheries Science Center, PIFSC Special Publication SP-11-002. Available from: http://www.pifsc.noaa.gov/pubs/credpub.php via the Internet. Accessed 25 October, 2013.

Page M. 1998. The biology, community structure, growth and artisanal catch of parrotfishes of American Samoa. American Samoa: Department of Marine and Wildlife Resources.

Pandolfi JM, Bradbury RH, Sala E, Hughes TP, Bjorndal KA, Cooke RG, Jackson JB. 2003. Global trajectories of the long-term decline of coral reef ecosystems. Science. 301(5635):955–958.

Pastorok RA, Bilyard GR. 1985. Effects of sewage pollution on coral-reef communities. Mar Ecol Prog Ser. 21(1):175–189.

Ponwith BJ. 1991. The inshore fishery of American Samoa: A 12-year comparison. Biological Report Series No 22. American Samoa: Department of Marine and Wildlife Resources.

Restrepo VR, Thompson GG, Mace PM, Gabriel WL, Low LL, MacCall AD, Methot RD, Powers JE, Taylor BL, Wade PR, et al. 1998. Technical guidance on the use of precautionary approaches to implementing National Standard 1 of the Magnuson-Stevens Fishery Conservation and Management Act. US Department of Commerce, NOAA Technical Memorandum NMFS-F/SPO-31. Available from: http://www.nmfs.noaa.gov/sfa/NSGtkgd.pdf via the Internet. Accessed 25 October, 2013.

Roberts CM, Polunin NVC. 1993. Marine reserves: Simple solutions to managing complex fisheries. Ambio. 22(6):363–368.

Sabater MG, Carroll BP. 2009. Trends in reef fish population and associated fishery after three millennia of resource utilization and a century of socio-economic changes in American Samoa. Rev Fish Sci. 17(3):318–335.

Sabater MG, Tofaeono S. 2007. Effects of scale and benthic composition on biomass and trophic group distribution of reef fishes in American Samoa. Pac Sci. 61:503–520.

Sabater M, Tulafono R. 2011. American Samoa archipelagic fishery ecosystem report. *In*: Spalding S, editor. Pacific Island fisheries monograph. Honolulu: Western Pacific Regional Fishery Management Council.

Saucerman S. 1995a. The inshore fishery of American Samoa, 1991 to 1994. Biological Report No. 77. American Samoa: Department of Marine and Wildlife Resources.

Saucerman S. 1995b. Assessing the management need of a coral reef fishery in decline. Joint FFA/SPC Workshop on the Management of South Pacific Inshore Fisheries, Noumea, New Caledonia.

Schaefer MB. 1954. Some aspects of the dynamics of populations, important for the management of the commercial fisheries. Inter-Am Trop Tuna Comm Bull. 1(2):23–56.

Severance C, Franco R, Hamnett M, Anderson C, Aitaoto F. 2013. Effort triggers, fish flow, and customary exchange in American Samoa and the Northern Marians: Critical human dimensions of the Western Pacific fisheries. Pac Sci. 67(3):383–393.

Vert KA, Amoroso RO, Jensen OP, Hilborn R. 2013. The frequency and intensity of productivity regime shifts in marine fish stocks. Proc Natl Acad Sci U S A. 110(5):1779–1784. Available from: http://dx.doi.org/10.1073/pnas.1214879110 via the Internet.

Wetzel CR, Punt AE. 2011. Model performance for the determination of appropriate harvest levels in the case of data-poor stocks. Fish Res. 110:342–355.

Williams I. 2010. U.S. Pacific reef fish estimates based on visual survey data. NOAA Pacific Islands Fisheries Science Center Internal Report IR-10-024. Honolulu: NOAA Pacific Islands Fisheries Science Center.

Williams ID, Richards BL, Sandin SA, Baum JK, Schroeder RE, Nadon MO, Zgliczynski B, Craig P, McIlwain JL, Brainard RE, et al. 2011. Differences in reef fish assemblages between populated and remote reefs spanning multiple archipelagos across the central and Western Pacific. J Mar Biol. 2011:14. Available from: http://dx.doi.org/10.1155/2011/826234 via the Internet.

Worm B, Hilborn R, Baum JK, Branch TA, Collie JS, Costello C, Zeller, D. 2009. Rebuilding global fisheries. Science. 325(5940):578–585.

WPRFMC (Western Pacific Regional Fishery Management Council). 2011. Omnibus amendment for the Western Pacific Region to establish a process for specifying annual catch limits and accountability measures. Honolulu: Western Pacific Regional Fishery Management Council.

Zeller D, Booth S, Davis G, Pauly D. 2007. Re-estimation of small scale fishery catches for U.S. flag-associated island areas in the Western Pacific: The last 50 years. Fish Bull. 105:266–277.

chapter twelve

Predictive mapping of coral reef fish species and communities

Simon J. Pittman and Anders Knudby

Contents

Introduction

At the spatial scales relevant to the routine movements of many coral reef–associated fishes, habitats exist as a spatially heterogeneous terrain varying in structural attributes, biotic communities, and seascape context (Grober-Dunsmore et al. 2009, Boström et al. 2011). Seascape studies, focusing on the ecological consequences of spatial patterning, have revealed that both the patch structure and the terrain morphology influence the geographical patterns of fish distributions and diversity at a range of spatial scales (Pittman et al. 2007a,b, Boström et al. 2011). In the past few decades, however, it has become evident that the physical structural complexity of shallow tropical seascapes is declining (Pandolfi et al. 2005). For instance, in the Caribbean Sea, multiple interacting stressors are associated with a decline in the abundance of live coral, particularly the large and architecturally complex branching species, and a measurable flattening of the topographic complexity over the past 60 years (Gardner et al. 2003, Pandolfi et al. 2005, Alvarez-Filip et al. 2009, 2011). In many locations, the biotic assemblage composition has also changed in what is often referred to as a "phase shift" from coral to algal dominance (Done 1992, Mumby 2009), with reduced-diversity coral communities comprised of stress-tolerant species (Green et al. 2008). These changes, combined with fishing pressure, have led to a region-wide decline in reef fish density (Paddack et al. 2009) and size composition, with the largest fishes becoming increasingly rare (Stallings 2009). A decline in habitat suitability for fishes of fishery value has negative effects on the sustainability of reef fisheries and human livelihoods (Souter and Linden 2000, Hughes et al. 2005). At broader spatial scales, the problem of declining quality of coral reefs is compounded by a loss of seagrasses (Orth et al. 2006) and mangroves (Polidoro et al. 2010) from the surrounding seascape. For multihabitat fishes, many of which are important food fish, adjacent vegetated habitat types provide complementary and supplementary resources that act synergistically to elevate local fish productivity and diversity in tropical seascapes (Parrish 1989, Nagelkerken et al. 2002, Mumby et al. 2004, Pittman et al. 2007b, Jones et al. 2010).

Degradation of fish habitat is geographically widespread and unevenly distributed, creating a significant logistical challenge for the prioritization of management actions. Much recent attention has been focused on developing spatial prioritization strategies to identify a subset of functionally important places, habitats, communities, and species that can be preferentially protected to ensure the sustainability of key ecosystem function and services (Roberts et al. 2003, Lourie and Vincent 2004, Klein et al. 2010). Coral reef ecosystems that are robust or resilient to stressors have also emerged as a criterion for prioritizing conservation investments (Maynard et al. 2010, McClanahan et al. 2012). As well as identifying areas of special concern, such as essential fish habitat, biodiversity and productivity hotspots, and robust and resilient reefs, managers implementing an ecosystem-based approach are increasingly interested in identifying key ecological relationships to better anticipate the consequences of environmental change (Crowder and Norse 2008, Foley et al. 2010). Place-based management strategies such as the setting up of marine protected areas (MPAs) and more comprehensive regional marine spatial management are increasingly applied worldwide, including in areas with tropical coral reef ecosystems (Mora et al. 2006, Agardy et al. 2011).

To be effective, prioritization strategies require reliable spatial data, with sufficient ecological detail to be meaningful for ecosystem-based management, but with sufficient geographical coverage to be operationally relevant for decision making. This is problematic because for many regions biological survey data are sparse and unevenly distributed and often highly clustered. Field sampling is typically conducted using sampling techniques with relatively small (from 10 cm^2 to 100 m^2) sample unit areas (line transects, quadrats, cores, etc.) providing useful detail, but at spatial scales that are too fine to support geographically comprehensive evaluations in spatial management.

Predictive mapping of biological distributions, sometimes referred to as species distribution modeling and ecological niche modeling, is now widely recognized as an effective analytical tool to address spatial information gaps in support of management (Guisan and Thuiller 2005, Elith and Leathwick 2009). Crucially important for the implementation of predictive mapping is the fact that most field samples now have an accurate and precise reference in time (time sample was collected) and space (geolocation of sample) that is becoming increasingly reliable with the widespread use of global positioning system (GPS) receivers. In addition, technological advances in high-resolution remote sensing and the increased availability of quantitative tools, such as spatial pattern metrics, allow us to measure and quantify the detailed patterning of coral reef ecosystems across multiple spatial and temporal scales (Knudby et al. 2007, Goodman et al. 2013). New, reliable, and cost-effective techniques and analytical tools, such as geostatistical modeling and machine-learning algorithms combined with geographical information systems (GIS), are being integrated to model and predictively map environmental features and biological distributions across broad geographical areas (Leathwick et al. 2008, Knudby et al. 2010a, Pittman and Brown 2011). These sophisticated modeling tools are capable of more than just providing missing data since they also have the potential to provide new ecological insights through analyses of complex macroecological relationships.

Here, we review the development of the predictive mapping of tropical marine fish distributions and then outline the rationale for a multiscale seascape ecology approach that draws on concepts and analytical tools associated with landscape ecology. We provide a short overview of the latest generation of machine-learning algorithms for predictive modeling and offer an operational framework for a statistically robust approach. We highlight the utility of the techniques for addressing a range of spatial information gaps in coral reef ecosystems with examples that focus on fish–seascape relationships of importance to the sustainability of reef fisheries.

History of predictive mapping in coral reef ecosystems

Early depictions of marine fish species distributions and habitats were typically drawn on paper maps guided by expert knowledge. It was not until the early 1990s, following the development of habitat suitability models in terrestrial systems by the U.S. Fish and Wildlife Service, that computer-based statistical modeling for delineating marine fish distributions emerged (e.g., Rubec and O'Hop 1996, Christensen et al. 1997). In the United States, the Magnuson-Stevens Fishery Conservation and Management Act (NMFS 1996) was amended to require the description and identification of essential fish habitat, representing societal recognition that habitat quantity and quality are critical to the health and productivity of fish populations. The Magnuson-Stevens Act therefore became an important driver of progress in mapping fish–habitat relationships at a time when the mapping process was being revolutionized by GIS software, GPS receivers, and a general increase in the availability of marine spatial data. One of the first applications of predictive mapping for tropical marine species was the Florida Estuarine Living Marine Resources System (FELMER) project, which statistically linked information on the habitat requirements of marine species (at different life stages) to maps describing the habitat characteristics of the seafloor and water column to predict species distributions for key fishery species (Rubec et al. 1998a,b). Habitat preferences for each species' life stage were determined by fitting polynomial regressions to mean catch-per-unit-effort (CPUE) across environmental gradients, with higher mean CPUEs indicating higher abundance. Habitat Suitability Index (HSI) maps created by linking the geometric mean of the index values to each habitat layer, were used to create seasonal maps (spring, summer, fall, winter) depicting the spatial distribution of each species' life stage (Rubec et al. 1998a,b, Rubec et al. 1999). Hotspots of suitability were then used to identify the most important habitats for each species with potential to be designated Habitat Areas of Particular Concern. This HSI approach was also used to develop models that could be applied to unsurveyed areas in order to cost-effectively fill spatial information gaps.

Many of the early efforts (e.g., Christensen et al. 2003) used linear models with many statistical assumptions and limited ability to model nonlinear patterns including threshold effects and complex multiscale interactions between environmental variables, all typical of ecological relationships. Relationships between organisms and environmental variability were modeled at single and sometimes arbitrary spatial scales, and much of the ecologically meaningful spatial heterogeneity in seafloor structure was not explicitly incorporated. More recently, a conceptual and statistical leap has been made in predictive modeling by incorporating environmental predictors quantified at multiple spatial scales using spatial pattern metrics from landscape ecology and geomorphology and the latest generation of machine-learning algorithms to model complex nonlinear relationships.

Comparative multimodel analysis has demonstrated that this spatial ecoinformatics approach to ecological modeling can boost predictive performance and provide new insights into potentially important drivers of ecological patterns (Elith et al. 2006, Pittman et al. 2007a, Knudby et al. 2010a, Pittman and Brown 2011).

Why a multiscale approach?

Ecological theory tells us that interactions across scales are important (Allen and Starr 1982, Kotliar and Wiens 1990, Holling 1992, Levin 1992). From the landscape ecology perspective, which incorporates a hierarchical conceptual framework, we recognize that species and communities respond to their environment at a range of spatial scales (Schneider 2001, Pittman and McAlpine 2003). Even within a single species, individuals at different life stages or of different sexes may respond to the environment in different ways, and at different scales, although some generalities will exist. To add to the already complex scale effects, the organism's response will also be modified by locational differences, such as the local distribution of habitat structure, predators, and prey.

However, for most marine species, even those that have been well studied, the relevant scale range is either estimated or, in most cases, remains unknown. Movement patterns can guide scale selection, but rarely are the relevant spatial and temporal dimensions of movement patterns known. Lacking knowledge of a specific focal spatial scale, we recommend adoption of an exploratory multiscale approach. This not only is judicious as a bet-hedging strategy when faced with scale uncertainty, but also provides a technique with which to begin incorporating a wider range of structural heterogeneity into the analyses and an opportunity to examine scale-dependent responses.

A multiscale seascape approach is particularly relevant to reef fish ecology since many fishes use multiple habitat types through routine daily home range movements, ontogenetic habitat shifts, and seasonal or spawning migrations (Nagelkerken et al. 2002, Pittman and McAlpine 2003). For example, in the context of locating and measuring habitat suitability, a princess parrotfish (*Scarus taeniopterus*) may only persist in seascapes where diurnal foraging habitat is sufficiently close to suitable nocturnal resting habitat providing adequate protection from predators. If suitable resources are not located within a distance traversable by an individual fish, then the seascape will likely offer low habitat suitability. If similar habitat preference occurs among multiple herbivorous fishes, then the low habitat suitability areas will support less herbivory than high-suitability areas. This habitat may consequently be of lower value to fisheries. Complex fish–seascape interrelationships with implications for fisheries and reef resilience can be modeled and made spatially explicit with the predictive mapping techniques.

Spatial representations of seascape structure

The spatial patterning of seascapes can be represented in two-dimensional (i.e., benthic habitat maps) and three-dimensional (i.e., digital elevation) models. From these models, a wide range of derivative environmental predictors can be created for predictive mapping of fish–seascape relationships. Predictors can be classified as relatively stable or relatively dynamic along a continuum of dynamism (Hyrenbach et al. 2000). Relatively stable predictors include seafloor features such as geology (geomorphological features/zones and bathymetry) and benthic communities or habitat types (seagrasses, colonized hardbottom, mangroves, etc.). Relatively dynamic predictors include water surface and water column properties such as salinity, temperature, ocean fronts, and phytoplankton concentration.

The majority of coral reef ecosystem applications have focused on the stable seafloor features as reliable representations of seascape structure with known ecological relevance to reef-associated fishes. Seafloor structure is usually represented as either thematic benthic habitat maps or three-dimensional digital terrain models. Benthic habitat maps are typically two-dimensional maps of internally homogeneous patches (habitat or biotope classes), with discrete boundaries represented as polygons. Benthic habitat maps can also be represented as a grid of cells (i.e., raster data) in which a habitat type is assigned a unique value. Predictors can be based on individual habitat types or on patch context by quantifying the seascape composition (amount and variety of habitat types) and configuration (spatial arrangement of habitat patches) in the surrounding seascape (Wedding et al. 2011). Advances in high resolution multispectral and hyperspectral remote sensing now make it more likely that habitat maps can incorporate reliable information on the biotic composition of patches such as the amount of live coral and macroalgae (Hedley 2013, Wozencraft and Park 2013). This could further improve model performance.

In addition to discrete patchiness, the marine environment exhibits spatial variability in the form of continuous, multidimensional gradients. For example, hydrodynamic interactions with coastline geomorphology, riverine influences, and seafloor topography create a wide range of spatial gradients in marine environmental conditions (e.g., salinity, wave action, depth, temperature, nutrients). The continuum model recognizes that a range of ecological processes may affect habitat suitability for different species through time, in a spatially continuous and potentially complex way. The premise is that individuals within a species are likely to respond to spatial gradients in resources such as food and refuge or other environmental conditions (Austin and Smith 1989, Fischer and Lindenmayer 2006). An additional benefit of the gradient model is that it retains the captured heterogeneity and avoids the subjectivity associated with boundary and thematic designations associated with habitat maps. High resolution bathymetric data and associate backscatter collected using techniques such as airborne laser altimetry (e.g., light detection and ranging or LiDAR) and sonar (e.g., multibeam or side-scan sonar) can be used to construct continuously varying surfaces exhibiting complex vertical, horizontal, and compositional structure (Wedding et al. 2008, Pittman et al. 2009).

In the absence of continuous sampling, interpolated or predictive surfaces can be modeled using point data from intensive georeferenced field samples. A cost-effective way to create predictor variables is to identify useful surrogate or proxy variables which describe heterogeneous ecological patterns that are difficult to map directly, such as biological diversity, community composition, and some biotic distributions. Useful surrogate variables have been found that can be either biotic or abiotic features providing they can be reliably mapped (Ward et al. 1999, Beger and Possingham 2008, Harborne et al. 2008, McArthur et al. 2010, Mellin et al. 2011). Seafloor maps that depict ecosystem structure, such as benthic habitat maps, are spatial representations that typically integrate both biotic and abiotic features of the seafloor (Brown et al. 2011), although some maps now also have incorporated both benthic and pelagic features into seascape classes (e.g., Roff et al. 2003).

Bathymetry is probably the single most useful predictor of marine biotic distributions due to its importance for marine ecological patterns and processes. Although not always available at sufficient spatial resolution for coral reef ecosystems, where bathymetry data have been examined, the three-dimensional arrangement of structural features over the seafloor surface (i.e., topographic complexity) has been an important driver of fish distributions and diversity (Pittman et al. 2009, Pittman and Brown 2011). Topographic complexity at a range of spatial scales influences a wide range of the biological, chemical, and physical aspects of a coral reef system such as hydrodynamics, nutrient pathways, predator–prey interactions, and larval settlement (Zawada 2011). In predictive models of habitat suitability for reef fish, higher species richness is associated with higher topographic complexity (Pittman et al. 2007a, 2009). Interestingly, the highest species richness is not associated with the highest topographic complexity, suggesting that the relationship is nonlinear or the relationship is affected by other variables (Pittman and Brown 2011). A wide range of metrics are available for quantifying complex structure on continuously varying surfaces (McGarigal and Cushman 2005) and we refer to these as terrain morphometrics because they measure the surface morphology (e.g., slope, aspect, curvature, rugosity, etc.) from a digital terrain model of bathymetry.

Geographical position metrics such as distance to shore, distance to shelf edge, and distance to river mouth have proved useful through interaction with other predictor variables because they can capture unobserved, unknown, or unmapped patterns such as inshore–offshore gradients in physical and chemical conditions and proximity to ecologically relevant features (e.g., nearshore nursery areas, shelf-edge spawning sites, freshwater outflow). Although rarely quantified, distance to shore has made a major contribution to models of reef fish distributions in examples from Australia (Mellin et al. 2010) and the Caribbean (Pittman and Brown 2011).

Machine-learning algorithms

Traditional statistical methods such as linear, logarithmic, and logistic regression have commonly been used to examine complex ecological relationships, leading to erroneous and misleading interpretations of ecological relationships when the assumptions underpinning the chosen regression model are not justified by the data (Jones and Syms 1998, Breiman 2001). An alternative and more flexible approach is provided by machine-learning or statistical-learning algorithms, a family of nonparametric models where the structure of the relationships between variables is not defined *a priori* but rather derived from iterative training and testing using random subsets of the available data (Hastie et al. 2009). Machine-learning algorithms have been shown to outperform parametric models in many multimodel comparisons in ecology, and have become the default tools for today's ecological modeling of all but the simplest ecosystems (Elith et al. 2006, Phillips et al. 2006, Pittman et al. 2007a, Knudby et al. 2010b). For example, Elith et al. (2006) compared 16 predictive modeling techniques, including both conventional and machine-learning algorithms using presence-only data for 226 species from six regions, and showed that machine-learning algorithms consistently outperformed other algorithms. For modeling coral reef ecosystems, Knudby et al. (2010b) showed that machine-learning algorithms provided significant increases in performance over more conventional modeling techniques such as generalized additive models and linear regression. Examples gaining popularity in ecology are artificial neural networks, support vector machines, and various forms of decision tree ensembles such as boosted regression trees (BRT) and random forest. Furthermore, much evidence suggests that predictions based on an ensemble of models of varying structure outperform those based on a single model type (Aruajo and New 2007, Marmion et al. 2009), although both the models and the method used to combine their predictions influence performance (Džeroski and Ženko 2004). Implementation through free software such as R (R Core Development Team 2012) has improved accessibility and some of the commonly perceived "black box" mechanics have now been explained to ecologists (De'ath 2007, Elith et al. 2008, Olden et al. 2008). For example, measures of relative variable importance and partial dependence plots can provide much insight into the model structure. Nevertheless, interpretation of model structure remains complicated when the value of one predictor variable influences the relationship between the response variable and another predictor variable (e.g., shelter space may correlate positively with fish biomass at shallow depths but not at deeper depths). Similarly, complications occur when multiple collinear predictors are used to train a predictive model, as is the case when a multiscale approach is used and in exploratory research where different but related independent variables (e.g., turbidity and distance from shore) are included (Knudby et al. 2010b).

A multiscale and multialgorithm comparative approach

We advocate a multiscale and multialgorithm comparative approach (Figure 12.1) in situations where there are no specific reasons to select a single scale or single modeling algorithm. This exploratory spatial ecoinformatics approach incorporates lessons learned from data mining applications in a wide range of disciplines. Comparative studies indicate that model results and subsequent interpretation of predictor relevance are often dependent on the type of model algorithm used (Knudby et al. 2010b). Furthermore, the evaluation of model performance can be biased by the choice of metrics and the spatial scale used to measure model accuracy (Aguirre-Gutiérrez et al. 2013). We recommend using multiple metrics because each metric quantifies different aspects of model performance and the literature includes reasonable arguments both for and against most of the commonly used model performance metrics (Fielding and Bell 1997, Pearce and Ferrier 2000). A multialgorithm and multimetric approach to model evaluation and selection therefore provides an opportunity to examine cross-model agreement (and disagreement) that should result in a more robust analysis with more complete quantification of errors leading to reduced uncertainty and greater realism in the ecological interpretation. It is popularly quoted

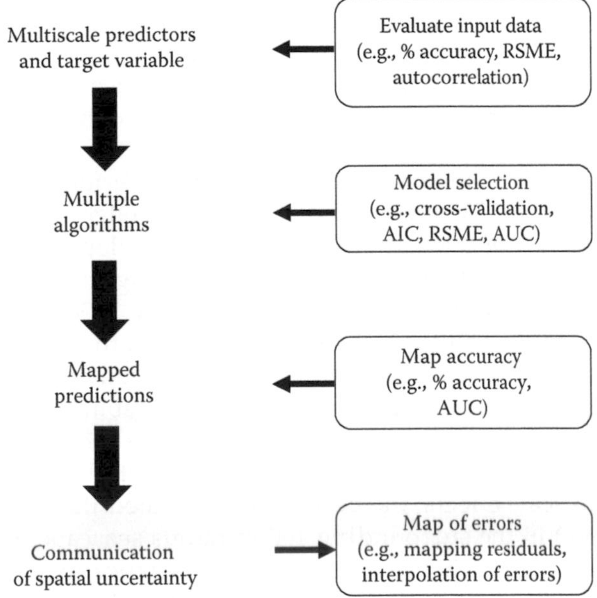

Figure 12.1 Multiscale and multimodel approach including three steps of model validation: (1) assessment of errors in the input data, (2) model validation and selection of best model or model ensemble, and (3) map accuracy assessment using independent data. AIC, Akaike's information criterion; RSME, root mean square error.

that "all maps are lies" and "all models are wrong," but we argue that by taking careful account of all sources of model bias and error throughout the process, we can produce predictive models that are of great value to marine ecologists and managers. Although rarely covered in the modeling literature, greater attention is needed so that sources of error can be accounted for and explained and techniques can be developed to enhance the communication of data caveats, errors, and spatial uncertainty (Hunsaker et al. 2001, Wedding et al. 2011).

Case studies of predictive mapping using fish–seascape relationships

Mapping habitat suitability for a harvested grouper species

Knowing the distribution of key species of ecological and economic concern across spatially complex seascapes supports management decision making, including siting of MPAs, evaluation of threats, zoning, and delineation of essential fish habitat. Pittman and Brown (2011) predicted suitable habitat for selected coral reef-associated fish species recorded in underwater visual surveys across the insular shelf of southwestern Puerto Rico. Species presence/absence association coefficients were calculated for a range of fish species including *Cephalopholis fulva* (coney), a common grouper species targeted by local fisheries. During the fish survey period, the bathymetry of the study area was mapped with airborne LiDAR and processed into a 4×4 m spatial resolution digital terrain model (<1 to 70 m depth). Seventy-nine spatial predictor variables were developed, including 11 derivatives of seafloor bathymetry quantified at seven different spatial scales (5, 15, 25, 50, 100, 200, 300 m radii) using a "moving window" technique within a GIS (Pittman and Brown 2011). Geographical setting was quantified by creating a grid of distance to shelf edge and distance to coastline measurement. Species distributions were modeled with BRT and MaxEnt to compare model performance. A high probability of occurrence indicated that the grid cell had suitable environmental conditions for the species. BRT provided "outstanding" model predictions (area under the curve [AUC] > 0.9) for three of five species and MaxEnt for two of five species, including *C. fulva* models that were the highest performing according to standard interpretation of the AUC metric. Distance to shelf edge was the most influential single predictor variable. Of the bathymetric derivatives, slope of the slope, a measure of topographic complexity, contributed most to explaining the response variables. For *C. fulva*, a threshold effect was evident at approximately 2000 m from the shelf edge, where species occurrence abruptly increased. For *C. fulva*, the predictive map from the MaxEnt model was more accurate than the BRT map based on assessment with independent species sightings data. Mapped predictions of habitat suitability showed that although almost all *C. fulva* were observed close to the shelf edge, not all shelf-edge areas offered high suitability due to variability in the spatial heterogeneity of the seafloor. With regard to scale effects, the best model indicated that topographic complexity (slope of the slope) in the surrounding 100 m radius seascape was most influential in defining suitable habitat when interacting with distance to shelf edge. The study demonstrated that useful distribution models of important reef fishery species can be developed using a wide range of multiscale explanatory variables derived from a single airborne sensor. Also demonstrated is the utility of applying multimodel algorithms that are capable of modeling nonlinear relationships with interactions among explanatory variables. Little is known, however, about the influence of spatial heterogeneity in fishing effort on the performance of models to accurately predict species distributions. Comparison between models developed for grouper in fished versus neighboring unfished reefs could be informative in determining the proportion of suitable habitat that is actually occupied by the species.

Mapping large-bodied fishes to identify essential fish habitat

Identifying and quantifying areas that concentrate biomass in coral reef ecosystems, such as fish spawning aggregation sites and other essential fish habitat, is a high priority for fisheries management. Costa et al. (2013) used a ship-based, splitbeam echosounder to map fish position in the water column and their body size along the shelf edge of the U.S. Virgin Islands simultaneously with a multibeam echosounder to map the seafloor bathymetry. Multiscale fish–seascape relationships were then modeled using BRT to identify important environmental drivers of fish occurrence and density and to predictively map locations of high fish productivity (Figure 12.2). Models for large fish performed well enough to provide useful maps of distributions (80.4%–86.1% accuracies), and highlighted areas with low fish density. Water depth, topographic complexity, and proximity to the shelf edge were the most influential explanatory variables. The integration of acoustic sensors and multiscale predictive mapping techniques offered a novel application of geospatial technologies to rapidly identify fish aggregations and associated habitats to support spatial prioritization in marine management. The study demonstrated the utility of integrating data from two different acoustic sensors capturing fish–seascape relationships in four dimensions (three in space and one in time). These data were collected at night and from water depths where ecological observations are rarely conducted due to low visibility. A limitation of the technique, however, is that species cannot be easily identified without additional underwater observations (divers/submersibles). Future research is now required to develop pattern recognition software that will automate the identification of fish species in the acoustic data based on body size combined with behavioral characteristics such as schooling, swim speed, and position in the water column.

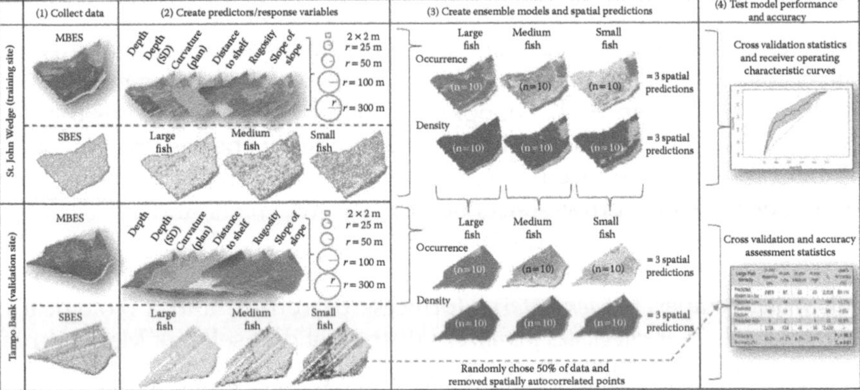

Figure 12.2 **(See color insert)** Example of analytical steps and data types in the multiscale seascape ecology approach to spatial predictive modeling. MBES, multibeam echosounder for mapping the seafloor; SBES, singlebeam echosounder for mapping location and body size of fish. (Adapted from Costa, B., Taylor, C.J., Kracker, L., Battista, T., and Pittman, S.J., *PLoS ONE*, 9(1), e85555, p1–7, 2013.)

Mapping indicators of coral reef resilience

The resilience of an ecosystem can be defined as the ecosystem's ability to maintain and restore structure and function in the face of disturbance (Pimm 1984, West and Salm 2003). Resilient areas are increasingly a focus for inclusion in MPAs and MPA networks, yet little is known about which areas are the most resilient to disturbances. Indicators quantifying ecosystem properties thought to confer resilience (McClanahan et al. 2012) can be used as proxies to guide identification of resilient reefs. Commonly accepted indicators of coral reef resilience include (but are not limited to) high coral cover and diversity, and low nutrient and sediment levels. For fish communities, high herbivore biomass and diversity is associated with resilience because herbivory will maintain algal biomass at lower levels, thus facilitating coral recovery following a disturbance. These indicators can be mapped with predictive modeling, but the work is in its infancy and map accuracies for many biological indicators are relatively low when predicted using satellite-based predictors (Rowlands et al. 2012, Knudby et al. 2013). Knudby et al. (2013) predicted the distribution of functional richness in herbivorous fish across coral reef ecosystems in the Kubulau traditional fisheries management area of Fiji. The authors found that random forest models, a machine-learning ensemble technique, were more accurate than geostatistical interpolation for mapping distributions of fish herbivore biomass and functional group richness (Figure 12.3).

Important future challenges include mapping and modeling connectivity, as well as the intelligent combination of resilience indicators and auxiliary information into more direct measures of resilience. Validation of resilience maps using time series of ecological observations will be the ultimate test of the success of predictive mapping efforts.

Forecasting the impact of declining reef complexity on fish distributions

High resolution digital terrain models of seafloor complexity provide a novel cost-effective tool for forecasting (and hindcasting) impacts on fish from changes to the topographic complexity of coral reef ecosystems. Coral reef rugosity, a measure of topographic complexity, is estimated to have decreased by more than 50% since the 1960s (Alvarez-Filip et al. 2009). Predictive mapping has recently been used as a proof of concept to develop impact scenarios by mapping the effect of reducing terrain topographic complexity on fish habitat suitability in southwestern Puerto Rico (Pittman et al. 2011b).

Fish species occurrence, from underwater visual surveys, was statistically linked to a suite of spatial predictors derived from airborne laser–derived bathymetry (LiDAR) using MaxEnt (Pittman and Brown 2011). Topographic complexity, measured as the slope of the slope, was used to spatially model and map probabilities of species presence. Slope of the slope was then uniformly reduced across the entire terrain by 25% to indicate the estimated decadal decline for southwestern Puerto Rican coral reefs, and by 50%, approximating Caribbean-wide declines since the 1960s. Predictions of high habitat suitability (using a consistent probability threshold for each scenario) were then remapped for an abundant herbivorous scraper, *S. taeniopterus* (princess parrotfish) and a biological indicator species of live coral cover, *Stegastes planifrons* (threespot damselfish). Mapped predictions were overlain and examined for differences in spatial patterning. Areas of suitable habitat, under different reef "flattening" scenarios, were quantified and compared to measure the change. Suitable habitat for *S. taeniopterus* reduced by 30% with a 25% reduction in terrain complexity, and reduced by as much as 66% when terrain was reduced by 50%. With a 25% reduction in topographic complexity, habitat was lost from the edges of large contiguous

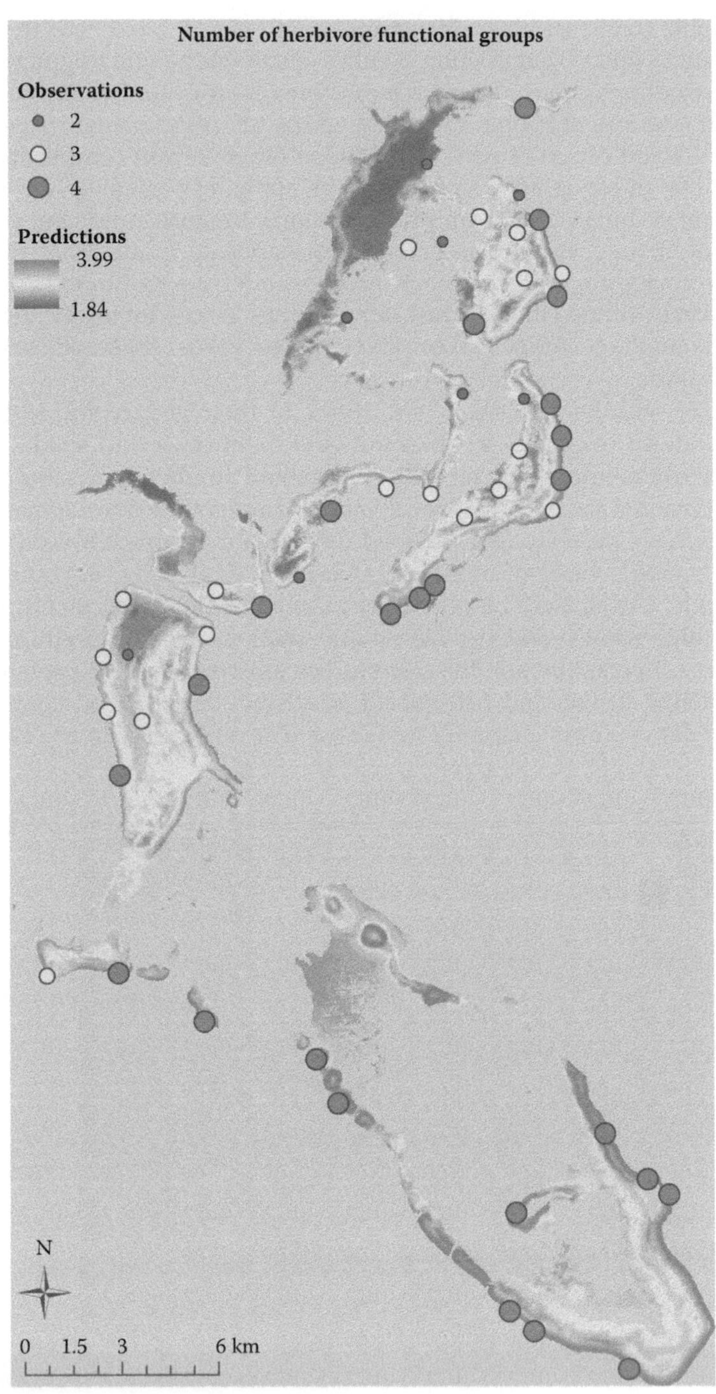

Figure 12.3 **(See color insert)** Predictive map of richness in herbivorous fish functional groups across coral reef ecosystems in Fiji as one indicator in a suite of resilience indicators. Predictive models were developed using random forest on the basis of high-resolution satellite data. (Adapted from Knudby, A., Jupiter, S., Roelfsema, C., Lyons, M., and Phinn, S., *Remote Sens*, 5, 1311–1334, 2013.)

patches of suitable coral reef where structure had already existed near the lower thresholds of suitability. With a 50% reduction, patches of suitable habitat fragment reduced even more and few large contiguous patches remain, whereas the number of small patches with relatively small interiors of habitat increased across the seascape. For *S. planifrons*, a 56% loss of suitable habitat occurred with a 25% reduction in terrain complexity (Figure 12.4). Simulating decline of topographic complexity by applying a spatially uniform reduction scenario is likely to be an overly simplistic scenario because it ignores the true spatial heterogeneity of change. In the absence of information to guide more realistic simulations, predictive mapping offers a flexible and cost-effective tool to examine a variety of scenarios that can be refined in the future, for example, to account for the different impacts of future storms on deep versus shallow and exposed versus sheltered reefs and impacts of ocean acidification.

The four case studies described here differ in their geographic setting, choice of algorithm(s), modeled response variable, and both the metrics and spatial scales used to quantify the environmental predictors. Given the small number of existing studies to date, such multidimensional variability between studies hinders the development of synthesis, and few studies have individually provided direct comparisons between aspects of this variability beyond the choice of modeling algorithm. Replication of comparison studies across geographic settings will be necessary to provide more informed guidance on the choice of both algorithm(s) and the metric and spatial scales of environmental predictors. Nevertheless, in combination the case studies illustrate the wide potential for spatial predictive modeling of fish and fish-related ecosystem variables on coral reefs, as well as the range of data sources currently in use for that purpose. The use of a multimodel

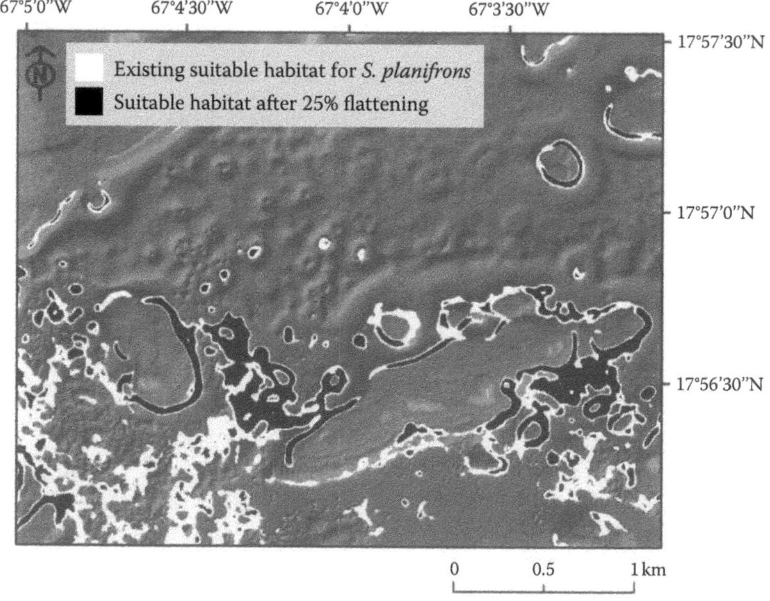

Figure 12.4 Predicted habitat suitability for threespot damselfish (*S. planifrons*) across a subset of the southwestern Puerto Rico study area using unaltered LiDAR-derived topographic complexity and numerically flattened topographic complexity to simulate 10-year declines for coral reefs in southwestern Puerto Rico. MaxEnt was used for modeling predictions. (Pittman, S.J., Costa, B., Jeffrey, C.F.G., and Caldow C., 2011. Proceedings of the 63rd Gulf and Caribbean Fisheries Institute 1–5 November, 2010 San Juan, Puerto Rico. p. 420–426.)

and multiscale approach is implicitly or explicitly adopted in both these and other studies (Kendall et al. 2003, Knudby et al. 2010a,b, Wedding et al. 2008, 2011, Pittman et al. 2009). This is a consequence of both our limited knowledge of optimal (focal) spatial scales at which specific ecological processes operate and the increasing realization that species respond to their environment at multiple scales.

Future research

In modern marine management, spatial modeling plays a key role in decision support and is central to the planning and implementation of ecosystem-based management (Pittman et al. 2011a). Predictive mapping is rapidly emerging as a viable tool for rapid, reliable, and cost-effective spatial ecological data analysis for marine management, particularly for site prioritization. Modeling techniques are increasingly adept at untangling large and complex ecological data and knowledge is growing about the errors and biases impacting the various stages of model development (samples size and distribution, scale of analysis, etc.). Future research is needed to apply the multiscale seascape approach to a wider range of marine ecosystems. Although model parsimony is desirable, it is essential to develop and test a wide range of remote sensing data to identify useful predictors. The techniques presented in this chapter can also be applied equally to map human use patterns and threats to ecosystem health. Little work has focused on predictive mapping of ecosystem function and services that could provide a cost-effective method to determine ecosystem value across broad spatial scales. For coral reef ecosystems, there is great potential for using models to link spatial patterns to ecological processes and the emerging work on mapping resilience indicators is beginning to make the linkage between ecological patterns and process. For some regions of the earth, it is now feasible to map the geography of herbivory by mapping individual herbivore species and groups of herbivores. With an increase in the number of multiscale ecological studies being conducted, it is conceivable that, at some point in the near future, there will be sufficient knowledge to begin the search for generalities that could lead to the formulation of new ecological laws explaining the distribution and biodiversity patterns in the seas.

Acknowledgments

We thank the organizers of the workshop "Inter-relationships between coral reef and fisheries" held in Tampa, Florida, for inviting this contribution. We are grateful to all of our scientific divers for many years of data collection in the Caribbean region. Funding for research conducted by SJP was provided by NOAA's Coral Reef Conservation Program and NOAA's Biogeography Branch. AK was supported by the Wildlife Conservation Society and Simon Fraser University, Canada.

References

Agardy T, di Sciara GN, Christie P. 2011. Mind the gap: Addressing the shortcomings of marine protected areas through large scale marine spatial planning. Mar Policy. 35:226–232.

Aguirre-Gutiérrez J, Carvalheiro LG, Polce C, van Loon EE, Raes N, Reemer M, Biesmeijer JC. 2013. Fit-for-purpose: Species distribution model performance depends on evaluation criteria—Dutch hoverflies as a case study. PLoS ONE. 8(5):e63708.

Allen TFH, Starr TB. 1982. Hierarchy: Perspectives for ecological complexity. Chicago: University of Chicago Press.

Alvarez-Filip L, Dulvy NK, Gill JA, Cote IM, Watkinson AR. 2009. Flattening of Caribbean coral reefs: Region-wide declines in architectural complexity. Proc Roy Soc Lond B Bio. 276:3019–3025.

Alvarez-Filip L, Gill JA, Dulvy NK, Perry AL, Watkinson AR, Côté IM. 2011. Drivers of region-wide declines in architectural complexity on Caribbean reefs. Coral Reefs. 30(4):1051–1060. Available from: http://dx.doi.org/10.1007/s00338-011-0795-6 via the Internet.

Aruajo MB, New M. 2007. Ensemble forecasting of species distributions. Trends Ecol Evol. 22:42–47.

Austin MP, Smith TM. 1989. A new model for the continuum concept. Vegetatio. 83:35–47.

Beger M, Possingham HP. 2008. Environmental factors that influence the distribution of coral reef fishes: Modeling occurrence data for broad-scale conservation and management. Mar Ecol Prog Ser. 361:1–13.

Boström C, Pittman SJ, Simenstad C, Kneib RT. 2011. Seascape ecology of coastal biogenic habitats: Advances, gaps and challenges. Mar Ecol Prog Ser. 427:191–217.

Breiman L. 2001. Statistical modeling: The two cultures. Stat Sci. 16(3):199–215.

Brown CJ, Smith SJ, Lawton P, Anderson JT. 2011. Benthic habitat mapping: A review of progress towards improved understanding of the spatial ecology of the seafloor using acoustic techniques. Estuar Coast Shelf Sci. 92:502–520.

Christensen JD, Battista TA, Monaco ME, Klein CJ. 1997. Habitat suitability index modeling and GIS technology to support habitat management: Pensacola Bay, Florida, case study. Report submitted to the Gulf of Mexico Program by National Oceanic and Atmospheric Administration, Strategic Environmental Assessments Division, Silver Spring, MD.

Christensen JD, Jeffrey C, Monaco ME, Caldow C, Kendall MS, Appledoorn RS. 2003. Cross-shelf habitat utilization patterns of reef fishes in southwestern Puerto Rico. Gulf Carib Res. 14:9–27.

Costa B, Taylor CJ, Kracker L, Battista T, Pittman SJ. 2013. Mapping reef fish and the seascape: Using acoustics and spatial modeling to guide coastal management. PLoS ONE, 9(1): e85555, p1–7.

Crowder L, Norse E. 2008. Essential ecological insights for marine ecosystem-based management and marine spatial planning. Mar Policy. 32:772–778.

De'ath G. 2007. Boosted trees for ecological modeling and prediction. Ecology. 81:3178–3192.

Done TJ. 1992. Phase-shifts in coral-reef communities and their ecological significance. Hydrobiologia. 247:121–132.

Džeroski S, Ženko B. 2004. Is combining classifiers better than selecting the best one? Mach Learn. 54:255–273.

Elith J, Graham CH, Anderson RP, Dudik M, Ferrier S, Guisan A, Hijmans RJ, Huettmann F, Leathwick JR, Lehmann A, et al. 2006. Novel methods improve prediction of species' distributions from occurrence data. Ecography. 29:129–151.

Elith J, Leathwick JR. 2009. Species distribution models: Ecological explanation and prediction across space and time. Annu Rev Ecol Evol Syst. 40:677–697.

Elith J, Leathwick JR, Hastie T. 2008. A working guide to boosted regression trees. J Anim Ecol. 77:802–813.

Fielding AH, Bell JF. 1997. A review of methods for the assessment of prediction errors in conservation presence/absence models. Environ Conserv. 24:38–49.

Fischer J, Lindenmayer DB. 2006. Beyond fragmentation: The continuum model for fauna research and conservation in human-modified landscapes. Oikos. 112:473–480.

Foley MM, Halpern BS, Micheli F, Armsby MH, Caldwell MR, Crain CM, Prahler E, Rohr N, Sivas D, Beck MW, et al. 2010. Guiding ecological principles for marine spatial planning. Mar Policy. 34:955–966.

Gardner TA, Cote IM, Gill JA, Grant A, Watkinson AR. 2003. Long-term region-wide declines in Caribbean corals. Science. 301:958–960.

Goodman JA, Purkis SJ, Phinn SR. 2013. Coral reef remote sensing: A guide for mapping, monitoring and management. Springer, Dordrecht, The Netherlands.

Green DH, Edmunds PJ, Carpenter RC. 2008. Increasing relative abundance of *Porites astreoides* on Caribbean reefs mediated by an overall decline in coral cover. Mar Ecol Prog Ser. 359:1–10.

Grober-Dunsmore R, Pittman SJ, Caldow C, Kendall MA, Fraser T. 2009. Chapter 14: A landscape ecology approach for the study of ecological connectivity across tropical marine seascapes. *In*: Nagelkerken I, editor. Ecological connectivity of coral reef ecosystems. Springer, Dordrecht, The Netherlands. p. 493–530.

Guisan A, Thuiller W. 2005. Predicting species distribution: Offering more than simple habitat models. Ecol Lett. 8:993–1009.

Harborne AR, Mumby PJ, Kappel CV, Dahlgren CP, Micheli F, Holmes KE, Brumbaugh DR. 2008. Tropical coastal habitats as surrogates of fish community structure, grazing, and fisheries value. Ecol Appl. 18:1689–1701.

Hastie T, Tibshirani R, Friedman J. 2009. The elements of statistical learning. 2nd ed. Dordrecht: Springer.

Hedley JD. 2013. Chapter 4: Hyperspectral applications. *In*: Goodman JA, Purkis SJ, Phinn SR, editors. Remote sensing of coral reefs: A guide for mapping, monitoring and management. Springer, Dordrecht, The Netherlands. p. 79–112.

Holling CS. 1992. Cross-scale morphology, geometry, and dynamics of ecosystems. Ecol Monogr. 62(4):447–502.

Hughes TP, Bellwood DR, Folke C, Steneck RS, Wilson J. 2005. New paradigms for supporting the resilience of marine ecosystems. Trends Ecol Evol. 20:380–386.

Hunsaker CT, Friedl MA, Goodchild MF, Case TJ. 2001. Spatial uncertainty in ecology: Implications for remote sensing and GIS applications. New York: Springer.

Hyrenbach KD, Forney KA, Dayton PK. 2000. Marine protected areas and ocean basin management. Aquat Conserv. 10:437–458.

Jones GP, Syms C. 1998. Disturbance, habitat structure and the ecology of fishes on coral reefs. Aust J Ecol. 23:287–297.

Jones DL, Walter JF, Brooks EN, Serafy JE. 2010. Connectivity through ontogeny: Fish population linkages among mangrove and coral reef habitats. Mar Ecol Prog Ser. 401:245–258.

Kendall MS, Christensen JD, Hillis-Starr Z. 2003. Multi-scale data used to analyze the spatial distribution of French grunts, *Haemulon flavolineatum*, relative to hard and soft bottom in a benthic landscape. Environ Biol Fish. 66:19–26.

Klein CJ, Ban NC, Halpern BS, Beger M, Game ET, Grantham HS, Green A, Klein TJ, Kininmonth S, Treml E, et al. 2010. Prioritizing land and sea conservation investments to protect coral reefs. PLoS ONE. 5(8):e12431. Available from http://dx.doi.org/10.1371/journal.pone.0012431 via the Internet.

Knudby A, Brenning A, LeDrew E. 2010a. Predictive mapping of reef fish species richness, diversity and biomass in Zanzibar using IKONOS imagery and machine-learning techniques. Remote Sens Environ. 114:1230–1241.

Knudby A, Brenning A, LeDrew E. 2010b. New approaches to modelling fish-habitat relationships. Ecol Mod. 221:503–511.

Knudby A, Jupiter S, Roelfsema C, Lyons M, Phinn S. 2013. Mapping coral reef resilience indicators using field and remotely sensed data. Remote Sens. 5:1311–1334.

Knudby A, LeDrew E, Newman C. 2007. Progress in the use of remote sensing for coral reef biodiversity studies. Prog Phys Geog. 31:421–434.

Kotliar NB, Wiens JA. 1990. Multiple scales of patchiness and patch structure: A hierarchical framework for the study of heterogeneity. Oikos. 59:253–260.

Leathwick J, Moilanen A, Francis M, Elith J, Taylor P, Julian K, Hastie T. 2008. Novel methods for the design and evaluation of marine protected areas in offshore waters. Conserv Lett. 1:91–102.

Levin SA. 1992. The problem of pattern and scale in ecology. Ecology. 73(6):1943–1967.

Lourie SA, Vincent ACJ. 2004. Using biogeography to help set priorities in marine conservation. Conserv Biol. 18:1004–1020.

Marmion M, Parviainen M, Luoto M, Heikkinen RK, Thuiller W. 2009. Evaluation of consensus methods in predictive species distribution modelling. Divers Distrib. 15:59–69.

Maynard JA, Marshall PA, Johnson JE, Harman S. 2010. Building resilience into practical conservation: Identifying local management responses to global climate change in the Southern Great Barrier Reef. Coral Reefs. 29:381–391.

McArthur MA, Brooke BP, Przeslawski R, Ryan DA, Lucieer VL, Nichol S, McCallum AW, Mellin C, Cresswell ID, Radke LC, et al. 2010. On the use of abiotic surrogates to describe marine benthic biodiversity. Estuar Coast Shelf Sci. 88:21–32.

McClanahan TR, Donner SD, Maynard JA, MacNeil MA, Graham NAJ, Maina J, Baker A, Jahson BAI, Beger M, Campbell SJ, et al. 2012. Prioritizing key resilience indicators to support coral reef management in a changing climate. PLoS ONE. 7(8):e42884. Available from http://dx.doi.org/10.1371/journal.pone.0042884 via the Internet.

McGarigal K, Cushman SA. 2005. The gradient concept of landscape structure. *In*: Wiens J, Moss M, editors. Issues and perspectives in landscape ecology. Cambridge: Cambridge University Press. p. 112–119.

Mellin C, Bradshaw CJA, Meekan MG, Caley MJ. 2010. Environmental and spatial predictors of species richness and abundance in coral reef fishes. Global Ecol Biogeogr. 19:212–222.

Mellin C, Delean S, Caley J, Edgar G, Meekan M, Pitcher R, Przeslawski R, Williams A, Bradshaw C. 2011. Effectiveness of biological surrogates for predicting patterns of marine biodiversity: A global meta-analysis. PLoS ONE. 6:e20141. Available from http://dx.doi.org/10.1371/journal.pone.0020141 via the Internet.

Mellin C, Parrott L, Andrefouet S, Bradshaw CJA, MacNeil MA, Caley MJ. 2012. Multi-scale marine biodiversity patterns inferred efficiently from habitat image processing. Ecol Appl. 22:792–803.

Mora C, Andrefouet S, Costello MJ, Kranenburg C, Rollo A, Veron J, Gaston KJ, Myers RA. 2006. Coral reefs and the global network of marine protected areas. Science. 312:1750–1751.

Mumby PJ. 2009. Phase shifts and the stability of macroalgal communities on Caribbean coral reefs. Coral Reefs. 28:761–773.

Mumby PJ, Edwards AJ, Arias-Gonzalez JE, Lindeman KC, Blackwell PG, Gall A, Gorczynska MI, Harborne AR, Pescod CL, Renken H, et al. 2004. Mangroves enhance the biomass of coral reef fish communities in the Caribbean. Nature. 427:533–536.

Nagelkerken I, Roberts CM, van der Velde G, Dorenbosch M, van Riel MC, de la Morinere EC, Nienhuis PH. 2002. How important are mangroves and seagrass beds for coral-reef fish? The nursery hypothesis tested on an island scale. Mar Ecol Prog Ser. 244:299–305.

NMFS (National Marine Fisheries Service). 1996. Magnuson-Stevens Fishery Conservation and Management Act: As amended through October 11, 1996. U.S. Department of Commerce, National Oceanic and Atmospheric Administration, National Marine Fisheries Service, NOAA Technical Memorandum NMFS-F/SPO-23. Silver Spring, MD: NOAA.

Olden JD, Lawler JJ, Poff NL. 2008. Machine learning methods without tears: A primer for ecologists. Quart Rev Biol. 83:171–193.

Orth RJ, Carruthers TJB, Dennison WC, Duarte CM, Fourqurean JW, Heck KL, Hughes AR, Kendrick GA, Kenworthy WJ, Olyarnik S, et al. 2006. A global crisis for seagrass ecosystems. Bioscience. 56:987–996.

Paddack MJ, Reynolds JD, Aguilar C, Appeldoorn RS, Beets J, Burkett EW, Chittaro PM, Clarke K, Esteves R, Fonseca AC, et al. 2009. Recent region-wide declines in Caribbean reef fish abundance. Curr Biol. 19:590–595.

Pandolfi JM, Jackson JBC, Baron N, Bradbury RH, Guzman H, Hughes TP, Micheli F, Ogden J, Possingham H, Kappel CV, et al. 2005. Are U.S. coral reefs on the slippery slope to slime? Science. 307:1725–1726.

Parrish JD. 1989. Fish communities of interacting shallow water habitats in tropical oceanic regions. Mar Ecol Prog Ser. 58:143–160.

Pearce J, Ferrier S. 2000. Evaluating the predictive performance of habitat models developed using logistic regression. Ecol Model. 133:225–245.

Phillips SJ, Anderson RP, Schapire RE. 2006. Maximum entropy modeling of species geographic distributions. Ecol Model. 190:231–259.

Pimm SL. 1984. The complexity and stability of ecosystems. Nature. 307:321–326.

Pittman SJ, Brown KA. 2011. Multi-scale approach for predicting fish species distributions across coral reef seascapes. PLoS ONE. 6(5):e20583. Available from http://dx.doi.org/10.1371/journal.pone.0020583 via the Internet.

Pittman SJ, Christensen J, Caldow C, Menza C, Monaco M. 2007a. Predictive mapping of fish species richness across shallow-water seascapes of the U.S. Caribbean. Ecol Model. 204:9–21.

Pittman SJ, Hile SD, Caldow C, Monaco ME. 2007b. Using seascape types to explain the spatial patterns of fish using mangroves in Puerto Rico. Mar Ecol Prog Ser. 348:273–284.

Pittman SJ, Costa B, Battista TA. 2009. Using Lidar bathymetry and boosted regression trees to predict the diversity and abundance of fish and corals. J Coastal Res. S(53):27–38.

Pittman SJ, Connor DW, Radke L, Wright DJ. 2011a. Application of estuarine and coastal classifications in marine spatial management. *In*: Wolanski E, McLusky DS, editors. Treatise on estuarine and coastal science. Vol. 1. Waltham, MA: Academic Press. p. 163–205.

Pittman SJ, Costa B, Jeffrey CFG, Caldow C. 2011b. Importance of seascape complexity for resilient fish habitat and sustainable fisheries. *In*: Proceedings of the 63rd Gulf and Caribbean Fisheries Institute 1–5 November, 2010 San Juan, Puerto Rico. p. 420–426.

Pittman SJ, Costa B, Wedding LM. 2013. Chapter 6: LiDAR applications. *In*: Goodman JA, Purkis SJ, Phinn SR, editors. Remote sensing of coral reefs: A guide for mapping, monitoring and management. Springer, Dordrecht, The Netherlands. p. 145–174.

Pittman SJ, McAlpine CA. 2003. Movements of marine fish and decapod crustaceans: Process, theory and application. Adv Mar Biol. 44:205–294.

Polidoro BA, Carpenter KE, Lorna C, Duke NC, Ellison AM, Ellison JC, Farnsworth EJ, Fernando ES, Kathiresan K, Koedam NE, et al. 2010. The loss of species: Mangrove extinction risk and geographic areas of global concern. PLoS ONE. 5:e10095. Available from http://dx.doi.org/10.1371/journal.pone.0010095 via the Internet.

R Core Development Team. 2012. R: A language and environment for statistical computing. Vienna, Austria: R Foundation for Statistical Computing. Available from http://www.R-project.org/ via the Internet.

Roberts CM, Andelman S, Branch G, Bustamante RH, Castilla JC, Dugan J, Halpern BS, Lafferty KD, Leslie H, Lubchenco J, et al. 2003. Ecological criteria for evaluating candidate sites for marine reserves. Ecol Appl. 13:S199–S214.

Roff JC, Taylor ME, Laughren J. 2003. Geophysical approaches to the classification, delineation and monitoring of marine habitats and their communities. Aquatic Conserv. 13:77–90.

Rowlands G, Purkis S, Riegl B, Metsamaa L, Bruckner A, Renaud P. 2012. Satellite imaging coral reef resilience at regional scale. A case-study from Saudi Arabia. Mar Pollut Bull. 64:1222–1237.

Rubec PJ, Bexley JCW, Norris H, Coyne MS, Monaco ME, Smith SG, Ault JS. 1999. Suitability modeling to delineate habitat essential to sustainable fisheries. *In:* Benaka LR, editor. Fish habitat: Essential fish habitat and restoration. Am Fish Soc Symp. 22:108–133.

Rubec PJ, Christensen JD, Arnold WS, Norris H, Steele P, Monaco ME. 1998a. GIS and modeling: Coupling habitats to Florida fisheries. J Shellfish Res. 17(5):1451–1457.

Rubec PJ, Coyne MS, McMichael Jr RH, Monaco ME. 1998b. Spatial methods being developed in Florida to determine essential fish habitat. Fisheries. 23(7):21–25.

Rubec PJ, O'Hop J. 1996. GIS applications for fisheries and coastal resources management. Gulf States Marine Fisheries Commission Publication No. 43. Ocean Springs, MS: Gulf States Marine Fisheries Commission.

Schneider DC. 2001. The rise of the concept of scale in ecology. BioScience. 51(7):545–553.

Souter DW, Linden O. 2000. The health and future of coral reef ecosystems. Ocean Coast Manage. 43:657–688.

Stallings CD. 2009. Fishery-independent data reveal negative effect of human population density on Caribbean predatory fish communities. PLoS ONE. 4:e5333. Available from http://dx.doi.org/10.1371/journal.pone.0005333 via the Internet.

Ward TJ, Vanderklift MA, Nicholls AO, Kenchington RA. 1999. Selecting marine reserves using habitats and species assemblages as surrogates for biological diversity. Ecol Appl. 9:691–698.

Wedding LM, Friedlander AM, McGranaghan M, Yost RS, Monaco ME. 2008. Using bathymetric lidar to define nearshore benthic habitat complexity: Implications for management of reef fish assemblages in Hawaii. Remote Sens Environ. 112:4159–4165.

Wedding L, Lepczyk C, Pittman SJ, Friedlander A, Jorgensen S. 2011 Quantifying seascape structure: Extending terrestrial spatial pattern metrics to the marine realm. Mar Ecol Prog Ser. 427:219–232.

West J, Salm R. 2003. Resistance and resilience to coral bleaching: Implications for coral reef conservation and management. Conserv Biol. 17:956–967.

Wozencraft JM, Park JY. 2013. Integrated LiDAR and hyperspectral. *In*: Goodman JA, Purkis SJ, Phinn SR, editors. Coral reef remote sensing: A guide for mapping, monitoring and management. Dordrecht: Springer. p. 175–191.

Zawada DG. 2011. Reef topographic complexity. *In*: Hopley D, editor. Encyclopedia of modern coral reefs: Structure, form, and process. Encyclopedia of Earth Sciences. Dordrecht: Springer. p. 902–906.

chapter thirteen

Progressing from data to information

Incorporating GIS into coral and fisheries management

Mark Mueller, John Froeschke, and David Naar

Contents

Introduction

According to the Magnuson-Stevens Fishery Conservation and Management Act (MSFCMA 2006), the US Gulf of Mexico Fishery Management Council (Gulf Council) is responsible for the conservation and management of coral and fishery resources in the federal waters of the Gulf of Mexico. With over 142 species in the Gulf Council's coral fishery management plan (GMFMC 2012a), including seven listed in the Endangered Species Act (assuming proposed rules in 77 FR 73220 are finalized), the trends and population status of Gulf coral species are themselves of direct concern. But beyond the regulatory obligation to directly manage coral as a natural resource, the Gulf Council recognizes the functional role of corals as habitat for a number of reef-associated fish species, including many of the most commercially significant species such as groupers and snappers (e.g., Szedlmayer and Lee 2004) on which many Gulf fishers depend for their livelihoods. For the most part, stock assessments and fishery management plans for these reef-associated species do not account for potential changes in the ecosystem services that corals and other types of habitat provide. This is, in part, because the mechanisms that affect coral health are complex and not fully understood. However, as explained elsewhere in this volume (e.g., Bortone, Chapter 1; Sale and Hixon, Chapter 2), changes in coral health and productivity may lead to unanticipated changes in associated fish populations, in turn requiring unplanned (and perhaps more restrictive) management measures. Fortunately, the degree of uncertainty in management decision making can be decreased by providing more information on the status of coral, including its known distribution, relative health, ecosystem interactions, and the various factors that impact these conditions.

Ecosystem-based management and the Gulf of Mexico Large Marine Ecosystem

As demonstrated elsewhere in this volume and in the literature (e.g., Beger and Possingham 2008, Crowder and Norse 2008, McClanahan et al. 2011), researchers (e.g., Pikitch et al. 2004, McClanahan et al. 2011) and resource managers at all levels (local, state, federal, and international) have shown increasing interest in implementing "ecosystem-based management" because of its potential to improve management outcomes through increased scientific understanding of complex systems. The term can be loosely defined as increased consideration of the interactive and ecosystem-wide effects and impacts between and among species, the habitats they depend on (including coral), and human activities. For example, in the United States, the Mid Atlantic Fishery Management Council is developing an Ecosystem Approach to Fisheries Management Guidance Document "to guide policy decisions as [it] transitions from single-species management toward an ecosystem-based approach" (MAFMC 2013). Although the Gulf Council does not have a directly equivalent document, it previously established a formal habitat policy recognizing that "all species are dependent on the quantity and quality of their essential habitats" and pledging to "protect, restore and improve habitats upon which commercial and recreational marine fisheries depend and to improve their productive capacity for the benefit of present and future generations" (GMFMC 2002, 1).

The essential habitats used by Gulf Council–managed species are not limited to the federal waters of the US exclusive economic zone (from the boundary of state waters out to 200 nmi/370 km). Many fish species in the Gulf of Mexico are estuarine dependent and occupy nearshore environments as juveniles (Beck et al. 2001, Able et al. 2007). Thus, precise habitat data sets of inshore areas must also be available so that ecosystem health as well as the biomass and long-term sustainability of fisheries that are later harvested offshore can be assessed (Cerveny et al. 2010). Likewise, the Florida Keys is a connected ecosystem that encompasses multiple jurisdictional boundaries requiring close coordination among management agencies. The coastal and pelagic ecosystems off Mexico and Cuba are similarly connected as part of the wider Gulf of Mexico Large Marine Ecosystem (UNEP 2009). Because many of the threats that impact Gulf of Mexico corals and fisheries are regional (e.g., hypoxia, nutrient/toxin pollution, harmful algal blooms, tropical storms) and even global (e.g., ocean acidification and warming) in their extent, obtaining useful, regional-level data on those threats is also crucial (SEDAR 2012, GMFMC 2013a).

Progressing from decentralized data to readily available, relevant spatial information

Successfully implementing ecosystem management measures depends on managers and supporting researchers having access to the best available biological and socioeconomic information relevant to corals and fishery stocks and their habitats. Pittman and Knudby (Introduction to Chapter 12 in this volume) explained the need for "reliable spatial data, with sufficient ecological detail to be meaningful for ecosystem-based management, but with sufficient geographical coverage to be operationally relevant for decision making." This balance between data having sufficient ecological detail and being management-ready presents challenges.

To that end, the Gulf Council, with assistance from the US National Oceanographic and Atmospheric Administration's (NOAA) Coral Reef Conservation Program, is implementing a project to build a "spatial baseline" by identifying and compiling widely dispersed data sets on coral locations (shallow, mesophotic, and deep sea), associated fish and fisheries, habitat (including oceanographic variables, bathymetry, benthic communities and bottom sediments, artificial structures, etc.), and relevant human-use and jurisdictional layers at appropriate spatial and temporal scales for federal management objectives. Unfortunately, these data sets are often decentralized. Currently they are maintained by a variety of user groups throughout the Gulf of Mexico, such as state and federal agencies and university research laboratories, making acquisition a logistical challenge. Therefore, a critical part of this project is establishing collaborative relationships with fellow data managers and researchers, including other groups working toward achieving similar data acquisition and synthesis goals such as the Gulf of Mexico Alliance's Ecosystem Integration and Assessment Team, of which the Gulf Council is a member.

Once obtained, these spatial data sets are vetted, formatted, and organized into standardized geographic information system (GIS) databases according to five broad categorical data themes (Figure 13.1). Subsequently, the data sets can be used to examine the complex interrelationships between relevant environmental factors and species of interest. For example, water column properties such as temperature, salinity, dissolved oxygen, turbidity, and currents directly influence fish species distribution (e.g., Froeschke et al. 2010). Proximity to other geographic features such as estuarine inlets can serve to enhance habitat value for fish species (Whaley et al. 2007, Froeschke et al. 2013). The same may be true for coral reefs in proximity to other habitat types such as seagrasses, mangroves, and deeper waters, all of which may have varying importance for the different life stages of reef fish (Pittman et al. 2007, Grober-Dunsmore et al. 2009, Cerveny et al. 2010). These sorts of spatial relationships could be more easily examined with the aid of an analysis-ready geospatial database. Similarly, seafloor characteristics such as submerged vegetative cover,

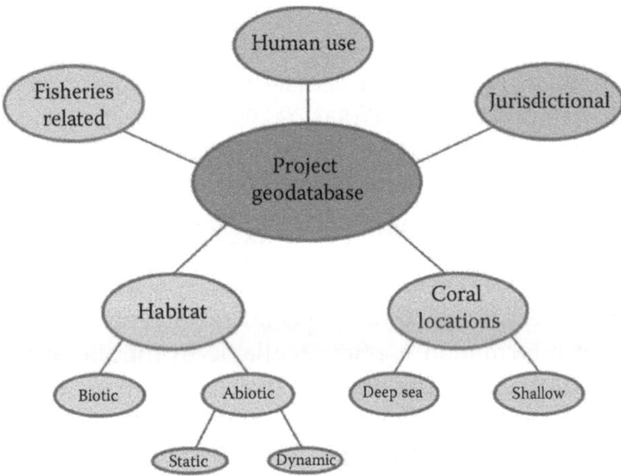

Figure 13.1 Schematic diagram of the spatial database being developed by the Gulf Council to support improved coral and reef fish management.

sediment type, rugosity, geomorphology (Harris and Baker 2011), and depth also influence the distribution of both mobile and sessile benthic species including invertebrates, sponges, shrimp (e.g., Clark et al. 2004), and coral reef fishes (Pittman and Brown 2011). The movements and activity patterns of heavily exploited reef fish species in and around marine protected areas (e.g., Farmer and Ault 2011) also merit management attention and the collection and analysis of relevant spatial data. A number of habitat-use relationships focused directly on corals and fisheries were explored in the workshop that inspired this volume (GMFMC 2013a). This chapter, partly guided by the workshop, is focused on compiling spatial data relevant to both corals and commercial fisheries, and making the most useful of those data sets available for stock assessments (e.g., SEDAR 2013) and other management uses, such as those described later in this chapter.

From data to information to management

Crisis response events clearly illustrate the need for readily available high-quality spatial information. During the massive 2010 Deepwater Horizon oil spill, the NOAA National Marine Fisheries Service (NMFS) had an immediate need for precise estimates of species distributions in order to determine which fish stocks were impacted and to identify species/areas that would require temporary harvest restrictions. This included several shrimp species, many of which are caught seasonally in the Gulf of Mexico. The Gulf Council and NMFS maintain GIS representations of areas known to be important to managed species. These representations, composed of multiple data layers, help define the essential fish habitat (EFH) for these managed species. Included in these characterizations is a multispecies generalized depiction of shrimp EFH (shown in crosshatched purple in Figure 13.2). While knowing the general extent of shrimp occurrence is useful for many purposes, not all species of shrimp are present in affected areas in equal numbers at the same time of year, and thus the information provided by the generalized EFH representation was somewhat limited. Decisions regarding specific area closures at specific times would be better informed by more precise spatially and temporally explicit information describing known and predicted species distribution and abundance patterns. For example, the species-habitat predictive model developed for brown shrimp spatial distribution (GMFMC 2010) shown in Figure 13.2 illustrates how certain areas within the EFH representation are likely of greater importance to brown shrimp than others; information that is highly relevant for managers. This work was based on empirical data from a multiyear, fisheries-independent trawl survey and used a generalized additive model (Wood 2006). Generalized additive models allow nonparametric smoothing of explanatory variables (Wood 2003) and are popular choices for fitting species distribution models. The field has been progressing rapidly due to enhanced computing power in combination with new statistical algorithms that are more robust to noisy data and can accommodate complex patterns and ecological interactions more efficiently than traditional approaches (e.g., multiple linear regression; Elith et al. 2006, 2008).

Developing spatial distribution models for many species requires empirical data on both the habitat and the records of known species occurrence from resource monitoring programs. This type of information is often available in the literature, especially at local scales (e.g., Clark et al. 2004), but is not always leveraged to assist in regional management decisions, in part because there is often a mismatch in the spatiotemporal scales of research products and the geographic area of responsibility for a particular management agency (Ogden et al., Chapter 14, this volume, Cerveny et al. 2010, Boström et al. 2011). But the benefits of having this type of predictive species-habitat distribution information available in crisis situations such as the Deepwater Horizon oil spill apply not only to brown shrimp

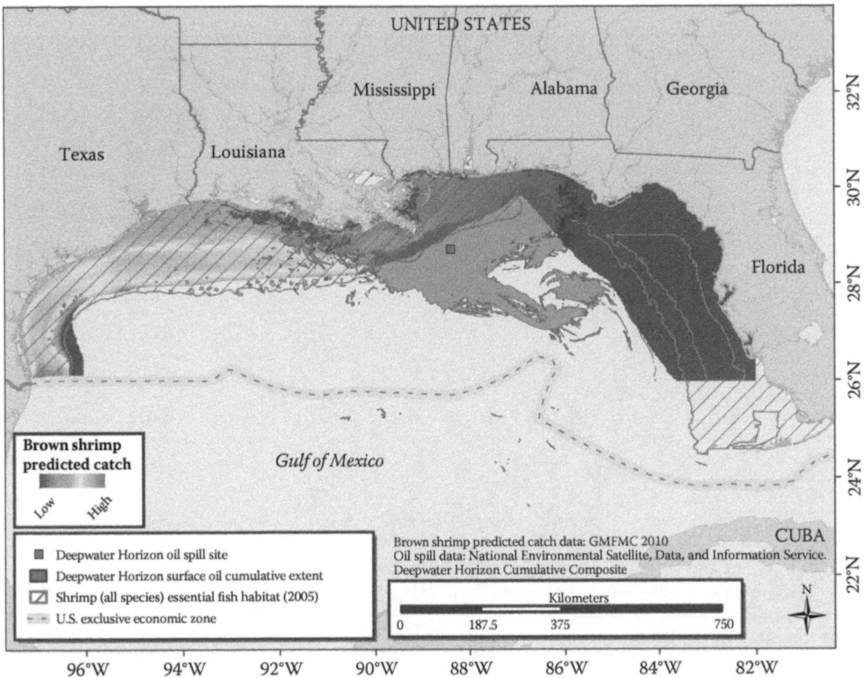

Figure 13.2 **(See color insert)** Shrimp essential fish habitat along with predicted relative abundance for brown shrimp, overlain with Deepwater Horizon oil spill maximum surface extent.

but to other potentially affected fisheries such as red snapper and coral. Direct harm to at least one deepwater coral colony was documented as a result of the spill (White et al. 2012). Such information is also critical for assessing the long-term impacts on various species and habitats through appropriate monitoring at a range of scales. Fortunately, groups such as the Gulf of Mexico Research Initiative and the NOAA Resources and Ecosystems Sustainability, Tourist, Opportunity, and Revived Economies (NOAA RESTORE) Act Science Program, charged with coordinating many postspill scientific efforts, recognize the need for such synthesized information and also the opportunity that expanded postspill research presents to help "integrate the disparate science efforts across the Gulf into something that would consider the connectivity and entirety of the Gulf of Mexico ecosystem and advance overall understanding as an integrated system" (NOAA RESTORE Act Science Program 2013, 2).

Making coral EFH more useful to resource managers

The formal definition of coral EFH is the "total distribution of coral species and life stages throughout the Gulf of Mexico" (GMFMC 2010, 15), meaning wherever coral are located is technically considered coral EFH. In practice, however, what is used by researchers and managers is the GIS layer representation (shown with blue hatching in Figure 13.3). This representation was last updated in 2005 and relied on bottom-type information dating back to the original 1985 Gulf of Mexico Atlas in determining that a large area of the West Florida Shelf was likely hard bottom, suitable for coral colonization. However, newer bottom-type information (Jenkins 2011) suggests that much of this area is dominated not by rocky surface sediments (Figure 13.3) but rather by sand. Sand is less suitable for long-term coral development and therefore not likely to be essential habitat. Conversely, other

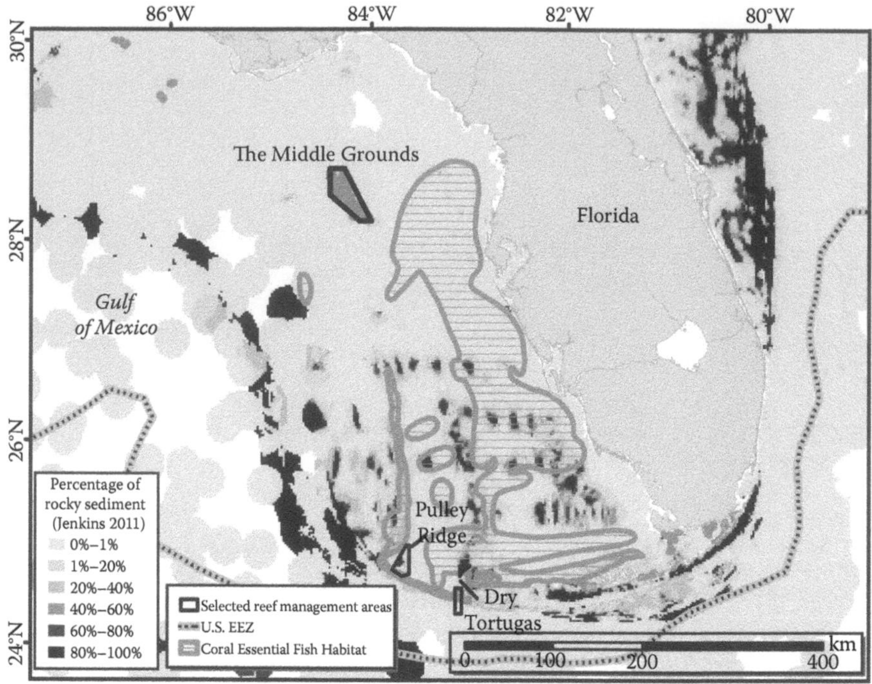

Figure 13.3 **(See color insert)** Coral essential fish habitat spatial representation (blue hatching) and percentage of rocky bottom sediments as described in Jenkins 2011. *Note*: white indicates no data.

areas where corals are now known to thrive are not included within the current GIS EFH boundary, such as most of Pulley Ridge and the southern Dry Tortugas Ecological Reserve—both established as coral Habitat Areas of Particular Concern (HAPCs) in 2006 (GMFMC 2005). Updating the coral EFH representation to depict areas that are truly essential to coral will enable more meaningful usage by both researchers and resource managers. Therefore, Gulf Council staff are actively compiling coral distribution information for as many locations throughout the Gulf of Mexico as possible, including shallow (Florida Keys reef tract, Dry Tortugas), shallow/mesophotic waters (Flower Garden Banks, South Texas banks, Pulley Ridge), and deeper waters (Viosca Knolls, West Florida Slope). Deepwater coral have been the subject of substantial field and modeling research in recent years (e.g., Reed et al. 2007, Ross and Quattrini 2007, 2009). Some recently discovered large deepwater coral aggregations were suggested as candidates for protection via creation of new HAPCs (GMFMC 2013a). This action was based on the potential threats from fishing gear interactions (e.g., Reed et al. 2007), requiring compilation of precise data on deepwater coral distributions. Updating the coral EFH representation with new information will also enable additional refinements to the EFH boundaries for other managed species such as shrimp and spiny lobster, all of which may be considered in the next 5-year EFH review in 2015.

Seafloor characteristics: At the foundation of marine ecosystems

Surficial sediments (Figure 13.3) such as those interpolated by Jenkins (2011) are important for coral settlement (e.g., Schroeder 2002) and for other Gulf Council–managed species such as red snapper (Szedlmayer and Lee 2004, SEDAR 2013) and shrimp (Clark et al. 2004,

Montero et al. forthcoming). Other seafloor characteristics such as canyons, seamounts, ridges, and so on were described in the recently produced Global Seafloor Geomorphology Map (Lawler 2013). These characteristics can also shape biological communities, including those associated with deepwater coral (Schroeder 2007). Of all of these characteristics, bathymetry is likely the most fundamental and important.

Over the past two decades, shallow seafloor bathymetric and backscatter surveys have become more frequent and most often used to characterize EFH. With the improvement and use of better ship-motion sensors and improved multibeam echo sounders at higher frequencies, high-resolution views (see Figure 13.4 insets), leading to a better understanding of the seafloor, are possible. In the 1980s, multibeam surveys were initiated by academic institutions and included deepwater multibeam surveys of the deep ocean, for example, midocean ridges (e.g., Lonsdale 1983, Macdonald and Fox 1983, Hey et al. 1986). However, the results of these surveys did not show an increase in precision due to the lower frequency (~12 kHz), narrower swath widths (45°), and the slow ping rate required in deep water. In addition, these surveys usually provided less than 100% coverage of the seafloor. Nevertheless, they yielded new views of seafloor spreading processes, including newly discovered features or better definitions of features such as overlapping spreading centers, propagating rifts, and microplates (see Naar and Hey 1991 and references therein, Theberge and Cherkis 2013).

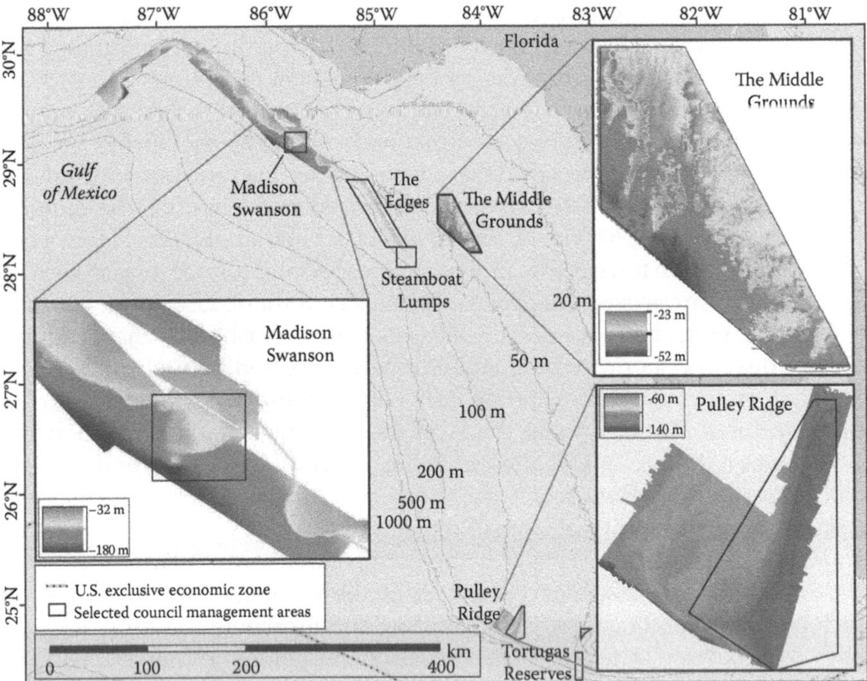

Figure 13.4 **(See color insert)** Overview map showing distribution of multibeam echo sounder imagery collected by multiple USF/USGS surveys, with selected Gulf Council fishery management areas in the eastern Gulf of Mexico. The Middle Grounds, Madison Swanson, and Pulley Ridge are expanded for detail. Depth range color ramps are specific to each inset area and cannot be directly compared.

The more recent shallowwater surveys provide much more detail due to the higher frequencies used (e.g., ~75 up to ~400 kHz). These surveys tend to provide complete coverage and a wealth of new products derived from the measured bathymetry and/or the backscatter intensity. Improvements to bathymetry and backscatter data processing methods have enhanced the overall quality of data being analyzed (Calder and Mayer 2003). The high-resolution bathymetry and backscatter data allow new derived variables to be determined, such as seafloor slope, rugosity, derivatives of the seafloor slope, and spatial groupings of different seafloor areas based on backscatter characteristics. When modern techniques are also used, improved and sometimes automated benthic habitat classifications are generated (e.g., Diaz 1999, Wright et al. 2002, Lundblad et al. 2006, Brown et al. 2011, Wall et al. 2011, Wright et al. 2012) that facilitate slope stability analysis (e.g., Lacono et al. 2012) or help to monitor temporal changes in the seafloor (e.g., Berman et al. 2005, Wolfson et al. 2007, Caress et al. 2012).

Rugosity, a measure of the "roughness" or complexity of the seafloor, is one of the more common of these products and has been successfully used as a substitute for estimating hardbottom habitat. For example, Dunn and Halpin (2009) achieved ~70% accuracy in an Atlantic Ocean regional model using relatively poor-resolution (90 m) bathymetry data (NOAA NGDC 2003). The less dramatic vertical relief prevalent throughout much of the Gulf of Mexico might reduce this prediction accuracy. Nevertheless, a Gulf of Mexico–wide rugosity estimate that makes use of the best available bathymetric data would be worth pursuing since the presence of hard bottom is of such great importance to a variety of commercially important fisheries (Parker et al. 1983, Szedlmayer and Lee 2004, SEDAR 2013).

Many shallow-water bathymetric and backscatter surveys are used to denote HAPCs and marine protected areas (e.g., Allee et al. 2012 and references therein). These data are valuable for planning purposes in regulating marine fisheries. The parameters estimated from these data also serve as important benchmarks for benthic habitat conditions and provide important GIS maps for high-resolution video and sampling surveys with remotely operated vehicles, autonomous underwater vehicles, and manned submersibles. During the recent Gulf of Mexico Deepwater Horizon oil spill, there were numerous requests for high-resolution, multibeam echo sounder data products from published and unpublished multibeam surveys collected in the Gulf of Mexico (Cross et al. 2005, Hine et al. 2008). However, locating the data in the appropriate format was problematic, especially for oceanographers who were not familiar with the surveys or the groups that collected them. Users had to visit multiple websites to find data sets; moreover, some data sets were published but not readily available online. A great deal of data processing and editing is required before multibeam data sets can be used, but some data archives retain only raw data and not the "processed" products in GIS-usable formats. Participants at the Workshop on Interrelationships between Corals Reefs and Fisheries (GMFMC 2013a) attested to the difficulty in locating and obtaining these high-resolution bathymetry data sets. Participants agreed that easier accessibility would be useful for the larger community of users. Problems associated with ease of data access and availability will increase as new types of bathymetric data sets are collected using additional techniques such as airborne laser surveys in very shallow waters. The challenge of the future will not be related to data collection, but to locating and incorporating these data into analyses after they have been collected (e.g., see discussion in Calder and Mayer 2003).

Artificial structures as EFH

In the Gulf of Mexico, fixed offshore petroleum platforms are used as habitat by several managed reef fish species and associated fauna. Recent Gulf Council discussions (GMFMC 2013b) and associated data compilation efforts have focused on the importance of artificial structures as they may be especially important to fisheries when these structures occur in areas otherwise devoid of natural hard bottom (Pickering and Whitmarsh 1997, Bortone et al. 2011). Sammarco et al. (2004, 2012) also demonstrated the importance of petroleum platforms as structure for coral colonization. The platforms serve as both sources and sinks for larval dispersal by hermatypic and ahermatypic corals, allowing important connections among metapopulations between the Flower Garden Banks and other pinnacle structures in the northern Gulf of Mexico (see Simmons et al., chapter 3, this volume). These structure-associated coral colonies have the potential to promote the long-term stability of natural coral communities. The presence of the many petroleum platforms in the northern Gulf of Mexico (3228 active and 777 nonproducing as of March 2012; BOEM 2012) has contributed to an expansion in the ranges of several corals. Sammarco et al. (2004) also noted statistically significant increases in total coral abundance with increasing platform age, the greatest community development beginning at >12–15 years, and a peak in abundance at depths between ~20 and 28 m.

Information on precise artificial reef locations is being collected by the Gulf Council from state agencies such as the Louisiana Department of Wildlife and Fisheries. Combining these data with bathymetry, other relevant habitat data, and estimated species distributions and fisheries effort data could help managers make decisions on the ideal siting of additional planned artificial structures.

Satellite imagery: An information-rich data source

Satellite imagery is used by groups such as the Gulf of Mexico Coastal Ocean Observing System and the University of South Florida's Institute for Marine Remote Sensing to develop data products that can be employed in fisheries science. Data collected on pelagic surface habitat variables and indicators (e.g., ocean color, temperature, sea surface height) can help end users derive information such as primary productivity and the direction of currents affecting larval dispersal. These variables can determine local species abundance. Better availability and leveraging of existing imagery-derived products could improve understanding of the impacts of threatening episodic events such as hurricanes, sea surface temperature changes (especially as related to coral bleaching events), oil spills, and harmful algal/dinoflagellate blooms (SEDAR 2012). Improved understanding could result in more accurate, ecosystem-level assessments of corals and fisheries stocks. High spatial and temporal resolution imagery offers great potential for improved habitat analysis by allowing more accurate differentiation of local-scale phenomena and short-term changes such as the subtle saline boundary shifts in dynamic estuarine environments (e.g., mangroves and salt marshes), where many managed fish species spend portions of their juvenile life stages. Imagery can also be used to directly map shallow-water coral reefs (Goodman et al. 2013) and even predict reef fish species richness, diversity, and biomass using detectable habitat variables such as depth, substrate complexity, and coral cover (Knudby et al. 2010).

Socioeconomic data matter

The Gulf Council's decision-making process considers the socioeconomic impacts of fisheries regulations; thus, the spatial database compilation effort also includes socioeconomic data sets. These data sets include information on the home ports or registered counties of commercial and recreational boats, and the locations of public launch sites such as boat ramps and marinas, which can be used to help estimate broad spatial trends in recreational fishing effort. The recent closure of 60 small areas in the Florida Keys to commercial trapping of spiny lobster (GMFMC 2012b) may provide a test case to examine the socioeconomic spatial effects resulting directly from coral conservation regulations. Of course, the GIS boundary layers and metadata of these and other fisheries jurisdictional and regulatory boundaries must be accurately maintained for any such analyses to be performed. There is also potential to leverage high-quality spatial demographic data. These data are currently available from the US Census Bureau and other sources such as state fishing license registration databases. One use of such data might be to inform the optimal siting of public hearing locations so that the maximum number of interested stakeholders can be reached while meeting costs are minimized.

Data portals: Making information accessible

The Gulf Council is developing a web-accessible data portal and map viewer to help disseminate compiled information on corals, reef-associated fisheries, related habitat and threats, human uses, and regulatory boundaries. These data will be available through the Gulf Council's website (www.gulfcouncil.org). Similar to other fishery management councils' web mapping applications (e.g., SAFMC 2013), this portal must cater to a diverse audience that includes state and federal fisheries managers, researchers, and stakeholders.

Consider one example of a data layer important to all of these audiences. Invasive lionfish (*Pterois volitans* and *P. miles*) threaten coral reef communities (Albins and Hixon 2008) and associated fisheries that depend on them. Estimates of the distribution and density of lionfish can improve our understanding of the extent and degree of impact. However, relying solely on intermittent updates provided via scientific literature (e.g., Schofield 2009, 2010, Côté et al. 2013) is inadequate since lionfish distribution is changing so rapidly and because such information outlets are not readily available to the general public. There is a need for up-to-date information that is continuously maintained and easily accessible online. In this case, the United States Geological Survey (curator of the best available regional lionfish database) recognized that need and added an attribute-enriched map of reported lionfish sightings to its real-time map viewer for nonindigenous aquatic species (USGS 2014).

One example of an application that is being used to improve accessibility is the Web Map Service (WMS). NOAA's Gulf Atlas (NOAA NCDDC 2013) is one authoritative data portal making use of WMS to help serve select data layers. Through WMS, a GIS layer can be viewed—but not downloaded—using desktop mapping software such as Google Earth. End users can still view the served layer in combination with other locally held layers of interest and produce customized maps as needed. This is a way for third party data facilitators (through their data portals) to promote data discovery while still requiring that end users contact the original data sources to obtain full access. This solution allows for some of the benefits of local viewing while preventing misuse or unauthorized redistribution.

The Gulf Council is certainly not alone in realizing the utility of data portals as a means of distributing compiled data sets, and a number of other institutions' portals both provide useful data and serve as models for the development of the Gulf Council's own data portal. In fact, there has been much discussion of centralizing portal content or developing "metaportals" to improve ease of access. While each data portal usually evolves to serve a slightly different audience or purpose (such as the Gulf Council's focus on corals and fisheries), the idea of sharing and integrating disparate data sets and resources has merit and might be developed further with Gulf of Mexico oil spill restoration funding.

An interactive data portal might also serve as a means for obtaining and sharing coral and coral-associated fisheries data collected by citizen scientists, such as scuba divers. Properly structured, such an endeavor could provide usable information on coral locations, health, and ecosystem interactions while also positively engaging the public and furthering education and outreach efforts.

Progressing forward

To fully leverage spatial tools such as GIS to improve coral and fisheries management, it is first necessary to identify, compile, and synthesize diverse spatial data sets and make that information widely available. Data analyses can inform specific ongoing research and management questions such as those described above, and more broadly aid the development of ecosystem-based management. It is recommended that researchers fill important data gaps and then design data collection efforts with an eye toward effective spatial analysis at scales meaningful for resource management. Investigators should take steps such as implementing data management plans and contributing to data portals and clearinghouses to aid ease of access, allowing others to build on their work with additional analyses and management applications. Research and management institutions should ensure adequate funding is provided to support such efforts, because leveraging spatial tools and analyses to their full potential can provide useful information to better understand the relationships between corals and fisheries and enhance their sustainable management.

Acknowledgments

The authors would like to acknowledge manuscript assistance from other Gulf Council staff, in particular C. Simmons, C. Schiaffo, and R. Rindone. Multibeam imagery was prepared by J. Brizzolara and B. Donahue at the University of South Florida. This work was funded by the Gulf of Mexico Fishery Management Council under award number NA11NMF4410063 from the NOAA, US Department of Commerce. The statements, findings, conclusions, and recommendations are those of the authors and do not necessarily reflect the views of the Gulf of Mexico Fishery Management Council, NOAA, or the Department of Commerce.

References

Able KW, Balletto JH, Hagan SM, Jivoff PR, Strait K. 2007. Linkages between salt marshes and other nekton habitats in Delaware Bay, USA. Rev Fish Sci. 15:1–61. Available from: http://www.tandfonline.com/doi/abs/10.1080/10641260600960995 via the Internet. Accessed 5 March, 2014.

Albins M, Hixon M. 2008. Invasive Indo-Pacific lionfish *Pterois volitans* reduce recruitment of Atlantic coral-reef fishes. Mar Ecol Prog Ser. 367:233–238. Available from: http://www.int-res.com/abstracts/meps/v367/p233-238/ via the Internet. Accessed 5 March, 2014.

Allee RJ, David AW, Naar DF. 2012. Two shelf edge marine protected areas in the Eastern Gulf of Mexico. *In:* Harris PT, Baker ED, editors. Seafloor geomorphology as benthic habitat: GeoHAB atlas of seafloor geomorphic features and benthic habitats. Amsterdam: Elsevier. p. 435–448.

Beck MW, Heck KL, Able KW, Childers DL, Eggleston DB, Gillanders BM, Halpern B, Hays CG, Hoshino K, Minello TJ. 2001. The identification, conservation, and management of estuarine and marine nurseries for fish and invertebrates. Bioscience. 51:633–641.

Beger M, Possingham H. 2008. Environmental factors that influence the distribution of coral reef fishes: Modeling occurrence data for broad-scale conservation and management. Mar Ecol Prog Ser. 361:1–13. Available from: http://www.int-res.com/abstracts/meps/v361/p1-13/ via the Internet. Accessed 5 March, 2014.

Berman G, Naar DF, Hine AC, Brooks G, Tebbens SF, Donahue BT, Wilson R. 2005. Geologic structure and hydrodynamics of Egmont Channel: An anomalous inlet at the mouth of Tampa Bay, Florida. J Coast Res. 21(2):331–357.

BOEM (Bureau of Ocean Energy Management). 2012. Gulf of Mexico Geographic Information System Data and Maps. Washington, DC: Bureau of Ocean Energy Management. Available from: http://www.boem.gov/Oil-and-Gas-Energy-Program/Mapping-and-Data/Maps-And-Spatial-Data.aspx via the Internet. Accessed 5 March, 2014.

Bortone SA, Brandini FP, Fabi G, Otake S. 2011. Artificial reefs in fisheries management. Boca Raton: CRC Press.

Boström C, Pittman S, Simenstad C, Kneib R. 2011. Seascape ecology of coastal biogenic habitats: Advances, gaps, and challenges. Mar Ecol Prog Ser. 427:191–217. Available from: http://www.int-res.com/abstracts/meps/v427/p191-217/ via the Internet. Accessed 5 March, 2014.

Brown CJ, Smith SJ, Lawton P, Anderson JT. 2011. Benthic habitat mapping: A review of progress towards improved understanding of the spatial ecology of the seafloor using acoustic techniques. Estuar Coast Shelf Sci. 92:502–520. Available from: http://linkinghub.elsevier.com/retrieve/pii/S0272771411000485 via the Internet. Accessed 5 March, 2014.

Calder BR, Mayer LA. 2003. Automatic processing of high-rate, high-density multibeam echosounder data. Geochem Geophys Geosyst. 4:6. Available from: http://dx.doi.org/10.1029/2002GC000486 via the Internet. Accessed 5 March, 2014.

Caress DW, Clague DA, Paduan JB, Martin JF, Dreyer BM, Chadwick WW, Denny A, Kelley DS. 2012. Repeat bathymetric surveys at 1-metre resolution of lava flows erupted at Axial Seamount in April 2011. Nat Geosci Lett. 5:483–488. Available from: http://dx.doi.org/10.1038/NGEO1496 via the Internet. Accessed 5 March, 2014.

Cerveny K, Appeldoorn RS, Recksiek CW. 2010. Managing habitat in coral reef ecosystems for fisheries: Just what is essential? Proceedings of the 63rd Gulf and Caribbean Fisheries Institute. San Juan, Puerto Rico, November 1–5, 2010.

Clark RD, Christensen JD, Monaco ME, Caldwell PA, Matthews GA, Minello TJ. 2004. A habitat-use model to determine essential fish habitat for juvenile brown shrimp (*Farfantepenaeus aztecus*) in Galveston Bay, Texas. Fish Bull. 102:264–277.

Côté IM, Green SJ, Hixon MA. 2013. Predatory fish invaders: Insights from Indo-Pacific lionfish in the Western Atlantic and Caribbean. Biol Conserv. 164:50–61. Available from: http://linkinghub.elsevier.com/retrieve/pii/S0006320713001171 via the Internet. Accessed 5 March, 2014.

Cross VA, Twichell DC, Halley RB, Ciembronowicz KT, Jarrett BD, Hammar-Klose ES, Hine AC, Locker SD, Naar DF. 2005. GIS compilation of data collected from the Pulley Ridge deep coral reef region. U.S. Geological Survey Open-File Report 2005-1089. Reston, VA: U.S. Geological Survey.

Crowder L, Norse E. 2008. Essential ecological insights for marine ecosystem-based management and marine spatial planning. Mar Policy. 32:772–778. Available from: http://linkinghub.elsevier.com/retrieve/pii/S0308597X08000651 via the Internet. Accessed 5 March, 2014.

Diaz JVM. 1999. Analysis of multibeam sonar data for characterization of seafloor habitats. Master's Thesis for the University of New Brunswick.

Dunn D, Halpin P. 2009. Rugosity-based regional modeling of hard-bottom habitat. Mar Ecol Prog Ser. 377:1–11. Available from: http://www.int-res.com/abstracts/meps/v377/p1-11/ via the Internet. Accessed 5 March, 2014.

Elith J, Graham CH, Anderson RP. 2006. Novel methods improve prediction of species' distributions from occurrence data. Ecography. 29:129–151.

Elith J, Leathwick JR, Hastie T. 2008. A working guide to boosted regression trees. J Animal Ecol. 77:802–813.

Farmer N, Ault J. 2011. Grouper and snapper movements and habitat use in Dry Tortugas, Florida. Mar Ecol Prog Ser. 433:169–184. Available from: http://www.int-res.com/abstracts/meps/v433/p169-184/ via the Internet. Accessed 5 March, 2014.

Froeschke JT, Stunz GW, Wildhaber M. 2010. Environmental influences on the occurrence of coastal sharks in estuarine waters. Mar Ecol Prog Ser. 407:279–292. Available from: http://www.int-res.com/abstracts/meps/v407/p279-292/ via the Internet. Accessed 5 March, 2014.

Froeschke BF, Tissot P, Stunz GW, Froeschke JT. 2013. Spatiotemporal predictive models for juvenile southern flounder in Texas estuaries. N Am J Fish Manage. 33(4):817–828.

GMFMC (Gulf of Mexico Fishery Management Council). 2002. Habitat policy. Tampa, FL: Gulf of Mexico Fishery Management Council. Available from: http://www.gulfcouncil.org/Beta/GMFMCWeb/downloads/habitat%20policy.pdf via the Internet. Accessed 5 March, 2014.

GMFMC (Gulf of Mexico Fishery Management Council). 2005. Final generic amendment number 3 for addressing essential fish habitat requirements, habitat areas of particular concern, and adverse effects of fishing in the following fishery management plans of the Gulf of Mexico: shrimp, red drum, reef fish, coastal migratory pelagics in the Gulf of Mexico and South Atlantic, stone crab, spiny lobster, and coral and coral reefs of the Gulf of Mexico. Tampa, FL: Gulf of Mexico Fishery Management Council. Available from: http://www.gulfcouncil.org/Beta/GMFMCWeb/downloads/FINAL3_EFH_Amendment.pdf via the Internet. Accessed 5 March, 2014.

GMFMC (Gulf of Mexico Fishery Management Council). 2010. 5-Year review of the final generic amendment number 3 addressing essential fish habitat requirements, habitat areas of particular concern, and adverse effects of fishing in the fishery management plans of the Gulf of Mexico. Tampa, FL: Gulf of Mexico Fishery Management Council. Available from: http://www.gulfcouncil.org/Beta/GMFMCWeb/downloads/EFH%205-Year%20Review%20Final%2010-10.pdf via the Internet. Accessed 5 March, 2014.

GMFMC (Gulf of Mexico Fishery Management Council). 2012a. Species listed in the fishery management plans of the Gulf of Mexico Fishery Management Council. Tampa, FL: Gulf of Mexico Fishery Management Council. Available from: http://www.gulfcouncil.org/Beta/GMFMCWeb/downloads/species%20managed.pdf via the Internet. Accessed 5 March, 2014.

GMFMC (Gulf of Mexico Fishery Management Council). 2012b. Final amendment 11 to the fishery management plan for spiny lobster in the Gulf of Mexico and South Atlantic. Tampa, FL: Gulf of Mexico Fishery Management Council. Available from: http://www.gulfcouncil.org/Beta/GMFMCWeb/downloads/Final_Spiny_Lobster_Amend_11_April_05_2012.pdf via the Internet. Accessed 5 March, 2014.

GMFMC (Gulf of Mexico Fishery Management Council). 2013a. Final summary report—Workshop on Interrelationships between Coral Reefs and Fisheries. Tampa, FL: Gulf of Mexico Fishery Management Council. Available from: http://www.gulfcouncil.org/docs/Coral%20Workshop%20Final%20Summary%20Report%209-26-13.pdf via the Internet. Accessed 5 March, 2014.

GMFMC (Gulf of Mexico Fishery Management Council). 2013b. Generic amendment 4: fixed petroleum platforms and artificial reefs and essential fish habitat: Options paper. Tampa, FL: Gulf of Mexico Fishery Management Council. Available from: http://www.gulfcouncil.org/docs/amendments/Artificial%20Reefs%20as%20EFH%20Amendment.pdf via the Internet. Accessed 5 March, 2014.

Goodman JA, Purkis SJ, Phinn SR. 2013. Coral reef remote sensing: A guide for mapping, monitoring and management. New York: Springer.

Grober-Dunsmore R, Pittman SJ, Caldow C, Kendall MS, Frazer TK. 2009. A landscape ecology approach for the study of ecological connectivity across tropical marine seascapes. *In:* Nagelkerken I, editor. Ecological connectivity among tropical coastal ecosystems. Dordrecht: Springer. p. 493–530.

Harris PT, Baker EK. 2011. Seafloor geomorphology as benthic habitat: GeoHAB atlas of seafloor geomorphic features and benthic habitats. Amsterdam: Elsevier.

Hey RN, Kleinrock MC, Miller SP, Atwater TM, Searle RC. 1986. Sea beam/deep-tow investigation of an active oceanic propagating rift system, Galapagos 95.5°W. J Geophys Res. 91:3369–3393. Available from: http://dx.doi.org/10.1029/JB091iB03p03369 via the Internet. Accessed 5 March, 2014.

Hine AC, Halley RB, Locker SD, Jarrett BD, Jaap WC, Mallinson DJ, Ciembronowicz KT, Ogden NB, Donahue BT, Naar DF, et al. 2008. Coral reefs, present and past, on the West Florida shelf and platform margin. *In:* Riegl BM, Dodge RE, editors. Coral reefs of the USA. Dordrecht: Springer. p. 127–173.

Jenkins C. 2011. Dominant bottom types and habitats in the Gulf of Mexico Data Atlas. Stennis Space Center, MS: National Coastal Data Development Center. Available from: http://gulfatlas.noaa.gov/ via the Internet. Accessed 5 March, 2014.

Knudby A, LeDrew E, Brenning A. 2010. Predictive mapping of reef fish species richness, diversity and biomass in Zanzibar using IKONOS imagery and machine-learning techniques. Remote Sens Environ. 114:1230–1241. Available from: http://linkinghub.elsevier.com/retrieve/pii/S0034425710000295 via the Internet. Accessed 5 March, 2014.

Lacono CL, Gracia E, Zaniboni F, Pagnoni G, Tinti S, Bartolome R, Masson DG, Wynn RB, Lourenco N, Pinto de Abreu M, et al. 2012. Large, deepwater slope failures: Implications for landslide-generated tsunamis. Geology. 40:931–934. Available from: http://geology.gsapubs.org/content/40/10/931 via the Internet. Accessed 5 March, 2014.

Lawler M. 2013. Global seafloor geomorphology map. Geoscience Australia/U.N. Environment Program/Conservation International. Available from: http://www.geoiq.grida.no/maps/1136 via the Internet. Accessed 5 March, 2014.

Lonsdale P. 1983. Overlapping rift zones at the 5.5°S offset of the East Pacific Rise. J Geophys Res. 88:9393–9406. Available from: http://dx.doi.org/10.1029/JB088iB11p09393 via the Internet. Accessed 5 March, 2014.

Lundblad ER, Wright DJ, Miller J, Larkin EM, Rinehart R, Naar DF, Donahue BT, Anderson SM, Battista T. 2006. A benthic terrain classification scheme for American Samoa. Mar Geod. 29:89–111.

Macdonald KC, Fox PJ. 1983. Overlapping spreading centres: New accretion geometry on the East Pacific Rise. Nature. 302:55–58. Available from: http://dx.doi.org/10.1038/302055a0 via the Internet. Accessed 5 March, 2014.

MAFMC (Mid Atlantic Fishery Management Council). 2013. Ecosystem Approach to Fisheries Management Guidance Document. Dover, DE: Mid Atlantic Fishery Management Council. Available from: http://www.mafmc.org/eafm/ via the Internet. Accessed 5 March, 2014.

McClanahan TR, Graham NAJ, MacNeil MA, Muthiga NA, Cinner JE, Bruggemann JH, Wilson SK. 2011. Critical thresholds and tangible targets for ecosystem-based management of coral reef fisheries. Proc Natl Acad Sci U S A. 108:17230–17233. Available from: http://www.pubmedcentral.nih.gov/articlerender.fcgi?artid=3193203&tool=pmcentrez&rendertype=abstract via the Internet. Accessed 5 March, 2014.

Montero JT, Chesney TA, Bauer JR, Graham J, Froeschke JT. forthcoming. Modeling brown shrimp (*Farfantepenaeus aztecus*) density distribution in the Northern Gulf of Mexico using boosted regression trees. ICES J Mar Sci.

MSFCMA (Magnuson-Stevens Fishery Conservation and Management Act). 2006. Public Law 94-265. As amended by the Magnuson-Stevens Fishery Conservation and Management Reauthorization Act (P.L. 109-479).

Naar DF, Hey RN. 1991. Tectonic evolution of the Easter microplate. J Geophys Res. 96:7961–7993.

NOAA NCDDC (NOAA National Coastal Data Development Center). 2013. Gulf of Mexico data atlas. Stennis Space Center, MS: NOAA NCDDC. Available from: http://gulfatlas.noaa.gov/ via the Internet. Accessed 5 March, 2014.

NOAA NGDC (NOAA National Geophysical Data Center). 2003. U.S. coastal relief model. Boulder, CO: NOAA NGDC. Available from: http://www.ngdc.noaa.gov/mgg/coastal/crm.html via the Internet. Accessed 5 March, 2014.

NOAA RESTORE (NOAA Resources and Ecosystems Sustainability, Tourist, Opportunity, and Revived Economies) Act Science Program. 2013. Gulf Coast Ecosystem Restoration Science, Observation, Monitoring, and Technology Program Science Framework. Available from: http://www.restoreactscienceprogram.noaa.gov/wp-content/uploads/2013/06/RESTOREScienceProgramFramework20130621.pdf via the Internet. Accessed 5 March, 2014.

Parker RO, Colby DR, Willis TD. 1983. Estimated amount of reef habitat on a portion of the U.S. South Atlantic and Gulf of Mexico continental shelf. Bull Mar Sci. 33:935–940.

Pickering H, Whitmarsh D. 1997. Artificial reefs and fisheries exploitation: A review of the "attraction versus production" debate, the influence of design and its significance for policy. Fish Res. 31:39–59. Available from: http://linkinghub.elsevier.com/retrieve/pii/S0165783697000192 via the Internet. Accessed 5 March, 2014.

Pikitch EK, Santora C, Babcock EA, Bakun A, Bonfil R, Conover DO, Dayton P, Doukakis P, Fluharty D, Heneman B, et al. 2004. Ecosystem-based fishery management. Science. 305(5682):346–347.

Pittman SJ, Brown K. 2011. Multi-scale approach for predicting fish species distributions across coral reef seascapes. PLoS ONE. 6:e20583. Available from: http://www.pubmedcentral.nih.gov/articlerender.fcgi?artid=3102744&tool=pmcentrez&rendertype=abstract via the Internet. Accessed 5 March, 2014.

Pittman SJ, Christensen JD, Caldow C, Menza C, Monaco ME. 2007. Predictive mapping of fish species richness across shallow-water seascapes in the Caribbean. Ecol Modell. 204:9–21. Available from: http://linkinghub.elsevier.com/retrieve/pii/S0304380006006338 via the Internet. Accessed 5 March, 2014.

Reed JK, Koenig CC, Shepard AN. 2007. Impacts of bottom trawling on a deep-water Oculina coral ecosystem off Florida. Bull Mar Sci. 81:481–496.

Ross SW, Quattrini AM. 2007. The fish fauna associated with deep coral banks off the Southeastern United States. Deep Sea Res Pt I: Oceanogr Res Pap. 54:975–1007. Available from: http://linkinghub.elsevier.com/retrieve/pii/S096706370700088X via the Internet. Accessed 5 March, 2014.

Ross SW, Quattrini AM. 2009. Deep-sea reef fish assemblage patterns on the Blake Plateau (Western North Atlantic Ocean). Mar Ecol. 30:74–92. Available from: http://doi.wiley.com/10.1111/j.1439-0485.2008.00260.x via the Internet. Accessed 5 March, 2014.

SAFMC (South Atlantic Fishery Management Council). 2013. Habitat and ecosystem atlas. Charleston, SC: South Atlantic Fishery Management Council. Available from: http://www.safmc.net/ecosystem-management/mapping-and-gis-data via the Internet. Accessed 5 March, 2014.

Sammarco PW, Atchison AD, Boland GS. 2004. Expansion of coral communities within the Northern Gulf of Mexico via offshore oil and gas platforms. Mar Ecol Prog Ser. 280:129–143.

Sammarco PW, Atchison AD, Boland GS, Sinclair J, Lirette A. 2012. Geographic expansion of hermatypic and ahermatypic corals in the Gulf of Mexico, and implications for dispersal and recruitment. J Exp Mar Biol Ecol. 436–437:36–49.

Schofield PJ. 2009. Geographic extent and chronology of the invasion of non-native lionfish (*Pterois volitans* [Linnaeus 1758] and *P. miles* [Bennett 1828]) in the Western North Atlantic and Caribbean Sea. Aquat Invasions. 4(3):473–479. Available from: http://www.aquaticinvasions.net/2009/AI_2009_4_3_Schofield.pdf via the Internet. Accessed 5 March, 2014.

Schofield PJ. 2010. Update of geographic spread of lionfishes (*Pterois volitans* [Linnaeus, 1758] and *P. miles* [Bennett, 1828]) in the Western North Atlantic Ocean, Caribbean Sea and Gulf of Mexico. Aquat Invasions. 5(Supplement 1): S117–122. Available from: http://www.aquaticinvasions.net/2010/Supplement/AI_2010_5_S1_Schofield.pdf via the Internet. Accessed 5 March, 2014.

Schroeder WW. 2002. Observations of *Lophelia pertusa* and the surficial geology at a deep-water site in the Northeastern Gulf of Mexico. Hydrobiologia. 471:29–33.

Schroeder WW. 2007. Seafloor characteristics and distribution patterns of *Lophelia pertusa* and other sessile megafauna at two upper-slope sites in the Northeastern Gulf of Mexico. OCS Study MMS 2007-035. New Orleans, LA: US Department of the Interior, Minerals Management Service, Gulf of Mexico OCS Region.

SEDAR (Southeast Data, Assessment, and Review). 2012. SEDAR episodic events workshop for Gulf of Mexico fisheries: workshop summary report. North Charleston, SC: SEDAR.

SEDAR (Southeast Data, Assessment, and Review). 2013. Stock assessment report for Gulf of Mexico red snapper. North Charleston, SC: Southeast Data, Assessment, and Review.

Szedlmayer ST, Lee JD. 2004. Diet shifts of juvenile red snapper (*Lutjanus campechanus*) with changes in habitat and fish size. Fish Bull. 102:366–375.

Theberge AE, Cherkis NZ. 2013. A note on fifty years of multi-beam. Hydro Int. 17(4):29–33. Available from: http://www.hydro-international.com via the Internet. Accessed 5 March, 2014.

UNEP (United Nations Environment Programme). 2009. The UNEP Large Marine Ecosystem Report: A perspective on changing conditions in LMEs of the world's regional seas. Nairobi: UNEP. Available from: http://www.lme.noaa.gov/index.php?option=com_content&view=article&id=178:unep-lme-report&catid=39:reports&Itemid=54 via the Internet. Accessed 5 March, 2014.

USGS (United States Geological Survey). 2014. Nonindigenous Aquatic Species Database. Gainesville, FL: USGS. Available from: http://nas2.er.usgs.gov/viewer/omap.aspx?SpeciesID=963 via the Internet. Accessed 5 March, 2014.

Wall CW, Donahue BT, Naar DF, Mann DA. 2011. Spatial and temporal variability of red grouper holes within Steamboat Lumps Marine Reserve, Gulf of Mexico. Mar Ecol Prog Ser. 431:243–254.

Whaley S, Burd JJ, Robertson B. 2007. Using estuarine landscape structure to model distribution patterns in nekton communities and in juveniles of fishery species. Mar Ecol Prog Ser. 330:83–99. Available from: http://www.int-res.com/abstracts/meps/v330/p83-99/ via the Internet. Accessed 5 March, 2014.

White HK, Hsing P-Y, Cho W, Shank TM, Cordes EE, Quattrini AM, Nelson RK, Camilli R, Demopoulos AWJ, German CR, et al. 2012. Impact of the Deepwater Horizon oil spill on a deep-water coral community in the Gulf of Mexico. Proc Natl Acad Sci U S A. 109:20303–20308. Available from: http://www.pubmedcentral.nih.gov/articlerender.fcgi?artid=3528508&tool=pmcentrez&rendertype=abstract via the Internet. Accessed 5 March, 2014.

Wolfson ML, Naar DF, Howd PA, Locker SD, Donahue BT, Friedrichs CT, Trembanis AC, Richardson MD, Wever TF. 2007. Multibeam observations of mine burial near Clearwater, Florida, including comparisons to predictions of wave-induced burial. IEEE J Oceanic Eng. 32(1):103–118.

Wood SN. 2003. Thin plate regression splines. J R Statist Soc Ser B. 65:95–114.

Wood SN. 2006. Generalized additive models: An introduction with R. Boca Raton, FL: Chapman & Hall/CRC.

Wright DJ, Donahue BT, Naar DF. 2002. Seafloor mapping and GIS coordination in America's remotest national marine sanctuary (American Samoa). *In:* Wright DJ, editor. Undersea with GIS. Redlands, CA: ESRI Press. p. 33–64.

Wright DJ, Roberts JT, Fenner D, Smith JR, Koppers AP, Naar DF, Hirsch ER, Clift LW, Hogrefe KR. 2012. Seamounts, ridges, and reef habitats of American Samoa. *In:* Harris PT, Baker ED, editors. Seafloor geomorphology as benthic habitat: GeoHAB atlas of seafloor geomorphic features and benthic habitats. Amsterdam: Elsevier.

chapter fourteen

Connectivity in the tropical coastal seascape

Implications for marine spatial planning and resource management

John C. Ogden, Ivan Nagelkerken, and Carole C. McIvor

Contents

Introduction

A major component of the high biological diversity of the tropics is the complex mosaic of interacting ecosystems, or "seascape," of the coastal marine zone. This relatively narrow band of shallow water, in intimate contact with terrestrial watersheds and open to the sea, is the source of much of the organic production and diversity of species in the tropical marine environment. It contributes to a significant proportion of global fisheries production and supplies other critical ecosystem services such as providing a wave-energy barrier to the coast and diluting, dispersing, and metabolizing the effluents of human society. A tropical seascape often is composed of some combination of coral reefs, seagrasses, and mangrove ecosystems. Their development and complexity of interaction largely depend upon the size of the adjacent land mass, the amount of terrestrial runoff, the configuration of the coast and offshore features such as islands and coral reefs, and water circulation patterns (Wolanski 2001). Of course, these ecosystems can thrive in isolation. Atolls, for example, have extensive coral reef development often in the absence of either seagrasses or mangroves. Despite the fact that the biogeographic realms of the Atlantic and the Indo-Pacific have distinct differences in species diversity and composition, the dominant biostructural component of each ecosystem—mangrove trees, seagrasses, and reef corals—have similar physiological and ecological roles (Birkeland 1987).

The component ecosystems of the tropical seascape, particularly coral reefs, have been studied for many years. Research in the last three decades has concentrated on ecosystem interactions, revealing interdependencies critical for increasingly urgent management concerns (Ogden 1988, Wolanski 2001, Ley and McIvor 2002, Nagelkerken et al. 2002). This chapter considers biological, biochemical, and physical interactions among tropical coastal ecosystems and their importance to coral reef structure and function and to the fisheries and other human uses they support.

Biological interactions in the seascape

Nurseries

The fish communities of mangroves and seagrass beds often overlap with those of adjacent systems as a result of short-term or ontogenetic migrations, and they may contain freshwater, estuarine, coral reef, or oceanic species (Whitfield 1998, Blaber 2000, Nagelkerken 2007). Mangroves and seagrass beds are considered important juvenile habitats for a variety of fish and invertebrate species that spend their adult life on coral reefs or offshore. The nursery function is particularly apparent in the Caribbean (see reviews by Parrish 1989, Blaber 2000, Nagelkerken 2009), but there is increasing evidence that this is the case in the Indo-Pacific as well (Nakamura et al. 2008, Olds et al. 2012a, 2013, Kimirei et al. 2013a). In areas where adjacent coral reefs are absent or limited in extent, the presence of juvenile reef fishes in mangroves and seagrass beds is less evident, and therefore, on some islands in the Indo-Pacific the value of these habitats as nurseries has been questioned (Thollot 1992) (Figure 14.1).

The salient features of mangroves and seagrass beds that make them valuable habitats for juvenile organisms are: (1) location away from the heavy predation characteristic of reefs (Grol et al. 2011); (2) protection afforded by the structural complexity of masses of leaves and roots (Cocheret de la Morinière et al. 2004); and (3) a rich food supply based on plant detritus, microorganisms, plankton, algae, and associated small invertebrates (Laegdsgaard and Johnson 2001). These features may significantly enhance the growth rates of juveniles and decrease postsettlement benthic mortality (Dahlgren and Eggleston 2000), which is typically very high for some marine fish species (Shulman and Ogden 1987). However, studies have recently shown that fish may trade growth for higher survival by using nurseries (Grol et al. 2008, Kimirei et al. 2013b). Furthermore, use of nonreef habitats by juvenile fish may temporarily alleviate competition for food and space between life stages within species and between congeneric species (Cocheret de la Morinière et al. 2002).

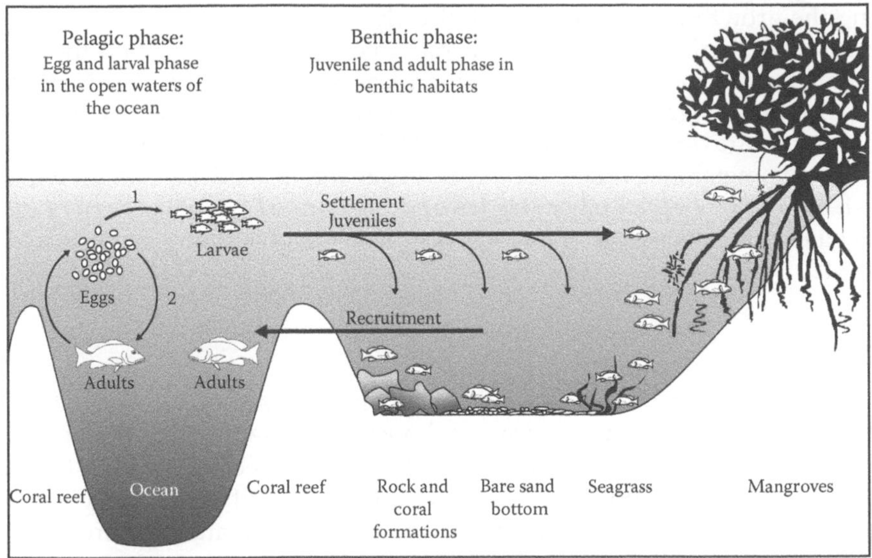

Figure 14.1 Overview of the life cycle of coral reef fishes that use off-reef areas such as seagrasses or mangroves as nurseries. Fishes either settle as pelagic larvae or recruit as early-stage juveniles into inshore benthic habitats. (Reprinted with permission from Nagelkerken I, editor. 2009. Ecological connectivity among tropical coastal ecosystems. Dordrecht: Springer.)

Fish larvae can settle directly from the oceanic plankton in a variety of shallow-water habitats (Pollux et al. 2007). Nevertheless, seagrass beds appear to be typical settlement areas, probably because of their relatively large surface areas that make them easier to target when moving inshore from the open ocean (Parrish 1989). In the Caribbean, species from a wide range of fish families (e.g., Lutjanidae, Haemulidae, Scaridae, Chaetodontidae, and Acanthuridae) use mangroves (Rooker and Dennis 1991), tidal channels (Nagelkerken et al. 2000b), or shallow patch reefs (Ogden and Ehrlich 1977) in lagoons as an intermediate life-stage habitat (Ley and McIvor 2002) before moving to deeper, offshore fringing or barrier reefs. This migration typically occurs when fishes have grown large enough that predation risk is similar in juvenile and adult habitats. Offshore migration is also likely to be driven by the need to reproduce or move to a habitat where higher growth rates can be sustained (Grol et al. 2008, Kimirei et al. 2013b).

Foraging movements and migrations

Diurnal feeding migrations among habitats are common features of fishes residing in mangroves and seagrass beds. Diurnally active fishes of common families such as Acanthuridae, Chaetodontidae, Gerreidae, Labridae, Scaridae, and Sparidae are found foraging on seagrass beds during daytime, but at night migrate to other complex habitats to shelter (Ogden and Buckman 1973, Beets et al. 2003, Nagelkerken et al. 2000a, 2008). Nocturnally active fishes of common families such as Apogonidae, Diodontidae, Haemulidae, Holocentridae, and Lutjanidae shelter in structurally complex habitats during daytime (Nagelkerken et al. 2000b) and migrate to seagrass beds or sand flats at night to feed (Ogden and Ehrlich 1977). The latter two habitats often contain higher densities of food items than coral reefs and are therefore more suitable as foraging habitats (Nagelkerken et al. 2000a). As a result of these migrations, fish can function as vectors of nutrients between seagrass beds and reefs, enhancing the growth rates of corals and directly adjacent seagrasses (Meyer et al. 1983, Allgeier et al. 2013), although with the current rate of coral degradation these nutrients may facilitate macroalgal growth, which suppresses coral survival and growth (Burkepile et al. 2013). Feeding migrations are linked to changing light levels (McFarland et al. 1979). These migrations are often precisely timed and take place along fixed migration pathways as much as a kilometer long that gradually bifurcate into a dendritic pattern with an individual fish at the tip of each branch (Ogden and Zieman 1977). This pattern of movement may allow individual fish to partition large nocturnal seagrass foraging areas (Quinn and Ogden 1984).

Edge effects

The boundary between coral reefs, seagrasses, and mangroves is an important component of each of the adjoining habitats. Diurnal and nocturnal foraging movements of herbivorous fishes and sea urchins from coral reefs into seagrass beds are known to create conspicuous grazed halos around Caribbean and some Indo-Pacific coral reefs (Randall 1965, Ogden et al. 1973). Adjoining habitats may also influence one another through exchange by larval, juvenile, and adult fish resulting in increased fish species richness and densities as compared to isolated habitats (Nagelkerken et al. 2001) (Figure 14.2).

In tropical areas influenced by large tidal ranges (e.g., many areas in the Indo-Pacific), tidal fish migrations also contribute to connectivity among shallow-water habitats (Forward and Tankersley 2001). Due to the large tidal amplitudes, many intertidal habitats are drained at low tide. With incoming tides, however, shoreline habitats such as mangroves, sand banks, and notches in fossil reef rock become temporarily available for juvenile fish that move in from adjacent seagrass beds and subtidal mud flats located at greater depths. At high tide, these intertidal habitats provide feeding and shelter areas when larger predatory fish appear with the incoming tide.

Dependencies on nonreef nursery habitats

Although mangroves and seagrass beds function as important shelter and foraging habitats for juvenile reef fish, the degree to which coral reef fish populations depend on these habitats remains unclear (Blaber 2000). The presence of juvenile fish in a habitat does not mean that the habitat functions as a nursery *per se*. Habitats are considered a nursery if their contribution in terms of biomass that is transferred to adult fish populations is larger than the average biomass production of all other habitats in which juveniles are found (Beck et al. 2001). This is closely related to the concept of "estuarine dependence" (Lenanton and Potter 1987), where estuarine-dependent species are defined as those for which estuaries are the principal environment for at least part of the life cycle and without which a viable population would cease to exist (Blaber et al. 1989).

Similarly, the concept of "bay habitat dependence" (i.e., the usage of shallow, inshore coastal areas by fishes) has been proposed for shallow-water habitats in nonestuarine environments (Nagelkerken and van der Velde 2002). Since mangroves are the principal habitat in many tropical estuaries and bays, the concepts of estuarine and bay habitat dependence are likely linked to dependence on mangroves (Blaber 2000). Studies in the western Atlantic suggest that at least 21–50 fish species (Lindeman et al. 2000, Nagelkerken and van der Velde 2003) may show ontogenetic movements from shallow-water inshore habitats (mangroves, seagrass beds, hardbottom substrata, shallow reefs) to adult deep/offshore coral reef habitat. In different areas, such as Australia, this is true for at least 14 fish

Figure 14.2 **(See color insert)** A Caribbean patch reef by day (a) and by night (b) showing surrounding seagrasses, a grazed "halo," and movements of day-active fishes including parrotfishes and night-active grunts (Haemulidae) and *Diadema* sea urchins.

species (Blaber et al. 1989). The dependence of fish on these habitats may be high in some cases. Caribbean reefs lacking access to these nursery habitats show significantly lower densities of adults of some fish species (e.g., species of Haemulidae, Scaridae, Lutjanidae, and Sphyraenidae) than reefs adjacent to mangrove and seagrass areas (Nagelkerken et al. 2002, Mumby et al. 2004).

Mangrove–reef connectivity has strong implications for fisheries and reef health: the movement of important species or guilds enhances the productivity and resilience of adjacent reefs, for example, through the enhanced biomass of herbivores that maintain reef macroalgae in a cropped state and improve coral settlement and survival (Mumby and Hastings 2008, Olds et al. 2012b). Furthermore, fringing reefs connected to nursery habitats by fish movements show an enhanced biomass of commercially important fish species as well as higher productivity of smaller prey species that may further fuel secondary or tertiary production (Nagelkerken et al. 2012, Olds et al. 2012a).

Fluxes of nutrients and organic material in the seascape

Mangroves

Mangrove forests, often highly productive, are commonly found along low-energy tropical and subtropical coastlines between 30° N and 30° S. Mangroves reach maximum extent in estuarine environments in river basins and coastal floodplains with abundant runoff of freshwater, sediments, and nutrients from the land. Nonestuarine mangroves occur on protected shores on islands and keys and lack direct channelized runoff (Kathiresan and Bingham 2001). Although we concentrate in this chapter on fluxes from mangroves to coral reefs, many of the same processes and ecosystem services also apply to seagrass beds in the tropical coastal seascape. A full review of the role of seagrasses is nonetheless beyond the scope of the present chapter, but they are mentioned where appropriate.

Mangroves as exporters of organic material

One of the ecosystem services attributed to mangroves is the exporting of organic matter to adjacent waters (e.g., Lee 2005, Hyndes et al. 2014). Bouillon and Connolly (2009) review three avenues for the export of organic matter from intertidal mangroves: (1) by the physical movement of dissolved or particulate matter; (2) through the migration of animals from intertidal to subtidal waters; and (3) via a series of predator–prey interactions known as a trophic relay (Kneib 1997).

The magnitude of export of mangrove-derived organic matter shows considerable geographic variation. Estimates of export of particulate organic carbon range from 16 g C m^{-2} y^{-1} for two high intertidal basin forests in subtropical southwest Florida (Twilley 1985) to 274 g C m^{-2} y^{-1} for a mountainous island catchment in wet tropical Australia (Robertson et al. 1992). In some cases, export of carbon may be primarily in the dissolved phase, and may be pulsed due to seasonal patterns of forest inundation (Twilley 1985) or to unpredictable but large storm events (Childers et al. 2000). Export via physical movement appears to increase with greater tidal amplitude, more runoff, increasing age of the wetland system, extent of mangroves, and lack of geomorphic features (e.g., shallow sills) that slow or trap ebb-tide entrained materials (Twilley et al. 1996, Childers et al. 2000).

The second avenue of export, tidal migration of fish between subtidal and intertidal waters, is a common phenomenon in tidal forests (Krumme 2009). Fish too large to reside permanently in the forest nonetheless follow the flood tide into the forest to feed. Material gathered in the forest during flood tides is processed as energy, converted to living biomass, or discharged as waste material into subtidal waters on the ebb tide, thus exporting mangrove productivity.

The trophic relay, the third avenue of export, is similar to tidal migration except that it covers longer temporal and larger spatial scales, moving production stepwise across the landscape with successive predator–prey interactions. As an example, Sheaves and Molony (2000) describe a scenario in subtropical Queensland, Australia, in which subadults of a serranid (*Epinephelus malabaricus*) and a snapper (*Lutjanus argentimaculatus*) forage within estuarine mangrove forests on leaf-feeding sesarmine crabs. The two fish species subsequently migrate offshore at larger sizes, thereby exporting mangrove productivity.

Mangroves as trophic support

The evidence that mangrove-derived production is a substantial source of nutrition for mangrove-associated fauna derives primarily from stable isotope analysis of animals and plants characteristic of mangrove and adjacent environments. Signatures of carbon isotopes vary substantially among primary producers using different inorganic carbon sources and photosynthetic pathways (Peterson and Fry 1987, Bouillon et al. 2008). Sulfur isotopes of rooted plants (mangroves, seagrasses) are considerably more negative than those of free-living algae (phytoplankton and benthic microalgae) because of different sulfur sources, for example, reduced sulfides in anoxic sediments (approximately −24%) versus seawater sulfate (+20%) (Fry et al. 1982). The carbon and sulfur signatures of animals reflect those of their diet, with little fractionation between trophic levels (Peterson and Fry 1987). Used in conjunction, these isotopes can thus be used to estimate the relative contributions of different primary producers to consumers. Nitrogen stable isotopes are subject to 2.0‰–3.5‰ fractionation with each step in the food web and are thus used to very reliably estimate trophic level (McCutchan et al. 2003). To illustrate the range of producer values for carbon isotopes in the coastal marine seascape, the negative values of mangroves ($\delta^{13}C$ values between −30‰ and −24‰) are distinctive from the higher values of seagrasses ($\delta^{13}C$ greater than −15‰) (Fry and Sherr 1984, Bouillon et al. 2008). The $\delta^{13}C$ values of oceanic phytoplankton average −21‰ (Peterson and Fry 1987), whereas those of benthic microalgae from mangroves vary from −16.5‰ to −21‰ (Bouillon et al. 2002, Lee 2006, Vaslet et al. 2012, McIvor unpublished).

Using the stable isotope approach, varying degrees of incorporation of mangrove-derived organic matter have been found in mangrove environments. The greatest incorporation has been seen in settings that favor retention, as opposed to export, of mangrove organic matter as typified by lagoonal and basin forest sites in Sri Lanka (Bouillon et al. 2004). Areal coverage of mangroves may also be a factor in determining the relative contribution of organic matter to consumers: mud crab (*Scylla serrata*) in more extensive mangroves in Micronesia showed greater incorporation of mangrove organic matter (Demopoulos et al. 2008). Lugendo et al. (2007) identified duration of tidal access as a factor influencing fish isotopic values in two forest types in Tanzania: fish from the vicinity of a frequently available riverine forest showed more mangrove-like signatures than those from the vicinity of a nearby more intermittently available fringing forest. When fringing forest settings were studied, consumers often exhibited greater reliance on organic matter from an adjacent seagrass habitat (e.g., Kieckbusch et al. 2004, Nagelkerken and van der Velde 2004, Vaslet et al. 2012). Where adequately measured, benthic microalgae have often been found to make substantial contributions to mangrove food webs (e.g., Kon et al. 2007, Tue et al. 2012). Such contributions are often attributed to the high digestibility of algae as well as to its high renewal rates (Bouillon et al. 2008, Kristensen et al. 2008).

Most evidence for mangrove trophic support of food webs has come from animals within or near the mangrove forest (Fry and Ewel 2003). The mangrove isotopic signal appears to quickly diminish with distance away from mangrove forests (Robertson et al. 1992, Loneragan et al. 1997). Torgerson and Chivas (1985) determined that terrestrial (i.e., mangrove) carbon in subtidal sediments decreased by an order of magnitude within 10 km of the shore off Missionary Bay, Australia, owing to its wide dispersal. The observed decline in the mangrove isotopic signal in offshore migrants that resided in mangrove environments in young life-history stages is consistent with simple physical dilution of mangrove carbon relative to other sources with greater distance from shore, with the increasing incorporation of other more abundant carbon sources (benthic microalgae, seagrass, phytoplankton), and with the relatively rapid turnover of carbon in the tissues measured (Peterson and Fry 1987, Fry et al. 1999).

Where mangroves and seagrasses occur in proximity to coral reefs (e.g., Belize Barrier Reef, many Caribbean islands, Far North Queensland in Australia), there is a reasonable expectation that mangrove and seagrass organic matter reaches the reef. However, documenting incorporation of this material into the biomass of coral reef organisms has met with limited success. On the Caribbean coast of Panama, Granek et al. (2009) used the stable isotopes of carbon and sulfur (in conjunction with a concentration-dependent mixing model) to demonstrate that mangrove organic matter was incorporated into sessile invertebrate filter feeders within a 10 km distance. The percent mangrove contribution to the diet was 0%–57% for sponges, 7%–41% for file clams, and 18%–52% for feather duster worms. Mangrove contribution to corals (*Acropora cervicornis, Agaricia fragilis, Agaricia tenuifolia, Montastrea annularis,* and *Diploria* sp.) ranged from 0% to 44% depending on taxa and location. Suchanek et al. (1985) used the deep diving submersible ALVIN to document incorporation of detrital seagrass carbon into deep-sea organisms (2455–3950 m) off St. Croix in the U.S. Virgin Islands. Sea urchins derived a significant proportion of their nutrition from detrital seagrasses via direct consumption (gut contents analysis), whereas holothurians deposit fed on seagrass-enriched sediments (carbon isotope analysis).

Additional evidence for habitat coupling in the coastal seascape comes from otoliths—the ear bones of fish. Unlike the muscle or soft tissues of fish and invertebrates, which exhibit rapid turnover in fast-growing animals, organic and inorganic materials laid down in daily otolith rings are conserved for life, providing a natural tracer of environments previously inhabited by individuals (Elsdon et al. 2008). An example of this approach is the work of Mateo et al. (2010), who used the concentrations of 16 elements as well as the stable isotopes of oxygen and carbon in otoliths to determine the nursery habitat of French grunt subadults collected from the fore reef in St. Croix and Puerto Rico. They found that 40% of French grunts in 2006 and 68% in 2007 in St. Croix had resided in mangrove habitats as juveniles, whereas for Puerto Rico, these percentages were 70% and 74%. In contrast, almost all schoolmaster snapper in both islands had resided as juveniles in mangrove habitats in both years. The implication of such work for management is that nursery habitats (mangroves, seagrass beds) as well as adult habitats (coral reefs) need to be contained within marine protected areas (MPAs) to conserve fish stocks.

Mangroves as refuge

In many cases, mangroves have been found to make relatively little trophic contribution to resident animals, but provide refuge, with feeding taking place in adjacent habitats, often seagrass beds (Igulu et al. 2013). In Gazi Bay, Kenya, barracuda, houndfish, conger eel, and rabbitfish moved routinely between mangroves and seagrass beds, their stable carbon isotopic signal more closely matching that of seagrass. These data suggest that such animals in Gazi Bay used both systems, but that they derived refuge support from mangroves and most trophic support from seagrasses (Marguillier et al. 1997). Other places where mangroves seemed to act more as a refuge than as a food source for fish species include island settings in southeast Florida and the Bahamas (Kieckbusch et al. 2004) and in the Caribbean (Nagelkerken and van der Velde 2004).

Linkage of mangroves, seagrasses, and coral reefs

Unequivocal evidence of habitat linkages in the coastal seascape has recently increased in quantity and technological complexity. The use of compound-specific stable isotope tags in otoliths provides a robust measure of prior habitat use. For example, at a location in the Red Sea, McMahon et al. (2012) identified five potential nursery areas for blackspot snapper (*Lutjanus ehrenbergii*) that were resident as adults on coral reefs across a 50 km seascape. They used stable isotope analysis of five amino acids in otoliths to determine each site's relative contribution to adult populations, and to identify migration corridors. The results confirmed the nursery value of mangrove and seagrass systems to inshore fish populations. Over 70% of adult *L. ehrenbergii* at the 2 km reefs and 45% at the 16 km reefs had migrated from these coastal wetland habitats as juveniles. A number of individuals had also moved at least 30 km from inshore nurseries to reefs on the edge of the continental shelf.

In summary, the use of naturally occurring compounds (bulk stable isotopes in soft tissues, compound-specific stable isotopes in otoliths) as biochemical tracers or tags has greatly increased our understanding of the linkages among mangroves, seagrasses, and coral reefs in the coastal seascape (see also Verweij et al. 2008, Huijbers et al. 2013). The strength of such connections is both temporally and spatially variable, with site-specific and species-specific differences. Continued research is needed to unravel the mechanisms and forces controlling the observed differences. Nonetheless, the demonstrated linkages have important implications for the management of the coastal seascape.

Physical interactions in the seascape

Terrestrial runoff

Coral reefs require clear water for vigorous growth and they can be damaged by excessive sediments and nutrients in uncontrolled terrestrial runoff. Riverine discharge containing a heavy sediment load can destroy or severely restrict coral reef community development to only the most sediment-tolerant species (West and Van Woesik 2001). The clearing of watersheds for agriculture, industry, and tourism, and the destruction of coastal estuaries, seagrass beds, and mangrove forests, which act as sediment traps, are among the most damaging impacts on coral reefs around the world (Ogden and Gladfelter 1986, Petersen et al. 1987, Wolanski 2001).

Similarly, sediment loads to the Great Barrier Reef region have increased severalfold and control of runoff from poor land-use practices and agricultural development are critical management problems for this region (Furnas and Mitchell 2001, Wolanski 2001, Wooldridge and Done 2009). Runoff from residential and agricultural developments has also been implicated in the decline of the seascape of south Florida (Ogden et al. 1994). The coral reef of Cahuita National Park, on the Atlantic coast of Costa Rica, has been virtually destroyed by decades of siltation delivered by rivers originating in highlands destabilized by agriculture (Cortes and Risk 1985).

In spite of a solid understanding of the inimical impact of sediments and poor water quality on coral reefs, there are few quantitative studies conclusively linking sedimentation with coral reef decline (Rogers et al. 2008). This is important as recent studies have suggested that improved water quality can increase the resilience of reefs in an era of climate change (Wooldridge and Done 2009).

Buffers and sinks

At the land–sea boundary, upland coastal forests, mangroves, and seagrasses act as buffers that intercept freshwater runoff, stabilize salinity, trap and bind sediments, and reduce pollutants, promoting the growth of coral reefs offshore (Wiebe 1987). Damage to and destruction of coastal forests, mangroves, and seagrass beds with development can lead to a cascading decline in coastal water quality, finally lowering water quality on offshore coral reefs (Duke and Wolanski 2001, Rogers et al. 2008) (Figure 14.3).

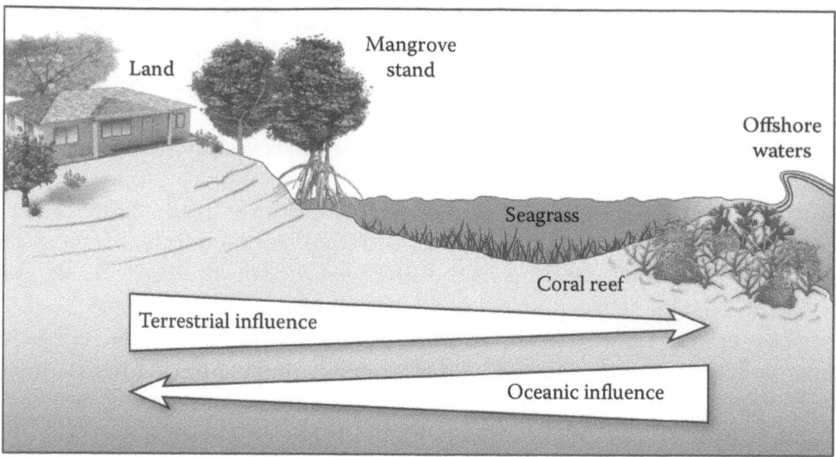

Figure 14.3 The coastal seascape showing the relative position of coastal ecosystems. Offshore reefs are buffered from the inimical impact of terrestrial influence by coastal forests, mangroves, and seagrass beds. Inshore areas are protected from the impacts of storms, waves, and tides by offshore coral reefs.

Periodic events such as hurricanes and heavy rains flush great quantities of accumulated material from mangrove and seagrass sinks. These events "reset" the sinks and have a potentially long-term effect on downstream ecosystems. In August 1992, Hurricane Andrew swept over the Florida peninsula, defoliating and eventually killing approximately 150 km^2 of mangrove trees (Ogden 1992). Suspension of organic-rich sediments within the mangroves caused large fish kills from anoxia and the long-term nutrient release associated with submerged mangrove detritus may have stimulated subsequent phytoplankton blooms. Phytoplankton and benthic algal blooms are also associated with runoff water and groundwater seepage (Marsh 1977, Johannes 1980).

Offshore healthy coral reefs buffer nearshore ecosystems from the destructive, erosive effects of storm waves. Reports from South Asia after the great tsunami of December 2004 indicated that regions which were protected by offshore coral reefs, intact coastal mangrove forests, and beach dunes suffered much less destruction (Pearson 2005).

Climate change and sea-level rise

There is little question that increases in atmospheric CO_2 from the burning of fossil fuels are causing sweeping environmental changes across the planet (IPCC 2013). Atmospheric and ocean temperatures are increasing and this is linked to increasing rates of sea-level rise through the thermal expansion of the oceans and the melting of glaciers. Seasonal abnormally high seawater temperatures are firmly linked to coral bleaching and possibly diseases that have decimated corals across the world. A more insidious and longer-term consequence of increasing CO_2 is ocean acidification, which has already slightly lowered ocean pH and may ultimately change the carbonate balance in seawater with disastrous consequences for corals, mollusks, and other marine organisms with carbonate shells (Doney et al. 2009). A review of these impacts is beyond the scope of this paper.

Sea-level rise has been a key architect of the physiography of coastal ecosystems over the past 10,000 years and, as a result of climate change, is now having an on-going impact on coastal ecosystems. The buffering capacity of coral reefs to protect coastal ecosystems is threatened by the projected rate of sea-level rise under scenarios of global warming. In the Florida Keys, the seaward movement of water inimical to coral reef growth from the shallow Florida Bay caused the demise of reefs in the central coral reef tract about 3000 years ago. Reefs to the north and south, protected from the influence of bay water, continued to thrive (Ginsburg and Shinn 1994, Neumann and Macintyre 1985).

Given an emissions "business as usual" estimate of global average sea-level rise of 52–98 cm by 2100, the vertical carbonate accretion rates of coral reefs are likely insufficient to keep up, disturbing the coastal protection function of coral reefs. Even with aggressive emissions reductions, a sea-level rise of 28–61 cm is predicted. Seagrass and mangrove communities will be eroded and moved shoreward, becoming less effective buffers, releasing nutrients, turbidity, and sediments further slowing offshore coral reef growth rate (Buddemeier and Smith 1988, IPCC 2013).

Florida Bay exerts a controlling influence on coral reef development in south Florida (Precht and Miller 2007). Decades of water management in the Everglades, and the draining of wetlands for development and agriculture, have severely decreased the amount of freshwater flowing into Florida Bay (McIvor et al. 1994). During hot, calm summer periods, salinities over twice that of normal seawater (70 parts per thousand) have been observed in the bay. In the summer, hot, heavy high-salinity water (over the coral bleaching threshold of 32°C) has been observed over the reef tract. The severe winters of 1976–1977 and 2009–2010 chilled Florida Bay water to levels lethal to corals, other invertebrates, and fishes. Intrusion of this water over the coral reefs of the Keys and the Dry Tortugas in 1977 killed many cold-susceptible corals, particularly elkhorn and staghorn corals (*Acropora* spp.) (Porter et al. 1982). Incursions over the Florida reef tract of potentially nutrient- and toxin-laden, low-salinity water carried by ocean currents from floods in the upper Mississippi River are periodically observed and may also negatively impact coral growth and reproduction (Ogden et al. 1994).

An overview of the future of coastal seascape management and conservation in the wider Caribbean

The connectivity and interdependence of coral reefs and seagrass and mangrove ecosystems over broad geographic scales documented in this chapter, makes a compelling case that change is needed in our approach to marine resources management and conservation of coastal ecosystems upon which our lives and economies depend. Nowhere is this more evident than in the wider Caribbean area, defined as the insular and coastal states and territories on the Caribbean Sea and Gulf of Mexico as well as adjacent waters of the Atlantic Ocean. The decline of coastal ecosystems has been in progress for at least several hundred years, keeping pace with the explosion of the human population, which began in the mid-nineteenth and early twentieth centuries (Jackson et al. 2001). The recent period of rapid decline was first noticed about five decades ago, coincident with the development of scuba technology and the expansion of field research on tropical coastal marine ecosystems. In spite of major long-term efforts in management and conservation and the attention of government and nongovernmental organizations, we have failed to arrest the decline (Pandolfi et al. 2005).

Marine protected areas

Supported by long-term observations and monitoring and the experience of local people, Caribbean pioneers of marine conservation implemented MPAs and communicated widely the ecological, political, and social considerations in their design. Starting from small beginnings, MPAs expanded across the region as the Great Barrier Reef Marine Park became the global icon of tropical marine management (Kelleher et al. 1995), followed in the United States in 1991 by the Florida Keys National Marine Sanctuary (National Marine Sanctuary Program 2005). A recent summary lists approximately 700 MPAs in Latin America and the Caribbean, however many are so-called "paper parks" with little or no protection or management (Guarderas et al. 2008).

Early studies of a class of MPAs called "no-take marine reserves," prohibiting all extractive use, documented the relatively rapid increase in size and number of demersal fishes after fishing prohibition. The Saba and Bonaire Marine Parks in the Dutch Antilles and the Hol Chan Marine Reserve in Belize were among the first and were influential. As elsewhere in the world, these reserves developed more and larger fishes after approximately 3–5 years of protection (Halpern 2003). However, longer-term studies of corals, for example, have shown that declines continue through a failure to implement kilometer-scale, no-take protection (Toth et al. forthcoming). Perhaps not surprisingly, fish, corals, and other major groups each react differently to no-take protection depending on the geographic scale of the ecological processes, particularly reproduction and recruitment, that drive their dynamics.

MPAs are necessary but insufficient

The implementation of numerous, small MPAs, even no-take marine reserves, has not been sufficient to protect coastal ecosystems from continued decline (Allison et al. 1998). Managers and scientists are approaching this problem in two different ways. Networks of existing MPAs have been proposed to unify management efforts and increase their virtual size. For example, in the United States, the "Islands in the Stream" concept linking seven existing MPAs in the Gulf of Mexico was proposed but not implemented at the end of the Bush administration in 2008 by National Oceanic and Atmospheric Administration (NOAA) Marine Sanctuaries. In 2009, NOAA's MPA Center (www.marineprotectedareas.noaa.gov) announced a national network of MPAs made up of 225 existing federal, state, and territorial protected areas, including Caribbean sites, but the work of making this a network truly representative of the nation's marine biodiversity, resources, and cultural heritage has just begun. A second approach has been to employ ecosystem-based management (EBM), acknowledging that more comprehensive ocean governance is needed that encompasses the geographic scales of marine biodiversity, human impacts, and ecological processes that sustain coral reefs and associated ecosystems (McLeod and Leslie 2009).

Ecosystem-based management

EBM is designed to support the long-term, sustainable use of marine resources and delivery of other ecosystem services while maintaining the resilience of ecosystems to natural and human disturbances. It is both a process for the development of policy actions as well as a framework for policy goals and objectives. EBM is place based and involves the people whose lives and economies are dependent upon the ecosystem services of that place. The size of the system under EBM may vary, but it is biologically, physically, and socially bound together. Most importantly, it is based on the latest scientific knowledge and supports management actions that are "information rich" to constantly increase understanding of ecosystem functioning (Rivera-Monroy et al. 2004, Spaulding et al. 2007, McLeod and Leslie 2009).

Marine spatial planning—an assessment and assembly of existing spatial scientific data and information in global information system (GIS) formats—is a key tool of EBM. It utilizes data on, for example, bathymetry, oceanography, sediment types, the distribution of key organisms including commercial species, key benthic habitats, and areas with high biological diversity or other unique or highly valued features. Human uses are also mapped, including fishing zones, aquaculture sites, shipping lanes, pipelines and cables, mineral leases, and protected areas, to name a few. The sources of this information include publications, databases, and local and traditional knowledge. The public meetings required to collect all of this information play an important role in building a sense of stewardship and a political constituency through an inclusive process (Crowder et al. 2006).

The GIS overlay maps show areas where information is abundant and areas where there are significant gaps. Continually updated maps from spatially organized databases allow assessments of changes and provide parameters for models to help predict the future under different scenarios of management and environmental change. The importance of maps in engaging the stakeholders, illuminating complex use problems, and informing solutions cannot be overemphasized (Carollo et al. 2009, Ogden 2010).

Similar to land-use planning, marine spatial planning concentrates on places of importance to human societies and provides a mapping and analysis framework for visualizing the finite nature of resources and the need for governance of human enterprises on the ocean through zoning, regulations, and permits (Day 2002, Crowder et al. 2006, Agardy 2010, Sanchirico et al. 2010). Young et al. (2007) outline four key principles for the implementation of EBM: (1) create governance policies that minimize geographic and temporal scale mismatches between biophysical systems and socioeconomic activities; (2) develop policies that recognize multiple uses of ocean areas and can mediate conflicts; (3) ensure that all interested parties have a voice in decision making from the beginning; and (4) monitor the results of management policies and change them as necessary as understanding of the dynamics of the ecosystem advances.

Marine spatial planning is an idea whose time has come (Norse 2010). It originated during the planning effort that established the Great Barrier Reef Marine Park in 1972. It has been used in Europe, notably in the extensively exploited North Sea, in the Mediterranean, and in several locations in Asia to balance economic and environmental objectives in ocean resource use (Douvere 2008). A recent step-by-step guide to marine spatial planning (http://www.unesco-ioc-marinesp.be/) highlights its importance and presents clear protocols (Ehler and Douvere 2009). This exemplary work shows that while the tools and approaches of marine spatial planning can be outlined, each location is unique in terms of the engagement of the stakeholders and the local and national political apparatus.

EBM and marine spatial planning may help to bring planning, clarity, and order to proposed human activities in the ocean. In many countries the ocean is a traditional commons, unevenly governed by sectors through local and national agencies with overlapping and conflicting legal mandates (Crowder et al. 2006). The U.S. Interagency Ocean Policy Task Force was charged by Executive Order in 2007 to develop a framework for EBM and marine spatial planning. Carrying this work further, the National Research Council (2013) released the National Ocean Policy Implementation Plan. Pending input from stakeholders and the outcome of the political process, there is now for the first time a roadmap toward a national ocean policy.

Bringing planning and order to managing the marine environment can be a contentious process, attracting strong opposition. However, the obvious need for more comprehensive governance of the oceans and the scientific and comprehensive approach of EBM may help us overcome short-term obstacles.

The Caribbean large marine ecosystem

Sherman (1994) defined a large marine ecosystem (LME) as a large region of the coastal ocean of 200,000 km^2 or larger characterized by distinct bathymetry, oceanography, and interdependent human and natural populations. There is abundant scientific evidence that the wider Caribbean functions as an LME (Sherman et al. 2005, Spaulding et al. 2007). The ocean currents of the wider Caribbean connect ecosystems over large areas through the planktonic transport of the larvae of many organisms (Cowen et al. 2000, Baums et al. 2006) as well as pollution from major population and agricultural centers. Among the best evidence for this connectivity is the startling mass mortality of the sea urchin *Diadema antillarum* in 1983–1984 caused by a presumed pathogen that spread throughout the whole region within a year, killing >90% of one of the most common invertebrates on coral reefs (Lessios et al. 1984). The spread of white band disease of acroporid corals was slower but similarly relentless throughout the region (Gladfelter 1982). These features have stimulated Caribbean countries to cooperate in the management and governance of marine resources and transboundary impacts and problems (Mahon et al. 2009).

In the United Nations, the World Bank (1991) defines governance as the manner in which power is exercised in the management of natural, economic, and social resources in development. There are several governance projects in the Caribbean that serve as examples. The Meso-American Barrier Reef System project with the support of the World Wildlife Fund and others pioneered a spatial planning approach to define biophysical characteristics, human uses, and potential conservation management measures within a four-country region of the western Caribbean (McField et al. 2002). The planning process was inclusive and thorough, but the social and environmental complexity of the region hampered implementation of internationally coordinated ecosystem-based management and governance. Nevertheless, the Meso-American Barrier Reef System serves as a heuristic example of how to initiate a marine planning process to serve the goal of large geographic scale marine governance.

The Caribbean Large Marine Ecosystem Project (CLME) http://www.clmeproject.org based at the Secretariat of the Intergovernmental Oceanographic Commission for the Caribbean (IOCARIBE) in Cartagena, Colombia, is a developing example of a multilevel governance network linking regional intergovernmental initiatives together with the cooperation of the Association of Caribbean States. The CLME approach has been top-down so far, concentrating on organizing Caribbean countries to develop conceptual designs including political considerations to approach comprehensive governance of the Caribbean LME. The CLME will ultimately use marine spatial planning to define international management concerns and identify use areas to implement governance (Fanning et al. 2007, Mahon et al. 2009).

Finally, there is developing interest in the public trust doctrine, applied currently in the United States only in state waters (shoreline to 3 nmi), which mandates that ocean resources be managed in the best interests of the citizens. Based on science and developing public concern, Turnipseed et al. (2009) advocated extension of the public trust doctrine to the exclusive economic zone. They stated that this will gather ocean assets under the same administration, avoid duplication and overlap, and finally help answer the important and vexing question: For whom and for what purpose should our ocean resources be managed?

Acknowledgments

JCO thanks Professor Charles P. McRoy of the University of Alaska, the visionary leader of the 1970–1980s Seagrass Ecosystem Study of the NSF International Decade of Ocean Exploration. He is grateful to his many Caribbean colleagues who are working toward comprehensive, ecosystem-based governance of the marine environment of the wider Caribbean. CCM was supported by the U.S. Geological Survey (USGS) Ecosystems Program. We thank Betsy Boynton of the USGS for graphics expertise. Any use of trade, product, or firm names is for descriptive purposes only and does not imply endorsement by the U.S. Government.

References

Agardy T. 2010. Ocean zoning: Making marine management more effective. Washington, DC: Earthscan.

Allgeier JE, Yeager LA, Layman CA. 2013. Consumers regulate nutrient limitation regimes and primary production in seagrass ecosystems. Ecology. 94:521–529.

Allison GW, Lubchenco J, Carr MH. 1998. Marine reserves are necessary but not sufficient for marine conservation. Ecol Appl. 8(1):S79–S92.

Baums IB, Miller MW, Hellberg ME. 2006. Geographic variation in clonal structure of a reef-building Caribbean coral, *Acropora palmata*. Ecol Monogr. 76:503–519.

Beck MW, Heck KL Jr, Able KW, Childers DL, Eggleston DB, Gillanders BM, Halpern B, Hays CG, Hoshino K, Minello TJ, et al. 2001. The identification, conservation, and management of estuarine and marine nurseries for fish and invertebrates. BioScience. 51:633–641.

Beets J, Muehlstein L, Haught K, Schmitges H. 2003. Habitat connectivity in coastal environments: Patterns and movements of Caribbean coral reef fishes with emphasis on blue striped grunt, *Haemulon sciurus*. Gulf Carib Res. 14:29–42.

Birkeland C. 1987. Nutrient availability as a major determinant of differences among coastal hard-bottom communities in different regions of the tropics. *In:* Birkeland C, editor. Comparison between Atlantic and Pacific tropical marine coastal ecosystems: Community structure, ecological processes, and productivity. UNESCO Reports in Marine Sciences 46. Norwich, UK: UNESCO. p. 45–97.

Blaber SJM. 2000. Tropical estuarine fishes. Ecology, exploitation and conservation. Fish and Aquatic Resources Series 7. Oxford: Blackwell Science.

Blaber SJM, Brewer DT, Salini JP. 1989. Species composition and biomasses of fishes in different habitats of a tropical Northern Australian estuary: Their occurrence in the adjoining sea and estuarine dependence. Estuar Coast Shelf Sci. 29:509–531.

Bouillon S, Connolly RM. 2009. Carbon exchange among tropical coastal ecosystems. *In*: Nagelkerken I, editor. Ecological connectivity among tropical coastal ecosystems. Dordrecht: Springer. p. 45–70.

Bouillon S, Connolly RM, Lee SY. 2008. Organic matter exchange and cycling in mangrove ecosystems: Recent insights from stable isotope studies. J Sea Res. 59:44–58.

Bouillon S, Koedam N, Raman AV, Dehairs F. 2002. Primary producers sustaining macro-invertebrate communities in intertidal mangrove forests. Oecologia. 130:441–448.

Bouillon ST, Moens ST, Overmeer I, Koedam N, Dehairs F. 2004. Resource utilization patterns of epifauna from mangrove forests with contrasting inputs of local versus imported organic matter. Mar Ecol Prog Ser. 278:77–88.

Buddemeier RW, Smith SV. 1988. Coral reef growth in an era of rapidly rising sea levels: Predictions and suggestions for long-term research. Coral Reefs. 7:51–56.

Burkepile DE, Allgeier JE, Shantz AA, Pritchard CE, Lemoine NP, Bhatti LH, Layman CA. 2013. Nutrient supply from fishes facilitates macroalgae and suppresses corals in a Caribbean coral reef ecosystem. Sci Rep. 3:1493.

Carollo C, Reed DJ, Ogden JC, Palandro D. 2009. The importance of data discovery and management in advancing ecosystem-based management. Mar Policy. 33:651–653.

Childers DL, Day JW Jr, McKellar HN Jr. 2000. Twenty more years of marsh and estuarine flux studies: Revisiting Nixon (1980). *In*: Weinstein MP, Kreeger DA, editors. Concepts and controversies in tidal marsh ecology. Boston: Kluwer Academic Publishers. p. 391–423.

Cocheret de la Morinière E, Nagelkerken I, van der Meij H, van der Velde G. 2004. What attracts juvenile coral reef fish to mangroves: Habitat complexity or shade? Mar Biol. 144:139–145.

Cocheret de la Morinière E, Pollux BJA, Nagelkerken I, van der Velde G. 2002. Post-settlement life cycle migration patterns and habitat preference of coral reef fish that use seagrass and mangrove habitats as nurseries. Estuar Coast Shelf Sci. 55:309–321.

Cortes J, Risk MJ. 1985. A reef under siltation stress: Cahuita, Costa Rica. Bull Mar Sci. 36:339–356.

Cowen RK, Kamazima M, Lwiza M, Sponaugle S, Paris CB, Olson DB. 2000. Connectivity of marine populations: Open or closed? Science. 287:857–859.

Crowder LB, Osherenko G, Young OR, Airame S, Norse EA, Baron N, Day JC, Douvere F, Ehler CN, Halpern BS, et al. 2006. Resolving mismatches in U.S. ocean governance. Science. 313:617–618.

Dahlgren CP, Eggleston DB. 2000. Ecological processes underlying ontogenetic habitat shifts in a coral reef fish. Ecology. 81:2227–2240.

Day JC. 2002. Zoning: Lessons from the Great Barrier Reef Marine Park. Ocean Coast Manage. 45:139–156.

Demopoulos AWJ, Cormier N, Ewel KC, Fry B. 2008. Use of multiple chemical tracers to define habitat use of Indo-Pacific mangrove crab, *Scylla serrata* (Decapoda: Portunidae). Estuar Coast. 31:371–381.

Doney SC, Fabry VJ, Feely RA, Klepas JA. 2009. Ocean acidfication: the other CO_2 problem. Ann Rev Mar Sci. 1:169–192.

Douvere F. 2008. The importance of marine spatial planning in advancing ecosystem-based sea use management. Mar Policy. 32:762–771.

Duke NC, Wolanski E. 2001. Muddy coastal waters and depleted mangrove coastlines: Depleted seagrass and coral reefs. *In*: Wolanski E, editor. Oceanographic processes of coral reefs: Physical and biological links in the Great Barrier Reef. Boca Raton: CRC Press. p. 53–76.

Ehler C, Douvere F. 2009. Marine spatial planning: A set-by-step approach toward ecosystem-based management. Intergovernmental Oceanographic Commission and Man and the Biosphere Program. IOC Manual and Guides No. 53, ICAM Dossier No. 6. Paris: UNESCO.

Elsdon TS, Wells BK, Campana SE, Gillanders BM, Jones CM, Limburg KE, Secor DE, Thorrold SR, Walther BD. 2008. Otolith chemistry to describe movements and life-history measurements of fishes: Hypotheses, assumptions, limitations, and inferences using five methods. Oceanogr Mar Biol Ann Rev. 46:297–330.

Fanning L, Mahon R, McConney P, Angulo J, Burrows F, Chakalall B, Gil D, Haughton M, Heileman S, Martinez S, et al. 2007. A large marine ecosystem governance framework. Mar Policy. 31:434–443.

Forward RB, Tankersley RA. 2001. Selective tidal-stream transport of marine animals. Oceanogr Mar Biol. 39:305–353.

Fry B, Ewel K. 2003. Using stable isotopes in mangrove fisheries research: A review and outlook. Isot Environ Health Stud. 39:191–196.

Fry B, Mumford PL, Robblee MB. 1999. Stable isotope studies of pink shrimp (*Farfantepenaeus duorarum* Burkenroad) migrations on the Southwestern Florida shelf. Bull Mar Sci. 65:419–430.

Fry B, Scalan R, Winters J, Parker P. 1982. Sulphur uptake by salt grasses, mangroves, and seagrasses in anaerobic sediments. Geochimica et Cosmochimica Acta. 46:1121–1124.

Fry B, Sherr EB. 1984. ^{13}C measurements as indicators of carbon flow in marine and freshwater ecosystems. Contrib Mar Sci. 27:13–47.

Furnas M, Mitchell A. 2001. Runoff of terrestrial sediment and nutrients into the Great Barrier Reef World Heritage Area. *In:* Wolanski E, editor. 2001. Oceanographic processes of coral reefs: Physical and biological links in the Great Barrier Reef. Boca Raton: CRC Press. p. 37–52.

Ginsburg RA, Shinn EA. 1994. Preferential distribution of reefs in the Florida reef tract: The past is a key to the present. *In:* Ginsburg RN, editor. Proceedings of the colloquium on global aspects of coral reefs: Health, hazards, and history. Rosenstiel School of Marine and Atmospheric Science. Miami: University of Miami. p. 21–26.

Gladfelter W. 1982. White-band disease in *Acropora palmata*: Implications for the structure and growth of shallow reefs. Bull Mar Sci. 32:639–643.

Granek EF, Compton JE, Phillips DL. 2009. Mangrove-exported nutrient incorporation by sessile coral reef invertebrates. Ecosystems. 12:462–472.

Grol MGG, Dorenbosch M, Kokkelmans EMG, Nagelkerken I. 2008. Mangroves and seagrass beds do not enhance growth of early juveniles of a coral reef fish. Mar Ecol Prog Ser. 366:137–146.

Grol MGG, Nagelkerken I, Rypel AL, Layman CA. 2011. Simple ecological trade-offs give rise to emergent cross-ecosystem distributions of a coral reef fish. Oecologia. 165:79–88.

Guarderas AP, Hacker SD, Lubchenco J. 2008. Current status of marine protected areas in Latin America and the Caribbean. Conserv Biol. 22:1630–1640.

Halpern BS. 2003. The impact of marine reserves: Do reserves work and does reserve size matter? Ecol Appl. 13:S117–S137.

Huijbers CM, Nagelkerken I, Debrot AO, Jongejans E. 2013. Geographic coupling of juvenile and adult habitat shapes spatial population dynamics of a coral reef fish. Ecology. 94:1859–1870.

Hyndes G, Nagelkerken I, McLeod R, Connolly R, Lavery P, Vanderklift MA. 2014. Mechanisms and ecological role of carbon transfer within coastal landscapes. Biol Rev. 89:232–254.

Igulu NM, Nagelkerken I, van der Velde G, Mgaya YD. 2013. Mangrove fish production is largely fuelled by external food sources: A stable isotope analysis of fishes at the individual, species, and community levels from across the globe. Ecosystems. 16:1336–1352.

IPCC. (Intergovernmental Panel on Climate Change) 2013. Climate Change 2013: The Physical Science Basis. Contribution of Working Group I to the Fifth Assessment Report of the Intergovernmental Panel on Climate Change [Stocker, T.F., D. Qin, G.-K. Plattner, M. Tignor, S.K. Allen, J. Boschung, A. Nauels, Y. Xia, V. Bex and P.M. Midgley (eds.)]. Cambridge University Press, Cambridge, United Kingdom. Available from https://www.ipcc.ch/report/ar5/wg1/ via the Internet.

Jackson JBC, Kirby MX, Berger WH, Bjorndal KA, Botsford LW, Bourque BJ, Bradbury RH, Cooke R, Erlandson J, Estes JA, et al. 2001. Historical overfishing and the recent collapse of coastal ecosystems. Science. 293(5530):629–638.

Johannes RE. 1980. The ecological significance of the submarine discharge of groundwater. Mar Ecol Prog Ser. 3:365–373.

Kathiresan K, Bingham BL. 2001. Biology of mangroves and mangrove ecosystems. Adv Mar Biol. 40:81–251.

Kelleher GC, Bleakley C, Wells S. 1995. A global representative system of marine protected areas. Volume II: Wider Caribbean, West Africa and South Atlantic. Great Barrier Reef Marine Park Authority, the World Bank, and The World Conservation Union (IUCN). Washington, DC: World Bank.

Kieckbusch DK, Koch MS, Serafy JE. 2004. Trophic linkages among primary producers and consumers in fringing mangroves of subtropical lagoons. Bull Mar Sci. 74:271–285.

Kimirei IA, Nagelkerken I, Mgaya YD, Huijbers CM. 2013a. The mangrove nursery paradigm revisited: Otolith stable isotopes support nursery-to-reef movements by Indo-Pacific fishes. PLoS ONE. 8(6):e66320.

Kimirei IA, Nagelkerken I, Trommelen M, Blankers P, van Hoytema N, Hoeijmakers D, Huijbers CM, Mgaya YD, Rypel AL. 2013b. What drives ontogenetic niche shifts of fishes in coral reef ecosystems? Ecosystems. 16:783–796.

Kneib RT. 1997. The role of tidal marshes in the ecology of estuarine nekton. Oceanogr Mar Biol Ann Rev. 35:163–220.

Kon K, Kurokura H, Hayashizaki K. 2007. Role of microhabitats in food webs of benthic communities in a mangrove forest. Mar Ecol Prog Ser. 340:55–62.

Kristensen E, Bouillon S, Dittmar T, Marchand C. 2008. Organic carbon dynamics in mangrove ecosystems: A review. Aquat Bot. 89:201–219.

Krumme U. 2009. Diel and tidal movements by fish and decapods linking tropical coastal ecosystems. *In:* Nagelkerken I, editor. Ecological connectivity among tropical coastal ecosystems. Dordrecht: Springer, p. 271–324.

Laegdsgaard P, Johnson C. 2001. Why do juvenile fish utilize mangrove habitats? J Exp Mar Biol Ecol. 257:229–253.

Lee RY. 2006. Primary production, nitrogen cycling and the ecosystem role of mangrove microbial mats on Twin Cays, Belize. PhD thesis, University of Georgia, Athens, GA.

Lee S. 2005. Exchange of organic matter and nutrients between mangroves and estuaries: Myths, methodological issues and missing links. Int J Ecol Environ Sci. 31:163–176.

Lenanton RCJ, Potter IC. 1987. Contribution of estuaries to commercial fisheries in temperate Western Australia and the concept of estuarine dependence. Estuaries. 10:28–35.

Lessios HA, Robertson R, Ross D, Cubit JD. 1984. Spread of *Diadema* mass mortality through the Caribbean. Science. 226:335–337.

Ley JA, McIvor CC. 2002. Linkages between estuarine and reef assemblages: Enhancement by the presence of well-developed mangrove shorelines. *In:* Porter JW, Porter KG, editors. The Everglades, Florida Bay, and coral reefs of the Florida Keys: An ecosystem sourcebook. Boca Raton, FL: CRC Press. p. 539–562.

Lindeman KC, Pugliese R, Waugh GT, Ault JS. 2000. Developmental patterns within a multispecies reef fishery: Management applications for essential fish habitats and protected areas. Bull Mar Sci. 66:929–956.

Loneragan NR, Bunn SE, Kellaway DM. 1997. Are mangroves and seagrasses sources of organic carbon for penaeid prawns in a tropical Australian estuary? A multiple stable-isotope study. Mar Biol. 130:289–300.

Lugendo BR, Nagelkerken I, Kruitwagen G, van der Velde G, Mgaya YD. 2007. Relative importance of mangroves as feeding habitats for fishes: A comparison between mangrove habitats with different settings. Bull Mar Sci. 80:497–512.

Mahon R, Fanning L, McConney P. 2009. A governance perspective on the large marine ecosystem approach. Mar Policy. 33:317–321.

Marguillier S, van der Velde G, Dehairs F, Hemminga MA, Rajagopal S. 1997. Trophic relationships in an interlinked mangrove-seagrass ecosystem as traced by $\delta^{13}C$ and $\delta^{15}N$. Mar Eco Prog Ser. 151:115–121.

Marsh JA Jr. 1977. Terrestrial inputs of nitrogen and phosphorous on fringing reefs of Guam. Proc Third Int Coral Reef Symp Miami Biol. I:331–336.

Mateo I, Durbin EG, Appeldoorn RS, Adams AJ, Juanes F, Kingsley R, Swart P, Durant D. 2010. The role of mangroves as nurseries for French grunt *Haemulon flavolineatum* and schoolmaster *Lutjanus apodus* assessed by otolith elemental fingerprints. Mar Ecol Prog Ser. 402:197–212.

McCutchan JJ Jr, Lewis WJ Jr, Kendall C, McGrath C. 2003. Variation in trophic shift for stable isotope ratios of carbon, nitrogen, and sulfur. Oikos. 102:378–390.

McFarland WN, Ogden JC, Lythgoe JN. 1979. The influence of light on the twilight migrations of grunts. Environ Biol Fish. 4:9–22.

McField M, Kramer PA, Kramer PR, editors. 2002. Ecoregional conservation planning for the Mesoamerican Coral Reef. Washington, DC: World Wildlife Fund.

McIvor CC, Ley JA, Bjork RD. 1994. Changes in freshwater inflow from the Everglades to Florida Bay including effects on biota and biotic processes: A review. *In:* Davis SM, Ogden JC, editors. Everglades: The ecosystem and its restoration. Deray Beach, FL: St. Lucie Press. p. 117–146.

McLeod K, Leslie H, editors. 2009. Ecosystem-based management for the oceans. Washington, DC: Island Press.

McMahon KW, Berumen ML, Thorrold SR. 2012. Linking habitat mosaics and connectivity in a coral reef seascape. Proc Natl Acad Sci U S A. 109:15372–15376.

Meyer JL, Schultz ET, Helfman GS. 1983. Fish schools: An asset to corals. Science. 220:1047–1049.

Mumby PJ, Edwards AJ, Arias-Gonzalez JE, Lindeman KC, Blackwell PG, Gall A, Gorczynska MI, Harborne AR, Pescod CL, Renken H, et al. 2004. Mangroves enhance the biomass of coral reef fish communities in the Caribbean. Nature. 427:533–536.

Mumby PJ, Hastings A. 2008. The impact of ecosystem connectivity on coral reef resilience. J Appl Ecol. 45:854–862.

Nagelkerken I. 2007. Are non-estuarine mangroves connected to coral reefs through fish migration? Bull Mar Sci. 80:595–607.

Nagelkerken I. 2009. Evaluation of nursery function of mangroves and seagrass beds for tropical decapods and reef fishes: Patterns and underlying mechanisms. *In*: Nagelkerken I, editor. Ecological connectivity among tropical coastal ecosystems. Dordrecht: Springer. p. 357–399.

Nagelkerken I, Bothwell J, Nemeth RS, Pitt JM, van der Velde G. 2008. Interlinkage between Caribbean coral reefs and seagrass beds through feeding migrations by grunts (Haemulidae) depends on habitat accessibility. Mar Ecol Prog Ser. 368:155–164.

Nagelkerken I, Dorenbosch M, Verberk WCEP, Cocheret de la Morinière E, van der Velde G. 2000a. Day-night shifts of fishes between shallow-water biotopes of a Caribbean Bay, with emphasis on the nocturnal feeding of Haemulidae and Lutjanidae. Mar Ecol Prog Ser. 194:55–64.

Nagelkerken I, Dorenbosch M, Verberk WCEP, Cocheret de la Morinière E, van der Velde G. 2000b. Importance of shallow-water biotopes of a Caribbean Bay for juvenile coral reef fishes: Patterns in biotope association, community structure and spatial distribution. Mar Ecol Prog Ser. 202:175–192.

Nagelkerken I, Grol MGG, Mumby PJ. 2012. Effects of marine reserves versus nursery habitat availability on structure of reef fish communities. PLoS ONE. 7:e36906.

Nagelkerken I, Kleijnen S, Klop T, van den Brand RACJ, Cocheret de la Morinière E, van der Velde G. 2001. Dependence of Caribbean reef fishes on mangroves and seagrass beds as nursery habitats: A comparison of fish faunas between bays with and without mangroves/seagrass beds. Mar Ecol Prog Ser. 214:225–235.

Nagelkerken I, Roberts CM, van der Velde G, Dorenbosch M, van Riel MC, Cocheret de la Morinière E, Nienhuis PH. 2002. How important are mangroves and seagrass beds for coral-reef fish? The nursery hypothesis tested on an island scale. Mar Ecol Prog Ser. 244:299–305.

Nagelkerken I, van der Velde G. 2002. Do non-estuarine mangroves harbour higher densities of juvenile fish than adjacent shallow-water and coral reef habitats in Curaçao (Netherlands Antilles)? Mar Ecol Prog Ser. 245:191–204.

Nagelkerken I, van der Velde G. 2003. Connectivity between coastal habitats of two oceanic Caribbean Islands as inferred from ontogenetic shifts by coral reef fishes. Gulf Carib Res. 14:43–59.

Nagelkerken I, van der Velde G. 2004. Are Caribbean mangroves important feeding grounds for juvenile reef fish from adjacent seagrass beds? Mar Ecol Prog Ser. 274:143–151.

Nakamura Y, Horinouchi M, Shibuno T, Tanaka Y, Miyajima T, Koike I, Kurokura H, Sano M. 2008. Evidence of ontogenetic migration from mangroves to coral reefs by black-tail snapper *Lutjanus fulvus*: Stable isotope approach. Mar Ecol Prog Ser. 355:257–266.

National Marine Sanctuary Program. 2005. Florida Keys National Marine Sanctuary Draft Revised Management Plan. U.S. Department of Commerce, National Oceanic and Atmospheric Administration, National Ocean Service, National Marine Sanctuary Program. Washington, DC: National Oceanic and Atmospheric Administration.

National Research Council. 2013. National Ocean Policy Implementation Plan. Washington, DC: National Academy Press.

Neumann AC, Macintyre I. 1985. Reef response to sea level rise: Keep-up, catch-up, or give up. Proc 5th Int Coral Reef Symp Tahiti. 3:105–110.

Norse E. 2010. Ecosystem-based spatial planning and management of marine fisheries: Why and how? Bull Mar Sci. 86:179–195.

Ogden JC. 1988. The influence of adjacent systems on the structure and function of coral reefs (status review). *In*: Choat JH, Barnes D, Borowitzka MA, Coll JC, Davies PJ, Flood P, Hatcher BG, Hopley D, Hutchings PA, Kinsey D, et al. editors. Proceedings of the 6th International Coral Reef Symposium. Vol. 1. Townsville, Australia: Plenary Addresses and Status Reviews. p. 123–130.

Ogden JC. 1992. The impact of Hurricane Andrew on the ecosystems of South Florida. Cons Biol. 6:488–490.

Ogden JC. 2010. Marine spatial planning (MSP): A first step to ecosystem-based management (EBM) in the wider Caribbean. Revista de Biologia Tropical. 58(Suppl. 3):71–79.

Ogden JC, Brown RA, Salesky N. 1973. Grazing by the echinoid *Diadema antillarum* Philippi: Formation of halos around West Indian patch reefs. Science. 182:715–717.

Ogden JC, Buckman NS. 1973. Movements, foraging groups, and diurnal migrations of the striped parrotfish *Scarus croicensis* Bloch (Scaridae). Ecology. 54:589–596.

Ogden JC, Ehrlich PR. 1977. The behavior of heterotypic resting schools of juvenile grunts (Pomadasyidae). Mar Biol. 42:273–280.

Ogden JC, Gladfelter EH, editors. 1986. Caribbean coastal marine productivity. UNESCO Rep Mar Sci. 41:59.

Ogden JC, Porter JW, Smith NP, Szmant AM, Jaap WC, Forcucci D. 1994. A long-term interdisciplinary study of the Florida Keys seascape. Bull Mar Sci. 54:1059–1071.

Ogden JC, Zieman JC. 1977. Ecological aspects of coral reef-seagrass bed contacts in the Caribbean. Proc 3rd Int Coral Reef Symp. 1:377–382.

Olds AD, Albert S, Maxwell PS, Pitt KA, Connolly RM. 2013. Mangrove-reef connectivity promotes the effectiveness of marine reserves across the western Pacific. Global Ecol Biogeogr. 22:1040–1049.

Olds AD, Connolly RM, Pitt KA, Maxwell PS. 2012a. Habitat connectivity improves reserve performance. Conserv Lett. 5:56–63.

Olds AD, Pitt KA, Maxwell PS, Connolly RM. 2012b. Synergistic effects of reserves and connectivity on ecological resilience. J Appl Ecol. 49:1195–1203.

Pandolfi JM, Jackson JBC, Baron N, Bradbury RH, Guzman HM, Hughes TP, Kappel CV, Micheli F, Ogden JC, Possingham HP, et al. 2005. Are U.S. coral reefs on the slippery slope to slime? Science. 307:1725–1726.

Parrish JD. 1989. Fish communities of interacting shallow-water habitats in tropical oceanic regions. Mar Ecol Prog Ser. 58:143–160.

Pearson H. 2005. Scientists seek action to fix Asia's ravaged ecosystems. Nature. 433:94.

Petersen RC Jr, Madsen BL, Wilzbach MA, Magadza CHD, Paarlberg A, Kullberg A, Cummins KW. 1987. Stream management: Emerging global similarities. Ambio. 16:166–179.

Peterson BJ, Fry B. 1987. Stable isotopes in ecosystem studies. Ann Rev Ecol Syst. 18:293–320.

Pollux BJA, Verberk WCEP, Dorenbosch M, Cocheret de la Morinière E, Nagelkerken I, van der Velde G. 2007. Habitat selection during settlement of three Caribbean coral reef fishes: Indications for directed settlement to seagrass beds and mangroves. Limnol Oceanogr. 52:903–907.

Porter JW, Battey JF, Smith GJ. 1982. Perturbation and change in coral reef communities. Proc Natl Acad Sci U S A. 79:1678–1681.

Precht WF, Miller SL. 2007. Ecological shifts along the Florida reef tract: The past is a key to the future. *In:* Aronson RB, editor. Geological approaches to coral reef ecology. New York: Springer. p. 237–312.

Quinn TP, Ogden JC. 1984. Field evidence of compass orientation in migrating juvenile grunts (Haemulidae). J Exp Mar Biol Ecol. 81:181–192.

Randall JE. 1965. Grazing effect on sea grasses by herbivorous reef fishes in the West Indies. Ecology. 46:255–260.

Rivera-Monroy VH, Twilley RR, Bone D, Childers DL, Coronado-Molina C, Feller IC, Herrera-Silveira J, Jaffe R, Mancera E, Rejmankova E, et al. 2004. A conceptual framework to develop long-term ecological research and management objectives in the wider Caribbean region. Bioscience. 54:843–856.

Robertson AI, Alongi DM, Boto KG. 1992. Food chains and carbon fluxes. *In:* Robertson AI, Alongi DM, editors. Tropical mangrove ecosystems. Washington, DC: American Geophysical Union. p. 293–326.

Rogers CA, Miller J, Muller EM, Edmunds P, Nemeth R, Beets J, Friedlander AM, Smith TB, Boulon R, Jeffrey CFG, et al. 2008. Ecology of coral reefs in the U.S. Virgin Islands. *In:* Reigel BM, Dodge RE, editors. Coral reefs of the USA. Berlin: Springer. p. 303–373.

Rooker JR, Dennis GD. 1991. Diel, lunar and seasonal changes in a mangrove fish assemblage off southwestern Puerto Rico. Bull Mar Sci. 49:684–698.

Sanchirico JN, Eagle J, Palumbi S, Thompson BH Jr. 2010. Comprehensive planning, dominant-use zones, and user rights: A new era in ocean governance. Bull Mar Sci. 86:273–285.

Sheaves M, Molony B. 2000. Short-circuit in the mangrove food chain. Mar Ecol Prog Ser. 199:97–109.

Sherman K. 1994. Sustainability, biomass yields, and health of coastal ecosystems. Mar Ecol Prog Ser. 112:277–301.

Sherman KM, Sissenwine M, Christensen V, Duda A, Hempel G, Ibe C, Levin S, Lluch-Belda D, Matishov G, McGlade J, et al. 2005. A global movement toward an ecosystem approach to management of marine resources. Mar Ecol Prog Ser. 300:275–279.

Shulman MJ, Ogden JC. 1987. What controls tropical reef fish populations: Recruitment or benthic mortality? An example in the Caribbean reef fish *Haemulon flavolineatum*. Mar Ecol Prog Ser. 39:233–242.

Spaulding MD, Fox HE, Allen GR, Davidson N, Ferdana ZA, Finlayson M, Halpern BS, Jorge MA, Lombana A, Lourie SA, et al. 2007. Marine ecoregions of the world: A bioregionalization of coastal and shelf areas. Bioscience. 57:573–583.

Suchanek T, Williams SL, Ogden JC, Hubbard DK, Gill IP. 1985. Utilization of shallow-water seagrass detritus by Caribbean deep-sea macrofauna: $\delta^{13}C$ evidence. Deep Sea Res. 32:201-214.

Thollot P. 1992. Importance of mangroves for Pacific reef fish species, myth or reality? Proc 7th Int Coral Reef Symp. 2:934–941.

Torgerson T, Chivas AR. 1985. Terrestrial organic carbon in marine sediment: A preliminary balance for a mangrove environment derived from ^{13}C. Chem Geol. 53:379–390.

Toth LT, Aronson RB, Smith SR, Murdoch TJT, Ogden JC, Precht WF, van Woesik R. Forthcoming. Do no-take marine reserves benefit corals? 14 years of change and stasis on Florida's reefs. Coral Reefs.

Tue NT, Hamaoka H, Sogabe A, Quy TD, Nhuan MT, Omori K. 2012. Food sources of macro-invertebrates in an important mangrove ecosystem of Vietnam determined by dual stable isotope signatures. J Sea Res. 72:14–21.

Turnipseed M, Roady S, Sagarin R, Crowder LB. 2009. The silver anniversary of the United States' exclusive economic zone: Twenty-five years of ocean use and abuse, and the possibility of a blue water public trust doctrine. Ecol Law Quart. 36:1–70.

Twilley RR. 1985. The exchange of organic carbon in basin mangrove forests in a southwest Florida estuary. Estuar Coast Shelf Sci. 20:543–557.

Twilley RR, Snedaker SC, Yanez-Arancibia A, Medina E. 1996. Biodiversity and ecosystem processes in tropical estuaries: Perspectives of mangrove ecosystems. *In:* Mooney HA, Cushman JH, Medina E, Sala OE, Schulze ED, editors. Functional roles of biodiversity: A global perspective. New York: Wiley. p. 327–370.

Vaslet A, Phillips DL, France C, Feller IC, Baldwin CC. 2012. The relative importance of mangroves and seagrass beds as feeding areas for resident and transient fishes among different mangrove habitats in Florida and Belize: Evidence from dietary and stable-isotope analyses. J Exp Mar Biol Ecol. 434–435:81–93.

Verweij MC, Nagelkerken I, Hans I, Ruseler SM, Mason PRD. 2008. Seagrass nurseries contribute to coral reef fish populations. Limnol Oceanogr. 53:1540–1547.

West K, van Woesik R. 2001. Spatial and temporal variance of river discharge on Okinawa (Japan): Inferring the temporal impact on adjacent coral reefs. Mar Poll Bull. 42:864–872.

Whitfield AK. 1998. Biology and ecology of fishes in southern African estuaries. Grahamstown, South Africa: J.L.B. Smith Institute of Ichthyology.

Wiebe WJ. 1987. Nutrient pools and dynamics in tropical, marine coastal environments, with special reference to the Caribbean and Indo-Pacific regions. *In:* Birkeland C, editor. Comparison between Atlantic and Pacific tropical marine coastal ecosystems: Community structure, ecological processes, and productivity. Paris: UNESCO Reports in Marine Science. Vol. 46. p. 19–42.

Wolanski E, editor. 2001. Oceanographic processes of coral reefs: Physical and biological links in the Great Barrier Reef. Boca Raton: CRC Press.

Wooldridge SA, Done TJ. 2009. Improved water quality can ameliorate effects of climate change on corals. Ecol Appl. 19(6):1492–1499.

World Bank. 1991. Managing development: The governance dimension. Discussion paper. Washington, DC: World Bank.

Young OR, Osherenko G, Ekstrom J, Crowder LB, Ogden JC, Wilson JA, Day JC, Douvere F, Ehler CN, McLeod K, et al. 2007. Solving the crisis in ocean governance: Place-based management of marine ecosystems. Environment. 49(4):20–32.

chapter fifteen

Sustaining the interrelationships between corals and fisheries

Managing for the future

Stephen A. Bortone

Contents

Objectives of management

In the United States, the Magnuson-Stevens Act provides a mandate to manage U.S. fisheries at sustainable levels for the overall benefit of the nation. This act also serves as a signpost to the world that sustainable fisheries management is of paramount importance for all humankind. The exact level of sustainability is, however, a moving target. The maximum production resulting from the activities of fisheries depends on many factors. The most important of these factors is the carrying capacity of our environment. Thus actions that help maintain or even increase the carrying capacity of our fisheries would be of benefit—as long as the increase in carrying capacity does not result in the detriment of something else (e.g., destroying mangroves to accommodate more shrimp mariculture farms).

The Regional Fisheries Management Councils, established by the Magnuson-Stevens Act, that have substantial coral habitat within their jurisdictions (i.e., the South Atlantic, Caribbean, Gulf of Mexico, and Western Pacific Councils) are actively involved with issues related to coral health and fisheries. Moreover, these councils have taken some initial steps to address the larger issue of these interrelationships. The interconnectedness of their approaches mirrors the similar ecological principles that seem to be governing these interactions as we currently know them. Continued interactions among management organizations of like thinking will go far in furthering our perspectives.

Reiterating the coral/fisheries premise

Fortunately, the United States through the Magnuson-Stevens Act has established a process to ensure that future fisheries are sustainable. However, the very real possibility that current levels of production may be decreasing due to seemingly insurmountable and ever-increasing obstacles is problematic. It is also obvious that the world human population and its accompanying environmental stressors are increasing. Accommodating for these stressors on natural resources is often fraught with problems, but through insight and cooperation most can be overcome. Larger, long-term global issues associated with climate change loom, however, as an "unstoppable force" influencing the environmental health of the biosphere. The harbinger of the ever-increasing influence that this may have on our biosphere's future is the worsening health and condition of corals and coral reefs. Concomitantly, as go the corals, so too go the organisms dependent upon them.

What we know about these interrelationships

Gilbert (1973) reviewed the relationships between habitat and general distributions of tropical/warm temperate fishes in the southeastern United States and Caribbean. His conclusion was that many reef fish, while not true obligates to a specific habitat, could be grouped as island species or continental species. The former have affinities for coral reefs, while the latter have affinities for hard structures, but not necessarily coral. Examples of species found in association with continental land masses are the red snapper (*Lutjanus campechanus*) and red grouper (*Epinephelus morio*), which are common off the warm temperate coastal areas of the southeastern United States and throughout the Gulf of Mexico but are rare from southernmost Florida to Caribbean and Bahamian waters. Conversely, many similar species found in these tropical Atlantic waters are rare in the warm temperate regions of the southeastern United States (e.g., the mutton snapper [*Lutjanus analis*] and tiger grouper [*Mycteroperca tigris*]) but are common in tropical regions where corals are the predominate hard structure. This distribution enabled by an association preference for substrate could be the result of temperature affinities or competition, but the repeated distribution patterns among many species indicates that this distribution pattern is the result of habitat affinity—coral reef versus no coral reef.

This observation by scientists is confirmed by fishers who target different cognate species depending on whether or not coral habitat is present. This suggests that fish affinities for coral reefs differ between species based on specific habitat preferences. This is not a remarkable idea in itself, but it does suggest that a "one size fits all" explanation for managing the interrelationships between corals and fishes may depend on larger climate- and earth-tuned forces that shape biogeographic patterns. Consequently, the importance of coral as habitat to species in more tropical areas may be thought of as an even more important, essential fish-habitat criterion. Also, fisheries associated with coral reefs may require special management considerations in light of anticipated future declines in fishery conditions (e.g., fish abundance, size, growth, or reproductive potential) if coral health and condition continue to deteriorate. Conversely, fishery managers responsible for fisheries in warm temperate regions may be afforded the luxury of determining essential life history requirements other than live coral and supplement these features, if necessary. Artificial reefs come to mind as a substrate feature that could be added to warm temperate waters if (and only if) it is determined that the necessary habitat for warm temperate fisheries is structure *per se* and not necessarily living structure as provided by coral reefs. Determining the essence of "essential fish habitat" that will suffice for any fishery's habitat requirements will be a critical area for further investigation in the near future lest we assume that warm temperate, hard-bottom-associated fisheries are doomed to decline because corals are declining.

Gilbert's (1973) argument that fishes have different habitat affinities based on whether or not they have continental or island habitat affinities, could offer a congruent perspective for deepwater versus shallow-water reef-associated fishes. In several of the chapters herein, it is noted that, while much less is known about the distribution and condition of deepwater corals as opposed to their shallow-water counterparts, a different complex of fishes is associated with corals by depth. Deepwater corals, to date, have apparently suffered less with regard to recent trends in environmental conditions and may at present reflect a "less spoiled" situation. Nevertheless, deepwater coral habitats deserve our attention before the message relayed by shallow-water coral habitats is lost. Managing fisheries associated with deepwater corals may require entirely different approaches than the management strategies for fisheries associated with shallow-water corals.

Where to from here?

Most of the chapters herein offered suggestions for the directions that future research should take regarding the interrelationships between corals and fisheries. These will not be reiterated here as they are provided and argued more eloquently in the individual chapters. I will, however, take a broader look at these future directions. This perspective should, hopefully, serve as "breadcrumbs in the forest" to lead future efforts.

It is beneficial, here, to reexamine the questions offered in the Preface that served as points of discussion for the workshop:

1. What are the "essential" features of the interrelationships between corals and fisheries?
2. If negative trends in coral condition continue, should we adjust management plans for coral-associated species?
3. What adjustments can be made or new techniques developed that will enhance our management approach to both corals and fisheries?
4. What more do we need to know to answer these questions?

Dealing with each of these questions demands further attention as they were only partially answered during the workshop and in this book. Although everyone did their best, it became obvious that there were far too many unknowns to allow complete resolutions. Importantly, however, directions are found by delving deeply to uncover solutions.

Conclusion

Pathways to a clearer perspective must include an understanding of the broad and varied cultural features of human societies. After all, it was mushrooming development among the world's economies that led to ever-increasing environmental stressors in the first place. Management always involves directly managing people and only indirectly involves managing resources. To help sharpen this facet of the approach, there is a need to improve the public's understanding and appreciation of the interrelationships between corals and fisheries.

Concomitant with improving society's appreciation of the interrelationships between corals and fisheries is the establishment of goals that further our understanding of these interrelationships. Above, I presented arguments that perhaps the goals for fisheries in tropical, warm temperate regions may be different from the goals for deepwater and shallow-water fisheries. In the great matrix of things, this could amount to at least four separate goals (shallow-tropical, shallow-warm temperate, deep-tropical, deep-warm temperate) for each faunal biogeographic province. Completing the matrix will help clarify the intricacies of the linkages between fisheries and corals and will go far in allowing adoption of prescriptive management measures that have a high probability of success.

Virtually all authors here argued for, at a minimum, the continuation of previously initiated monitoring studies, especially those directed toward elucidating the nature of the connectivity between corals and fisheries. The dynamic interrelationship between corals and fisheries demands broad approaches—both spatially and temporally. Short-term (less than a decade) and regional (<1000 km in extent) efforts will do little to provide full answers to the proposed questions as corals and their fisheries are truly global in nature. Most ecologists consider that unifying ecological principles are at work that transcend local areas and brief time periods, as illustrated by considering the Moran Effect (Moran 1953) where it was proffered that similar responses to stress should occur in dispersed populations. It has also been argued that this principle or effect is applicable to coral reef fish communities (Cheal et al. 2007). Thus, studies sufficiently broad in time and space that investigate the interrelationships between corals and fisheries in one area can add to and draw from similar studies carried out in distant areas. We can already appreciate this concept as the world's shallow-water coral populations seem to be reacting similarly across the expanse of our seas to global stressors. Moreover, it is critical to continue evaluating currently perceived trends in the health, condition, interrelationships, and relative abundances of both corals and fisheries. Understanding the long-term, broadly based trends and relationships is key to further advancement of our state of knowledge.

Questions regarding the degree of dependency (both obligative and facultative or shades in between) on interrelationships between corals and fisheries remain and it behooves the ecological research community to address this issue. This is especially true for questions whose answers could satisfy the needs of resource managers. To aid in this effort, consideration should be given to making use of marine protected areas as experimental and observational platforms from which much can be learned.

I suggested above that alternative management strategies for fisheries that often are associated with corals but are not dependent on the living features that corals provide, could benefit from artificial reef deployments. There is an increasing body of information on the potential benefits (and detriments) that such deployments may have on certain fisheries (see Bortone et al. 2011). However, it should be noted that we still do not have sufficient expertise to prescribe artificial reefs as a solution to diminishing habitat. Much more directed research, specifically geared to providing resource managers with practical artificial reef deployment strategies, is needed (Bortone 2011). This is especially true in areas like the Gulf of Mexico where an extensive array of over 3000 oil platforms are functioning as "reef" habitat over and above the habitat naturally provided by existing living and dead coral reefs.

Compounding our view here are the other interactive biological features of any aquatic ecosystem. Herein, the authors have made it clear that coral and fishery interactions with other nearshore biocenoses, such as mangroves, must also be considered in discussions on the relative significance of the interaction between corals and fisheries. Truly, all aspects of the ecosystem must be brought into the equation when managing from an ecosystem perspective.

The general use of global information systems (GIS) has been a "game changer" in ecosystem-based management. The ability to simultaneously incorporate multiple layers of varied data (i.e., biotic and abiotic) is revolutionary for the science of resource management. Along with the advent of any new technology, however, comes the realization of how much better our ability to manage resources would be if we only had information on...... (fill in the blank with your favorite variable!). This realization is not facetious. The use of GIS has raised our level of sophistication, but it has also increased our data requirements. Chapters herein have readily acknowledged the future importance of GIS for understanding coral and fisheries interrelationships. Simultaneously, the authors have also raised issues regarding the incomplete status of many of the data layers that are needed to make fuller use of GIS in the management process. While there are calls for more and better mapping of bottom features (as well as associated biotic community features), all should realize that these maps are not the end, but serve as a pathway to a future that incorporates ecosystem-based management into its assessments.

Along with the incorporation of GIS into the management "tool box," it should be noted that there is still a need to improve fishery stock assessment methods. While true for all fisheries, this is especially true for coral reef–associated fisheries. Because of their associated high species diversity, these communities of organisms on coral reefs are represented by a high number of species that are not particularly abundant. These species, because of their lack of frequency, are often characterized as data-poor species. It is important to realize that currently data-poor species are likely to always remain data poor. Polunin et al. (1996) argued that it was not the lack of fisheries' models *per se* that prevented us applying these models to reef fish fisheries but more a lack of data. This is due to the fact that the very essence of high species diversity as a fundamental characteristic of shallow-water tropical areas precludes that any, or even a few, species will have any great abundance. The Gulf of Mexico and South Atlantic Councils are each responsible for managing nearly 70 fish species, the Caribbean Fishery Management Council is responsible for managing over 200 species, and the Western Pacific Council is responsible for managing between 150 and 220 species depending upon the region of the Pacific Ocean. However, fishery management plans have only been developed for a few dozen of these many species. This is chiefly due to a lack of data and funding to conduct stock assessments. High numbers of species with a low frequency of occurrence and low predictability of capture are the hallmarks of tropical fisheries. Thus, collecting more fish, gathering more data, and intensifying standard analytical methods will still render these species data poor. So there is a dilemma when assessing coral reef–associated fisheries. Consequently, the analytical methods of stock assessment developed for low diversity, abundant cold temperate species (e.g., cod, halibut, etc.) are not particularly useful when applied to data-poor fish stocks. In this volume, consideration is given to "thinking outside the box" when conducting stock assessments, especially when developing new analytical techniques to examine data-poor fish stocks. While the directions may be apparent, the exact formulas may take considerable effort for complete resolution. Importantly, however, the effort to develop stock assessment techniques that are applicable to data-poor stocks associated with coral reef–associated fisheries has begun.

References

Bortone SA. 2011. A pathway to resolving an old dilemma: Lack of artificial reefs in fisheries management. *In*: Bortone SA, Brandini FP, Fabi G, Otake S, editors. Artificial reefs in fisheries management. Boca Raton, FL: CRC Press. p. 311–321.

Bortone SA, Brandini FP, Fabi G, Otake S, editors. 2011. Artificial reefs in fisheries management. Boca Raton, FL: CRC Press.

Cheal AJ, Delean S, Sweatman H, Thompson AA. 2007. Spatial synchrony in coral reef fish populations and the influence of climate. Ecology. 88(1):158–169.

Gilbert CR. 1973. Characteristics of the western Atlantic reef-fish fauna. Quart J Florida Acad Sci. 35(2–3):130–144.

Moran PAP. 1953. The statistical analysis of the Canadian lynx cycle. II. Synchronization and meteorology. Aust J Zool. 1:291–298.

Polunin NVC, Roberts CM, Pauly D. 1996. Developments in tropical fisheries science and management. *In*: Polunin NVC, Roberts CM, editors. Reef fisheries. London: Chapman & Hall. p. 361–377.

Index